THE BREAKDOWN
AND RESTORATION
OF ECOSYSTEMS

NATO CONFERENCE SERIES

I   Ecology
II   Systems Science
III Human Factors
IV   Marine Sciences
V   Air—Sea Interactions

I ECOLOGY

THE BREAKDOWN AND RESTORATION OF ECOSYSTEMS

Edited by
M. W. Holdgate
and M. J. Woodman
Institute of Terrestrial Ecology
Natural Environment Research Council
Cambridge, England

Published in coordination with NATO Scientific Affairs Division

PLENUM PRESS · NEW YORK AND LONDON

Library of Congress Cataloging in Publication Data

Conference on the Rehabilitation of Severely Damaged Land and Freshwater Ecosystems in the Temperate Zones Reykjavík, Iceland, 1976.
The breakdown and restoration of ecosystems.

(NATO conference series: I, Ecology; v. 3)
Includes index.
1. Ecology—Congresses. 2. Fresh-water ecology—Congresses. 3. Reclamation of
land—Congresses. 4. Nature conservation—Congresses. I. Holdgate, Martin W. II.
Woodman, M. J. III. North Atlantic Treaty Organization. Special Program Panel on
Eco-Sciences. IV. Title. V. Series.
QH540.C66 1976   574.5   77-18922
ISBN-13: 978-1-4613-4014-0   e-ISBN-13: 978-1-4613-4012-6
DOI: 10.1007/978-1-4613-4012-6

Proceedings of the Conference on the Rehabilitation of
Severely Damaged Land and Freshwater Ecosystems in Temperate
Zones held in Reykjavik, Iceland, July 4-10, 1976, sponsored
by the NATO Special Program Panel on Ecology

© 1978 Plenum Press, New York
Softcover reprint of the hardcover 1st edition 1978
A Division of Plenum Publishing Corporation
227 West 17th Street, New York, N.Y. 10011

Preface

This volume contains the papers presented at a conference on
"The rehabilitation of severely damaged land and freshwater ecosystems in temperate zones", held at Reykjavik, Iceland, from
4th to 11th July, 1976. The meeting was held under the auspices
of the Ecosciences Panel of the N.A.T.O. Science Committee, and
the organising expenses and greater part of the expenses of the
speakers and chairmen were provided by N.A.T.O. The scientific
programme was planned by M. W. Holdgate and M. J. Woodman, in
consultation with numerous colleagues, and especially with the
Administrative Director of the Conference in Iceland, Dr. Sturla
Fridriksson.

Iceland proved a particularly suitable location for such a
Conference. Geologically, it is one of the youngest countries in
the world, owing its origin to the up-welling of volcanic rock
along the spreading zone of the mid-Atlantic ridge within the
past 20 million years. Its structure, northern oceanic situation,
recent glaciation and continuing volcanic activity make it distinct
as a habitat and have given it a flora and fauna of especial
interest. It is also a land of great natural beauty with its ice
caps, waterfalls, volcanic landforms, geothermal features and
dramatic coasts. In addition, its ecosystems have proved exceptionally vulnerable to man's impact and it presents the kind of
problem with which the Conference was concerned in an acute form.

The symposium was of value to those who took part in it
through the exchange of ideas that it made possible. Those from
overseas hope that it has helped their Icelandic colleagues, just
as they benefited from hearing about the environment of Iceland
at first hand. Two excursions permitted us to inspect areas of
soil erosion, soil reclamation and afforestation as well as to
view volcanic landscapes and farmlands and see such world-famous
places as Hekla, Gullfoss, Langjokull, Geysir and Thingvellir.
It is our hope that the contacts so agreeably established will
endure, and that the meeting will consequently prove of value to
the people of Iceland by increasing the pool of ecological
expertise available for them to consult.

This volume brings together much information about the vulnerability, degradation and restoration of temperate ecosystems. It is not, however, a "cookery book" designed to provide a Government or an Agency confronted with an environmental problem with a formula to remedy their ills. Knowledge is still insufficient and the situations in the world too diverse for such an approach to be feasible even were it desirable. What this book does is provide a quarry for certain types of information and a directory to certain people whose knowledge may be of value to those facing the kinds of situation the papers published here describe.

As organisers and editors we would like to record our gratitude to all who, in many different ways, made the meeting and the book possible. Among them we would particularly name His Excellency, Mr. Einar Agustsson, Icelandic Minister for Foreign Affairs, who opened our proceedings; Dr. F. K. Hare, Professor J. W. Heslop Harrison and other members of the Ecosciences Panel, who conceived of the meeting in the first place; and Dr. T. D. Allan of N.A.T.O. who supported us in the work of organisation and through whom our funds came. Mr. M. Clark of Thomas Cook Limited in Cambridge and staff of Icelandair and of the Loftleidir Hotel in Reykjavik organised our travel and accommodation most capably, and we are especially grateful to Mrs. Erla Gunnarsdottir, Conference Officer in the Loftleidir Hotel and to Miss Svava Arasdottir who was our Conference Secretary. Our colleagues in the Institute of Terrestrial Ecology in Cambridge have given us constant support, especially Mr. R. T. Collins, Miss M. J. Moxham, Mrs. E. M. Chambers and Mrs. J. M. Bowen who provided stalwart help in managing finances and preparing papers. Finally, and above all, it is a pleasure to record our thanks to our colleague, Dr. Sturla Fridriksson, himself a member of the Ecosciences Panel, who proposed and arranged the substantial contribution from Icelandic scientists to the meeting, organised our memorable excursions and in many other ways made the Conference enjoyable as well as successful.

M. W. HOLDGATE

M. J. WOODMAN

Contents

PART III: THE RESTORATION OF DEGRADED ECOSYSTEMS

Nato and Science

T. D. Allan

Executive Officer
Eco-Sciences Panel

Most scientists of the Western world are aware of NATO's involvement in promoting scientific endeavour and co-operation throughout and beyond the Alliance - 'the smiling face of NATO'.

The Science Committee was brought into existence in 1958 in response to a widespread feeling that the Alliance must look beyond a resolve to unite in mutual military defence. It must also determine to maintain and strengthen the power of its science and technology. The standard of scientific excellence which characterises the Western democracies has been largely achieved through freedom of movement and exchange of ideas between professional scientists of like-thinking nations.

Since its creation 18 years ago the work of the Science Committee has evolved into five main activities. These are Research Grants Programme, Fellowship Programme, Senior Scientists Programme, Advanced Study Institutes and Special Programme Panels.

RESEARCH GRANTS PROGRAMME

Grants are made to individuals or teams of research workers, their applications being judged by internationally chosen referees. Particular emphasis is placed on international contact between scientists, and projects which do not fit easily into a national support system. In the last year 368 applications were made of which 259 were funded, but at only a fraction of the level requested.

---

* This paper is an edited version of an address of welcome given by Dr Allan at the opening session of the conference. In it, he conveyed the good wishes of Professor M. N. Ozdas, Assistant Secretary General for Scientific and Environmental Affairs, who was unfortunately unable to be present.

THE FELLOWSHIP PROGRAMME

This programme permits scientists to be exchanged between the
nations of the Alliance. The programme is operated by the various
nations but by being centrally monitored brings a cross fertilisation of the scientific work of the nations which would be too
complex to arrange on a bilateral basis. In 15 years this programme
has helped more than 10,000 scientists.

THE SENIOR SCIENTISTS PROGRAMME

This is a small but useful fund-lending support for senior scientists to work with colleagues in other nations of the Alliance.
Ten awards were made in 1975.

ADVANCED STUDY INSTITUTES

The Institutes have been a very successful activity of the Science
Committee. The scheme was created to enable scientists, particularly young people, to get together to hear leading workers lecture
in depth on their work and to contribute themselves to a major
reappraisal of a field, chosen because it lies in an area of new
and significant knowledge. About fifty Institutes are held each
year and the proceedings are published in an authoritative series
of books giving up-to-date 'state of the art' reviews. Many
visitors from non-NATO countries have taken part in the ASI's.

SPECIAL PROGRAMME PANELS

Whereas the Research Grants and Advanced Study Institutes Programmes
respond to requests initiated by investigators, there are special
areas of science which, as a result of careful study by the
Committee, have been deemed worthy of special effort. The Panels
of individual experts advising on these programmes can recommend
and develop research grants, symposia, seminars, exchange visits,
etc. The programmes are not intended to be permanent but strive
in a few years to produce a catalytic effect on the advancement of
a particular subject. When international collaboration is well-
established the programme may be terminated and another area taken
up. At present there are five programmes: Eco-Sciences, Marine
Sciences, Air-Sea Interaction, Systems Science and Human Factors.

The Conference in Reykjavik was sponsored by the Eco-Sciences
Panel. It took place almost three years after Panel Members first
proposed that a conference on Rehabilitation of Severely Damaged

Terrestrial Eco-Systems should be held. Professor Heslop Harrison,
then Director of the Royal Botanic Gardens at Kew, and Dr.
Fridriksson, Administrative Director of the Conference, were
particularly active as members of the Eco-Sciences Panel in promoting the idea of this conference and we are grateful for their
enthusiasm.

But, of course, Special Programme Panels represent bodies of
expertise which meet briefly no more than two or three times a
year. For the overall organisation and scientific direction of a
Conference such as this, the Panels must necessarily rely on a
recognised expert willing to devote a considerable amount of his
time and effort to the task. Here we have been particularly
fortunate in persuading Dr. Martin Holdgate and Mr. Michael Woodman,
of the Institute of Terrestrial Ecology in Cambridge, to apply
their versatile talents to the job. I should like to thank them
for the work they have put in.

As a comparative layman in a company of experts I have no
intention of trying to make any profound statement on the subject
of this Conference. However, it is apparent from the earlier
comments of various experts when the feasibility of holding the
Conference was discussed, that a degree of concern has existed for
some time. There are, apparently, many examples of ecosystems with
a rich diversity of organisms which have become severely degraded
or substantially reduced in area so that there is real doubt whether
any samples at all will survive. If, as some ecologists believe,
major efforts at restoration will become necessary in the next
century or so, then international conferences such as this appear
to be essential.

Introduction

The Theme of the Symposium

M. W. Holdgate*

The theme of this symposium is the rehabilitation of severely damaged land and freshwater ecosystems in temperate zones. The importance of this theme is obvious. We live in a world whose population must certainly double and will very probably quadruple before development, education and medical technology combine to bring stability. We know that at present we do not feed the world's population as well as we should, and the need to expand food supplies must grow as the numbers of people rise. This demand can be met in several ways: by bringing new land under cultivation, by cropping marine resources not used at present, by breeding new strains of plant and animal, by processing plant materials we and our livestock cannot at present use, and by changes in human diet so as to increase the efficiency with which we use the primary production of the earth.

None of these things is easy. The parts of the world that have predictable climates, plenty of water, soils of high fertility, and freedom from extremes of environmental stress — the temperate-zones - have mostly been settled for millennia. The new lands being brought into use today in the tropics are not too easy to use. They often suffer drought. In forest areas a high proportion of the essential nutrients are locked up in a standing crop of trees and are all too readily dissipated if the forest is cleared by fire. These areas are difficult to develop without loss of fertility. Education, energy and money may be needed on a scale unavailable in the developing countries. Today, the losses to agriculture through erosion, the expansion of deserts, and salination due to inadequate irrigation each year, cancel out the gains. The frontiersman's solution to demands for more produce — go out and clear the wilderness - is no solution now. The best wilderness has been cleared, and much of what remains is valued for its timber and control of water run-off. We have to manage what we have more efficiently, especially in the temperate zones where agriculture and productive forestry are most easy to sustain.

* The Conference in Reykjavik was opened by His Excellency Mr. Einar Agustsson, Minister for Foreign Affairs of Iceland, who delivered an address of welcome. This was followed by a keynote speech by Dr. Holdgate as Scientific Director of the Conference, which has been slightly amended for publication here.

There are certainly important new resources in the sea.
Probably 100 million tonnes a year of krill - enough to double
world fishery landings - could be taken from Antarctic waters
without endangering the balance of the ecosystem (though we do
not know this for certain, and need more research before we are
sure). Much can be done through plant and livestock breeding
if we conserve the genetic resources of wild relatives and primitive strains, so as to have the full range of diversity to draw
on). But let us recall a lesson of the 'green revolution' in
India: that the successes achieved have come when new plants and
animals have been made available to people educated to use them
and with the social infrastructure to harvest, store, transport,
and market the produce. A miracle plant is only useful if its
yield can be got to the consumer in a usable condition. Finally,
various people have made thoroughly novel suggestions for enchancing
food supplies - let us eat antelopes which crop wild scrublands,
rather than cattle, for which we have to grow grass. Let us get
protein out of the leaves of wild vegetation rather than disturb
the soil to grow crops. Let us eat earthworms. Let us have
hydroponics. All these may have a great future - in the right
places and in the right communities - but many demand advanced
technology and use much energy. For the present, crops or animals
people can eat, grown in soil or water people can cultivate
or fish, are bound to be our mainstay.

The Conference was on a theme that is in many ways central
to the human situation. How can we sustain the fertility and the
richness of the temperate lands and waters that are vital to the
world's food production? How can we restore the fertility of
those lands and waters that have been degraded by mismanagement
in the past? These are practical themes and the Conference had
practical objectives. In its simplest form, the aim was to
examine what ecological science could tell us about the factors
that determine the patterns and processes of ecosystems and
their responses to human perturbation, so that we can be guided to
avoid damage and to reverse the consequences of past errors. The
programme was shaped to attack this issue logically. First, attention was focussed on basic ecological principles. What is an
ecosystem? Tansley defined it as an assemblage of plants, together
with the animals directly dependent upon them and the immediate
components of the physical environment with which both plants and
animals interact. The first part of this book is concerned with
how far ecosytems function as if they are entities: organisms or
taxa with homeostasis at their own level; and how far their properties are simply the integral of the properties of all their
components. What determines their stability (if they have any),
and the survival of key species within them? Can we describe their
properties in strictly logical models? Can we use these and predict
the probabilities of various kinds of change if man acts in various
kinds of way?

This leads us back to the disciplines of the real world.   In
Part II there are reviews of damage to ecosystems in North West
Europe, Iceland, the Mediterranean and the freshwaters of the
Rhine.   The participants in the meeting were asked a basic question
- could they have predicted and prevented this damage with the
ecological understanding available today?   If not, where were the
gaps in understanding?   In Part III there follow success stories
of many kinds: examples of the restoration of freshwater systems,
the re-creation of soil fertility, and the re-establishment of
vegetation and fauna.   Again, the wider lessons of these examples
demand analysis - how will they help us to understand other problems
in the future?

Several papers in this part of the volume therefore lead us
forward towards the establishment of guide-lines.   These are at
several levels.   What mosaic of land use is best in a world whose
physical environment varies greatly?   How should we manage water
and soil, key to much of productivity?   How should we conserve the
genetic diversity of the world's life?   How do we judge priorities,
relating the potential in the environment to the infinitely variable
pattern of human needs, aspirations and traditions?

The Conference was, in a sense, about Conservation.   Not the
narrow sense of the word, as applied to the protection of wildlife,
much though that contributes to the quality of human experience,
but in the wider sense of safeguarding the integrity of the
environment - soil fertility, water balance, plant productivity,
genetic richness, and the working of the great natural cycles on
which all life, including man's, depends.   This broad conservationist
approach was stressed throughout the discussions.   It was also
emphasised that the participants' responsibility as scientists was
to state the options and present the consequences of policies that
must also take economic and social factors into account.

Part I: Basic Ecological Principles

Introduction

M. W. Holdgate

The first session of the Conference was concerned with basic principles of ecology. A group of ecologists who had specialised in the analysis and description of the behaviour of populations and ecosystems using mathematical language were invited to illustrate the kinds of insight this approach gives. The session was particularly directed towards improving the accuracy of predictions about the course of change in ecological systems, whether in response to natural factors or human influence. Such predictions are equally important to the management of productive land, which we wish to protect from degradation, or devastated land we wish to restore. Four basic questions, all relevant to this theme, were posed at the start of the session:

1. What makes certain species or populations vulnerable to particular kinds of change?

2. What critical environmental conditions have to be sustained to prevent unwelcome changes in ecosystems or populations?

3. How far are traditional human uses of systems liable to promote instability? (May's paper, for example, implies that cropping at maximum sustained yield appears bound to make a species less efficient at withstanding other environmental perturbations).

4. How far do ecosystems respond as entities, and how far, in contrast, is their response no more than the aggregation of the responses of their component species, reflecting the diverse strategies of those species?

1. FACTORS CONTROLLING THE STABILITY AND BREAKDOWN OF ECOSYSTEMS

R. M. May

Princeton University

Princeton, U.S.A.

INTRODUCTION

The current search for principles that may guide our understanding and management of terrestrial and marine ecosystems is amply illustrated by the growing number of conferences, journals and books addressed to this theme.

There are many different lines along which this search may fruitfully be advanced. One approach involves the construction of detailed multi-parameter models for specific systems, which may then be explored on a computer with the eventual aim of having a model which is sufficiently realistic and reliable to serve as a management tool. Work of this kind is described in later papers in this session. At the other end of the spectrum of possible approaches are more abstract models, which attempt to capture the essential features of natural systems, sacrificing detailed realism in order to reach for a general conceptual framework. Such models are at best metaphors for, or caricatures of, reality; they serve a useful purpose in organising data, sharpening discussion, or suggesting the key factors the empiricist should be focussing on. My paper deals mainly with this latter class of relatively general models.

The present paper offers no more than a brief trail guide to the contemporary literature; it is in the nature of a discursively annotated bibliography. With so much printed material being produced these days (May and May, 1976), it seems unpardonable to write repetitive reviews. In particular, I have been quite shameless in referring extensively to the multi-authored volume

11

Theoretical Ecology: Principles and Applications (May, 1976a),
hereafter referred to as TE.

SINGLE POPULATIONS

Theoretical Models and Metaphors

We begin by directing attention to the dynamical behaviour of a
single population, $N(t)$: here N is the number of animals (or sometimes their biomass) at time t. Truly single-species situations
can be realised in the controlled setting of a laboratory experiment, but are rarely met in the natural world. It is nevertheless
helpful to make a start by pretending that a population's interactions with its physical environment and with other biological
populations can be subsumed in passive parameters, and see what can
be learned from such models.

   If population growth is a continuous process, one of the simplest equations characterising a population with density dependent
regulation is the familiar logistic,

$$dN/dt = rN(1 - N/K). \tag{1}$$

The exact form of the right hand side of this equation is not to be
taken seriously; rather it is representative of a wide class of
population equations with nonlinear regulatory mechanisms (TE,
p. 5-6). This equation can be used to make an important and
paradigmatic distinction between the statics and the dynamics of
the behaviour of the population: the magnitude of the equilibrium
population, $N^* = K$, depends only on the parameter K; the time
taken to attain this equilibrium, or to recover to it following a
disturbance, depends on the rate parameter r. This point (which
underlies the metaphor of r and K selection developed below) can be
emphasised by introducing the dimensionless population variable
$N' = N/K$, and the dimensionless time variable $t' = rt$, to get the
parameter-free equation

$$dN'/dt' = N'(1 - n') \tag{2}$$

This sort of reasoning, to produce dimensionless variables and
combinations of parameters, is a technique that is commonplace in
engineering and physics, and should be more widely used in ecological studies than it is.

   In any such dynamical system, it is useful to identify a
"characteristic return time", T_R, which gives an order-of-magnitude
estimate of the time the population takes to return to equilibrium,

following a disturbance (for a more precise discussion, see May et.
al., 1974, and Beddington et. al., 1976a). From the above remarks,
it is clear that for equation (1) $T_R = 1/r$.

These features of equation (1) can be used to make some
generalisations about the life history strategies of species, in
terms of the deliberately oversimplified notions of r and K selection (see, e.g. MacArthur, 1972). A K-selected organism sees its
environment as relatively stable and predictable (and consequently
the population is usually around its equilibrium value, $N \simeq K$).
The evolutionary pressures on an organism in these circumstances
are, crudely, to be a good competitor; to increase its effective
value of K; to have fewer offspring but invest more time and energy
in raising them. The emphasis is on the population statics. Conversely, an r-selected organism sees its environment as relatively
unstable and unpredictable (and is usually at low population values,
growing exponentially, and undergoing episodes of boom and bust).
The evolutionary pressures here are for opportunism; for large r
to exploit the transient good times; to have many offspring, few of
which expect to mature. The emphasis is on the population dynamics.
Of course, reality is more complicated than this and the paradigms
of r- and K- selection are no more than the opposite ends of what
in fact is a continuum.

In equation (1), the regulatory effects (expressed by the
factor $[1 - N/K]$) operate instantaneously. To put it mildly, this
is unlikely to be the case in the real world, where such regulatory
effects are liable to operate with some built-in time delay, the
characteristic magnitude of which may be denoted by T. Such time
delays may be incorporated by generalising the logistic to read

$$dN/dt = rN \left[1 - N (t - T)/K\right]; \tag{3}$$

this equation is representative of a wider class of more realistic
equations with distributed delays (see, e.g., May, 1973a). The
dynamical behaviour of equation (3) now depends on the relative
magnitude of the two time scales, T and $T_R = 1/r$. If the time delay
T is short compared with the characteristic return time T_R (i.e.,
if rT is small), disturbances will be damped monotonically back to
the equilibrium point $N^* = K$, as in equation (1). As T begins to
approach T_R, there is a tendency for the regulatory mechanism to
produce overshoot and overcompensation. At first this leads to
oscillatory (rather than monotonic) return to the equilibrium point,
but as T becomes significantly larger than T_R (i.e., as rT becomes
larger than some number of order unity) this pattern of overshoot
and overcompensation leads to self-sustaining stable limit cycles.
Such stable cycles are an explicitly nonlinear phenomenon, in which
the population density N(t) oscillates up and down in a cycle whose
amplitude and period are determined uniquely by the parameters in
the equation.

In short, populations with time delayed regulatory mechanisms may exhibit monotonic damping, or damped oscillations, or sustained cyclic fluctuations, depending on the relative magnitude of the time delay T and their natural time scale T_R (see TE, pp. 6-9 and references therein).

Variations on this theme can arise in many different ways.

For many plant and animal species generations do not overlap, and population growth is a discrete rather than a continuous process. The appropriate equations are then difference equations, relating the populations at successive discrete intervals of time; that is relating N_{t+1} to N_t. Even if there are no explicit time delays in the nonlinear regulatory mechanisms in such equations, there is a one time step lag inherent in the structure of the difference equation. Depending on the relation between the populations' natural time scale or "characteristic return time" and the built-in time lag, there again can be monotonic or oscillatory damping to an equilibrium point, or self-sustained cycles: this analogy between difference equations and time-delayed differential equations is developed more carefully in May et al., (1974).

In addition, difference equations admit of a further remarkable complication. If the nonlinearities are sufficiently severe (corresponding in effect to the time delays being sufficiently great), the regular pattern of stable cycles can give way to apparently chaotic fluctuations. That is, a simple and deterministic population model can give rise to dynamical trajectories which are indistinguishable from random noise. This fact holds obvious and disturbing implications for the interpretation and analysis of real world population data: both the phenomenon itself (e.g. May, 1976b) and its biological implications (e.g. May and Oster, 1976) are reviewed more fully elsewhere.

As another variation on the theme of relative time scales, consider the logistic equation (1) with a time dependent carrying capacity, K (t):

$$dN/dt = rN \left[1 - N/K(t)\right] \qquad (4)$$

Suppose K(t) varies periodically (e.g., seasonally) with period τ. The population will now tend to "track" these variations, or to average them out, depending on whether the characteristic response time T_R (= 1/r) is short or long compared with τ. Similar considerations apply to the more general circumstance where K(t) exhibits random variability: the population will tend to "track" the low frequency (long time scale) components of the noise spectrum, and to average over the high frequency (short time scale) components, with the transition zone occurring for noise frequency components $\omega \sim r$ (time scales of the order of T_R). These points are developed more fully by Roughgarden (1976) and TE (pp. 20-23).

The above notions may also be applied very broadly to consider the dynamical behaviour of a population in a randomly varying environment. In this statistical situation, the population magnitude will be described by some probabilistic distribution function. Roughly speaking, this probability cloud is in tension between two countervailing tendencies: the population's regulatory mechanisms are trying to keep it around some average value (with departures therefrom tending to return in the characteristic time T_R), while the random environmental fluctuations are tending to make the population cloud diffuse (with a characteristic diffusion time t_D, roughly inversely proportional to the variance which characterises the environmental noise: $t_D \sim 1/\sigma^2$). The probability cloud will tend to remain compact, and the population to persist, if

$$T_R < t_D \qquad\qquad (5)$$

If this qualitative criterion is not fulfilled the population is liable to fluctuate to extinction under the influence of the random environmental noise. These qualitative remarks contain the essence of the more detailed work of Levins (1969), May (1973b), Roughgarden (1976), Abrams (1976), Ludwig (1975) and others.

In summary, we conclude that regardless of whether population growth is a continuous or a discrete process, simple density dependent models show that single populations may exhibit either a stable equilibrium point (damped in a monotonic or an oscillatory manner), or sustained fluctuations (as stable cycles or as chaos), depending on whether the natural periodicities or time delays in the regulatory mechanisms are short or long compared with the characteristic response time of the system.

Since a population's characteristic response is often of the order of $1/r$, these considerations introduce elements of self-consistency into the evolution of population parameters, and the concomitant notions of r- and K- selection. In an unpredictable environment, there will be an advantage in having a largish r, to recover from bad times and to exploit the good times. But large r (short T_R) condemns the population to track environmental fluctuations, and makes for the sort of overcompensation which is inimical to population regulation, thus exacerbating the perceived unpredictability of the environment. Conversely, relatively small r implies a long response time, with the advantage that the population may maintain steady values and may average over environmental variations, but with the disadvantage of slow recovery from traumatic disturbances.

We proceed to indicate practical applications of these ideas.

Applications

The above considerations enlarge the notions of r- and K- selection,
and allow a great deal of empirical data on the life history
strategies of organisms to be codified and understood. In particular, Southwood (Southwood et al., 1974; Southwood, 1975; TE, ch.3)
has recently attempted to give a synoptic account of the relation
between an organism's bionomic strategy (as manifested by size,
longevity, fecundity, range and migration habit, etc.) and its
environment: this synthesis applies to systematic trends both
within and between taxa.

One overall moral is clear. It is difficult to produce extinction in r- selected organisms, such as crop pests or the Fire ant,
for their primary adaptation is to bouncing back from low population
values. Conversely, K- selected organisms are geared to maintain
the population around its equilibrium value, and may have difficulty
in recovering from a severe disturbance which is outside their
evolutionary experience. The extinct Passenger pigeon is an
example of a once abundant organism of a K- selected kind; in general, it is K- selected species that present conservation problems.
Evaluation of an organism's general bionomic strategy should always
be the first step in any programme of rational conservation or
management: methods for regulating duck hunting will not be applicable to Sand-hill cranes.

It may also be noticed that, broadly speaking, the contrast
between the relatively predictable and stable tropical environment
and the relatively vagarious temperate one can be viewed as tending
to produce a K- selection/r- selection contrast. One manifestation
of this is that ecologically analogous animals tend to give birth to
fewer progeny in the tropics than in the temperate and boreal zones.
Examples are agouti and paca compared with rabbits, peccaries with
wild pigs, curassows and guans with pheasants and quails. Lack,
Cody and others have presented a wealth of evidence that birds
tend to have smaller clutch sizes in the tropics: this and other
such data have been reviewed by Southwood et al., (1974). The
overall effect is that tropical animals tend to have less capacity
for the sort of explosive "pioneer" population growth which
characterises their higher-latitude cousins (mythologised by lemmings). They are not so well adapted to recovery from bad times.

There is a corresponding tendency to select for competitive
ability, rather than weed-like opportunism and adaptability, in
tropical plants. One consequence is that seeds tend to have
little or no dormancy period, being adjusted to germination in the
sandy, moist, cool environment of the forest. This is in striking
contrast to the wide variety of dormancy periods and survival
strategies used by, say, desert plants.

Another useful application of these ideas is given by Conway and Southwood (TE, chs. 3 and 14), who make a rough classification of insect pests along a continuum whose extremes are "r pests" and "K pests". This classification then reveals systematic patterns in the successes and failures of pest control programmes, and suggests which of the principal control techniques (pesticides, biological control, plant and animal resistance, sterile mating, etc.) may be most appropriate against different categories of pests.

Single population models are also useful in laying bare some of the essential dynamical features of a harvested population. Again, the logistic equation (1) can serve to illustrate the general behaviour, even though practical applications will use more detailed and realistic models (see, e.g. Beverton and Holt, 1957, for fisheries models).

If the population is harvested with a constant effort, E (that is, by taking a constant proportion of the population), the dynamics may be caricatured by

$$dN/dt = rN (1 - N/K) - EN \qquad (6)$$

There is a large and classic literature dealing with this deterministic model (i.e., where r and K are constant), and with its more detailed realistic relatives: see, e.g., Clark (1976). The optimum sustained yield (i.e. with $dN/dt = 0$) is achieved with $E = r/2$, and is $Y = rK/4$. The above discussion, and equation (5) in particular, adds insight into the dynamical behaviour of such a harvested population in an environment which has elements of random variability (i.e., where r and K contain randomly varying components). It is easily seen that for equation (6) the characteristic return time T_R is

$$T_R = 1/(r - E) \qquad (7)$$

a result which is to be compared with the previous $T_R = 1/r$ for the pure logistic equation (1). It follows that, as a larger proportion of the population is harvested (as E increases), the characteristic time for it to recover from disturbances lengthens; the value of E which produces the optimum sustained yield leads to a T_R double that for the unharvested population, and yet larger values of E lead to larger T_R. Referring to equation (5), we recall that in a randomly varying environment a lengthening T_R means a diminished capacity to damp the population fluctuation caused by the fluctuating environment. For the example of a harvested fish population, this enhanced variability in N will show up as an increased relative variability (increased coefficient of variance) in the fish catch as E is increased; this goes some way towards a qualitative explanation of one of the most conspicuous features of many fisheries since World War II.

For the alternative strategy of harvesting for a constant yield, Y (constant quota), the simple model is

$$dN/dt = rN (1 - N/K) - Y \qquad\qquad (8)$$

The behaviour of this system in a deterministic environment is discussed, complete with numerical applications to some harvested populations, by Brauer and Sanchez (1975). The dynamical behaviour of the system (8) in a fluctuating environment is similar to that of equation (6), except more dramatic; as one tunes Y to obtain what would be the optimum sustained yield in a deterministic world, $T_R \rightarrow \infty$, and the population fluctuations become unboundedly large as time goes on.

These are, of course, grossly oversimplified models for harvested populations in a stochastic world. But, in conjunction with the earlier insights (embodied in equation (5)) as to relevant time scales, they give a good account of qualitative trends. For a more thorough discussion, see Beddington and May (1976).

Summary

I conclude this brief account of simple models for single populations by reiterating that the population's dynamics (monotonic damping, oscillatory damping, cycles, or chaotic fluctuations) are determined by the relative magnitudes of the ecologically relevant time scales in the system. Assigning rough values to these time scales should be one aim of management studies.

Although I have glossed over it here, the question of spatial scale will usually be equally important; some aspects of this question are discussed in TE (Chs. 3, 5, 9, 10 and references therein).

TWO POPULATIONS

This sort of analysis of simple models can be extended to elucidate some of the basic features of two populations interacting as prey-predator or as competitors or as mutualists. Of this, the work on prey-prediator systems (with plant-herbivore and human host-parasite systems being special cases), is probably the most extensive, and has made the most quantitative contact with field and laboratory experiments and observations. This is specially true of the growing body of work on arthropod prey-predator systems, where quite details correspondence between theory and experiment is emerging (see, e.g., Beddington et al., 1976b, Hassell et al., 1976 and the review in TE, ch. 5).

One important feature which arises in a natural way in many
such prey-predator models is the occurrence of two alternative
stable states for the system. Specific examples include: models
for the North American Spruce budworm studied by Holling and co-
workers (Dixon Jones, and Ludwig, Priv. comm.); simple graphical
models for the dynamics of grazing systems, where plants are the
prey and herbivores the predators (Noy-Meir, 1975); models for
many tropical human host-parasite diseases, such as schistosomiasis
(see, e.g., MacDonald, 1965, or Nasell and Hirsch, 1973). Although
these and other examples have each been discussed sui generis, the
2 alternate stable states arise in the same basic way in these
various prey-predator systems, and the phenomenon may properly be
regarded as generic. Nor need the 2 stable states both be points:
Guckenheimer et al., (1976) have observed that, in simple prey-
predator models obeying difference equations (discrete generations),
one can easily have alternate states in which one or both is a
cyclic oscillation rather than an equilibrium point.

These specific examples bear out a point earlier made in more
abstract form by Holling (1973): if there are 2 or more alternate
stable states, then a system may recover its original configuration
following a weak disturbance, but a large disturbance can carry it
into an entirely new and locally stable state. Such a system may
be thought of as "resilient" to weak disturbances, but altered in a
qualitative way by strong disturbances. All the above authors
stress the important fact that, once there are 2 or more alternate
stable states, continuous changes in some environmental or biological
parameter or in a population value can produce discontinuous
effects; the system may return to the original configuration following a perturbation by a factor of 2.0 to some population, but be
precipitated into an entirely new state (with different population
values, including possibly some extinctions) by a perturbation by
a factor of 2.1. Noy-Meir makes this point for his grazing systems,
and adduces evidence to suggest that such changes of state have
occurred in some managed grasslands and ranges.

It is convenient here to anticipate the next section, on multi-
species models, and to remark that in nonlinear systems of many
interacting populations it is usual to find a multiplicity of alternate stable states. An interesting study of this kind is that
reported recently by Gilpin and Case (1976), who examined the
typical number of alternate states, and the relative magnitude of
their domains of attraction, for model ecosystems in which the
interactions were of the quadratically nonlinear Lotka - Volterra
type. Although this study is much more metaphorical and abstract
than the budworm, grazing system and tropical disease ones described
above, it does underline the robustness of the phenomenon.

In a more empirical fashion, recent reappraisals of the theory of plant succession (e.g., Horn, TE ch. 10; Connell and Slatyer, priv. comm.) call into question the Clementsian vision of an orderly procession of successional states, and observe that there are various reasons why one may often find alternative stable states, and alternative successional trajectories. Partly for this reason, Horn (1975) and others have suggested that a system will, in the large, tend to realise its maximum potential diversity when exposed to some intermediate level of disturbance: too much or too little disturbance may channel the system into a particular single state, with consequently lowered diversity.

It is to be emphasised that linear models cannot exhibit multiple stable states; nor, in a linear model, can a continuously changing parameter produce a discontinuous effect. Insofar as alternate stable stages and discontinuous changes are a natural feature of minimally realistic prey-predator models, and even much more of multi-species models, I would be highly sceptical of any programme which sought to use a linear or quasi-linear model as a management tool. Whatever the merits and demerits of large but linear input-output models in economic analysis, I think they have little place in ecology.

MANY POPULATIONS

Once one is dealing with many interacting species, life becomes very complicated. A variety of different approaches have been employed to say useful things about multi-species situations. These include, inter alia, descriptions of the patterns of trophic organisation and energy flow; of the patterns of species, relative abundance and diversity in the community; and of the relation between the number of species in a region and its area and degree of isolation. Some of this material will be covered in later papers in this session.

Relatively abstract model ecosystems also have a role to play, in helping to clarify the relation between stability and complexity in ecological systems. (Careful analysis of the various definitions that can be attached to the term "stability" are to be found in Orians, 1975, Whittaker, 1975 and May, 1973b).

Earlier ecological theory often held that complexity, in the sense of many species and a rich web of interconnecting relation-ships, tended of itself to confer stability on an ecosystem. More recent studies both of theoretical models and of empirical evidence, suggest that complex communities are dynamically fragile. Although well adapted to persist in the relatively predictable environments in which they have typically evolved, complex ecosystems are likely to be much less resistant to the large scale disturbances wrought by man than are relatively simple and dynamically robust systems. Although these perceptions clearly have a bearing on the "factors controlling stability and breakdown in ecosystems", they do not provide the quantitative tools for analysis of the kind outlined earlier in this review. For this reason I shall not elaborate further; for a more full discussion, see TE (ch. 8, and references therein).

CONCLUSION

For the reasons given at the outset, this paper has been merely a sketchy and insubstantial outline. The bones are, however, amply given flesh in the references provided, although even here I have favoured works of review and interpretation rather than the original sources.

I am indebted to King's College, Cambridge and the Cambridge University Statistical Laboratory, for their hospitality during the time this paper was written. This work was supported in part by the NSF under grant BMS 75 - 10464.

REFERENCES

Abrams, P. E. 1976. Environmental variability and niche overlap. Math.Biosci., 28, 357-372.

Beddington, J. R. and May, R. M. 1976. The dynamical effects of harvesting in a randomly varying environment (in preparation).

Beddington, J. R., Free, C. A. and Lawton, J. H. 1976. Concepts of stability and resilience in predator-prey models. (In press). (Referred to as Beddington et al., 1976a).

Beddington, J. R., Hassell, M. P. and Lawton, J. H. 1976. The components of arthropod predation. II. The predator rate of increase. J. Anim. Ecol., 45, 165-85. (Referred to as Beddington et al., 1976b).

Beverton, R. J. H. and Holt, S. J. 1957. On the dynamics of exploited fish populations. Fishery Invest., Lond., 19, 7-533.

Brauer, F. and Sanchez, D. A. 1975. Constant rate population harvesting: equilibrium and stability. Theor. Populat. Biol., 8, 12-30.

Clark, C. W. 1976. Mathematical Bioeconomics. New York:
    Academic Press.

Gilpin, M. E. and Case, T. J. 1976. Multiple domains of attraction
    in competition communities. Nature, Lond., 261, 40-42.

Guckenheimer, J., Oster, G. F. and Ipaktchi, A. 1976. The dynamics
    of density dependent population models. Theor. Populat. Biol.,
    (in press).

Hassell, M. P., Lawton, J. H. and Beddington, H. R. 1976. The
    components of arthropod predation. I. The prey death-rate.
    J. Anim. Ecol., 45, 135-64½

Holling, C. S. 1973. Resilience and stability of ecological systems.
    Annu. Rev. Ecol. & Syst., 4, 1-24.

Horn, H. S. 1975. Markovian properties of forest succession. In:
    Ecology and Evolution of Communities, edited by M. L. Cody and
    J. M. Diamond, 196-211. Cambridge, Mass.:Harvard Univ. Press.

Levins, R. 1969. Some demographic and genetic consequences of
    environmental heterogeneity for biological control. Bull. ent.
    Soc. Am., 15, 237-240.

Ludwig, D., 1975. Persistence of dynamical systems under random
    perturbations. SIAM Rev., 17, 605,40.

MacArthur, R. H. 1972. Geographical Ecology. New York: Harper and
    Row.

Macdonald, G. 1965. The dynamics of helminth infections, with
    special references to schistosomes. Trans. R. Soc. trop. Med.
    Hyg., 59, 489-506.

May, R. M. 1973a. Time-delay versus stability in population models
    with two and three trophic levels. Ecology, 54, 315-25.

May, R. M. 1973b. Stability and Complexity in Model Ecosystems.
    Princeton: Princeton Univ. Press.

May, R. M. (ed.) 1976a. Theoretical Ecology: Principles and
    Applications. Oxford: Blackwell Scientific Pubs.

May, R. M. 1976b. Simple mathematical models with very complicated
    dynamics. Nature, Lond., 261, 459-467.

May, R. M., Conway, G. R., Hassell, M. P. and Southwood, T. R. E. 1974. Time delays, density dependence, and single species oscillations. J. Anim. Ecol., 43, 747-70.

May, R. M. and Oster, G. F. 1976. Bifurcations and dynamic complexity in simple ecological models. Am. Nat., 110, (in press).

May, J. and May, R. M. 1976. The ecology of the ecological literature. Nature, Lond., 259, 446-447.

Nasell, I. and Hirsch, W. M. 1973. The transmission dynamics of schistosomiasis. Communs pure appl. Math., 26, 395-453.

Noy-Meir, I. 1975. Stability of grazing systems: an application of predator-prey graphs. J. Ecol., 63, 459-81.

Orians, G. H. 1975. Diversity, stability and maturity in natural ecosystems. In: Unifying Concepts in Ecology, edited by W. H. Van Dobben and R. H. Lowe-McConnell, The Hague:Junk.

Roughgarden, J. 1976, Resource partitioning among competing species: a coevolutionary approach. Theor. Populat. Biol., (in press).

Southwood, T. R. E. 1975. The dynamics of insect populations. In: Insects, Science and Society, edited by D. Pimental, 151-199. New York: Academic Press, 151-99.

Southwood, T. R. E., May, R. M., Hassell, M. P. and Conway, G. R. 1974. Ecological strategies and population parameters. Am. Nat., 108, 791-804.

Whittaker, R. H. 1975. Communities and Ecosystems. (2nd edition). New York; London: Macmillan.

DISCUSSION: PAPER 1

L. M. TALBOT   I appreciated your approach to sustained yield in a harvested population, and particularly your use of "optimum sustained yield" rather than "maximum sustained yield (MSY)". The concept of MSY as it is being applied has the real danger of causing breakdown of the ecosystem involved and loss of the species being harvested. Last year, it appeared that the concept was going to be enshrined in international law through the Law of the Sea

negotiations. Consequently, I organised a
series of workshops bringing together international experts in the productivity and
management of wild populations. This distinguished group concluded that neither MSY nor
any other ecologically simplistic concept was
adequate as a basis for – or goal of – management. Instead, any specific management regime
should be guided by a set of principles, which
they defined.

MSY, as applied, was bound to fail because
it neither takes into account the many factors
which affect the species being harvested nor
those involved with that species' inter-
relationships with the ecosystem. This is
essentially one of the problems you addressed
in your paper. This whole field illustrates
the general difficulty of translating ecology
into policy: interim working concepts may be
translated into inflexible legal dogma.

R. M. MAY

I am a fence-sitter: I did not set out to
attack notions about MSY, but rather to emphasise that classical models need enlargement
for application in a world exhibiting
stochastic environmental fluctuations. Models
for the harvesting of a population in a
fluctuating environment differ in some
qualitative ways from the conventional deterministic models. Moreover, as indicated in
more detail above, these qualitative differences are more significant if the harvesting
is done for constant yield (constant quotas)
than for constant proportions (constant effort).
The next generation of models should acknowledge these consequences of environmental
stochasticity if they are to be useful policy
instruments.

H. REGIER

Observation of fish populations in the
Great Lakes suggests that the steady state
approach to fish dynamics is not useful. It is
an empirical inference that oscillations
increase with stress from opportunistic fishing.
In Lake Erie, Blue pike rose to a peak of
abundance in the mid-1950s, but did not

reproduce and is now extinct. We have now
turned our backs on conventional population
statistics, and our empirical findings are now
matching up with our models!

J. N. R. JEFFERS   Professor May edged near to, but did not
refer to, catastrophe theory. Does he consider
the topological models of catastrophe theory
to be useful?

R. M. MAY   I have reservations as to the usefulness
of Thom's catastrophe theory in ecology; I
think the technique has been oversold by
enthusiasts.

It is true that catastrophe theory
provides an elegant way of bringing out the
essential similarities in, e.g., the Spruce
budworm or schistosomiasis or other prey-
predator situations, where there can exist
alternate stable states and where discontinuous
change can be produced in response to continuous change in a control variable (as when
water freezes in response to a smooth and
continuous temperature change). But I think
these phenomena can be well understood - and
were well understood - without recourse to the
sophistications of Thom's theory.

What _is_ important, and is not generally
appreciated, is that catastrophe theory is
simply not applicable unless there is some
underlying "gradient field". This means the
theory is totally inapplicable to systems with
time delays, or to systems which obey difference
equations. But essentially all ecologically
interesting systems involve one or both of
these factors, and thus are completely outside the domain where the mathematical
techniques of "catastrophe theory" apply.

2. CRITICAL AREAS FOR MAINTAINING VIABLE POPULATIONS OF SPECIES

J. M. Diamond

UCLA Medical Center

Los Angeles, U.S.A.

INTRODUCTION

Different species require different areas of habitat in order to maintain viable populations. Thus, before one can assess whether preservation or rehabilitation of some piece of habitat will be adequate to maintain the population of a particular species, one must specify the area as well as the quality of the piece. This paper serves three purposes: to offer a method of depicting such area requirements quantitatively; to illustrate the problem of area requirements by briefly summarising 13 studies on various taxa and geographic regions; and to interpret the existence of area requirements.

HOW TO DEPICT AREA REQUIREMENTS

The phrase "area requirement" should not be taken to mean that, for species S_i, there is some area A_i below which it has no chance of survival and above which its survival is ensured. Rather, probability of occurrence or survival will in general increase smoothly with area. Fig. 1 illustrates a convenient method of depicting this relation. Suppose that one has surveyed numerous census plots or islands with similar habitat but with different areas. Group the plots into sets which share similar values of area within a modest range — e.g., 1 - 5, 6 - 20, 21 - 50, 51 - 200, 201 - 1000 ha, etc. For a particular species and for each set of plots, calculate the fraction of the plots of a given set that are actually found to support a population of the species. Plot this fraction, which is termed the incidence of the species and abbreviated by J, against area A (Diamond 1975a).

Fig. 1 gives such plots for distributions of six bird species
on islands of the New Hebrides Archipelago in the Southwest Pacific.
The Black duck (Anas superciliosa) is on all islands larger than
ca. 2000 km^2, is on a decreasing fraction of islands with areas
from 2000 to 100 km^2, and is on no island smaller than 100 km^2.
The Rainbow loukeet (Trichoglossus haematodus) reaches much smaller
islands: it is present on all islands larger than 30 km^2, and is
still on half of the islands of area ca. 2 km^2. The Thicket warbler
(Cichlornis whitneyi) requires even larger areas than Anas
superciliosa (absent on all islands of 2000 km^2 or less area),
while at the opposite extreme the Cardinal honeyeater (Myzomela
cardinalis) occurs on all surveyed islands larger than 2 km^2.

For many species in many faunas, the incidence functions have
the general form of Fig. 1, increasing from 0 at low A to 1 at high
A. $A_{0.5}$, the value of A at which J = 0.5, provides a rough, single-
parameter description of the empirical curve and is also simply
related to a theoretical interpretation in terms of immigration-
extinction equilibria (see Fig. 3 later). Other forms of incidence
functions observed in nature are that J may apparently reach an
asymptote at a value less than 1 at high A, or that (for so-called
supertramp species in some rich tropical faunas) J may be maximal
at an intermediate A, declining to zero at both low and high A.

EXAMPLES OF AREA REQUIREMENTS IN NATURE

Our first six examples are drawn from studies of birds, mammals,
lizards, and ants on real islands:

1. Incidence functions were determined for all land and freshwater bird species of the Bismarck (Diamond 1975a), Solomon (Mayr
and Diamond 1976), and New Hebrides (Diamond and Marshall 1977)
archipelagoes in the Southwest Pacific, from surveys of 28 - 52
islands in each archipelago. Values of $A_{0.5}$ ranged from ca. 1 ha,
as for the sunbird (Necharinia jugularis) and the mound-builder
(Megapodius freycinet), to more than 10,000 km^2, as for the eagle
(Harpyopsis novaeguineae) and 31 other species.

2. Jones (personal communication) has determined incidence
functions from distributions of bird species on British islands.
Values of $A_{0.5}$ ranged from several ha, as for Rock pipit and
oystercatcher, to 6 km^2, as for Common snipe, and higher values for
some species.

3. Morse (1971) noted distributions of five forest parulid
warbler species in three consecutive breeding seasons on seven

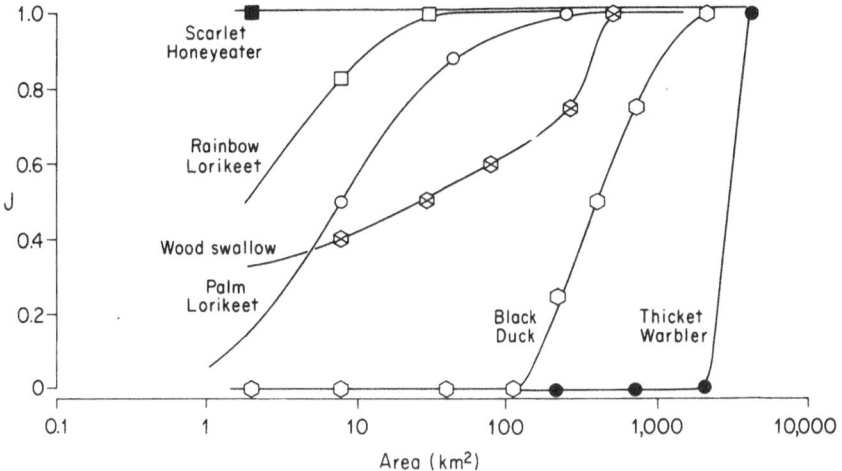

Figure 1

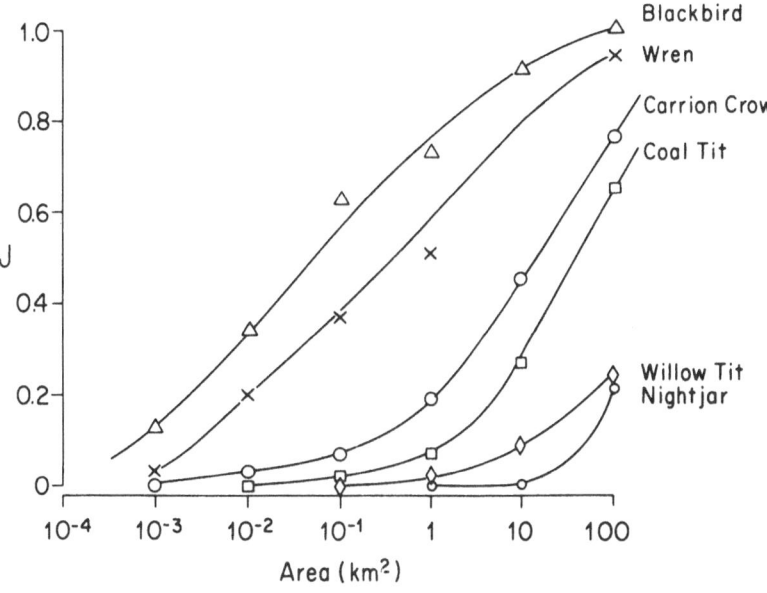

Figure 2

islands of different sizes off the coast of northeastern North
America. Parula warbler bred in almost all years on all islands,
including five islands with only 0.16 - 0.69 ha of forest; Myrtle
warbler bred in all years on all islands larger than 0.35 ha but
never bred on the 0.16-ha island; Black-throated green warbler
never bred on islands of 0.39 ha or less, sometimes bred on 0.49-
and 0.54-ha islands, and always bred on 0.69- and 1.50-ha islands;
and Magnolia warbler and Blackburnian warblers never bred on islands
of 1.50 ha or less but did breed on an island of 134 ha.

4. Hope (1973) reported distributions of mammals on 26 islands
of Bass Straits between Tasmania and Australia. Among 10 species
of herbivorous marsupials, none occurred on any island smaller than
1.4 km^2, the wallaby (Thylogale billardierii) was the most widespread
($A_{0.5} \sim 3$ km^2), and several species such as the Grey kangaroo
(Macropus giganteus) were confined to the largest island, Tasmania
(67,900 km^2). Carnivores had higher values of $A_{0.5}$ than similar-
sized herbivores, and large carnivores higher values than small
carnivores.

5. Ford (1963) studied reptile distributions on 33 islands
off western Australia. Islands smaller than 0.16 ha had no reptiles,
and islands of 0.20 - 32 ha usually had geckos but no other reptiles.
Among geckos, Egernia pulchra had a relatively high $A_{0.5}$ (~ 5 ha)
and occurred on no island smaller than 3 ha, while Ablepharus lineo-
ocellatus was on several islands smaller than 1 ha.

6. Goldstein (1975) surveyed ants on nine islands in Long
Island Sound off northeastern North America. The area of the largest island was 4 ha, while the other islands were 0.016 - 0.4 ha.
Of 34 ant species found on the adjacent mainland, nine reached no
island; eight reached only the largest island (4 ha); 13 more
reached some islands of 0.3 - 0.4 ha, but no smaller island; and
only four species reached any island in the size range 0.016 - 0.3
ha.

Our remaining seven examples of area requirements are drawn
from habitat patches on mainlands. Such patches function as
virtual islands for species that can live in the patch habitat, but
not in the surrounding habitat type.

7. Moore and Hooper (1975) recorded bird species in 433 British
woodlands during the breeding season. The percentages listed in
their Table 2 may be plotted directly to obtain incidence functions
for each of 53 species (see Fig. 2). Values of $A_{0.5}$ ranged from
$\leq$ 0.1 ha for blackbird and Wood pigeon to ca. 100 ha for Coal tit
and treecreeper and at least several hundred ha for Lesser spotted
woodpecker and nightjar. A similar study by the same authors for
water bird species on ponds and lakes of various sizes yielded
$A_{0.5} \sim 0.5$ ha for coot, 1 ha for Great crested grebe.

8. Whitcomb et al. (1976) censused bird species during the
breeding season in 25 isolated woodlands of eastern North America.
Eight species, termed "edge species", occurred regularly in small
(1 - 4 ha) woods and on the edge of medium-sized (6 - 14 ha) woods
but were virtually absent from large (30 - 800) woods. Twenty-two
species occurred in woods of all sizes studied (1 - 800 ha). Fourteen species, most of them long-distance migrants, were virtually
absent in small woods, occurred at low density in medium-sized woods,
and at high density in large woods.

9. Galli et al. (1976) censused birds in eastern North America
(New Jersey) on 31 forest plots ranging in area from 0.01 to 44 ha.
Incidence increased with area for most insectivorous species and
for the sole hawk species observed. Minimum areas below which J
was 0 varied among species from 0.8 to 10 ha and were larger for
nonpasserines than for passerines.

10. Fleming (1975) and Diamond (unpublished) examined distributions of native New Zealand bird species in native forests of
different sizes. The smallest or most disturbed or most isolated
forest fragments contained only species that invaded New Zealand
from Australia in evolutionarily recent times, as judged by having
differentiated not at all or only at the subspecies or allospecies
level (Rhipidura fuliginosa, Gerygone igata, Zosterops lateralis).
Medium-sized forests contained in addition some forms endemic at
the species or even genus level (Petroica macrocephala, Anthornis
melanura). Members of old endemic genera and families, such as
Mohoua ochrocephala, Acanthisitta chloris, Prosthemadera novaesee-
landiae, Callaeas cinerea, and Nestor meridionalis, were mainly
confined to large tracts of native forest.

11. Diamond (unpublished) tabulated distributions of New
Guinea's 194 montane bird superspecies on seven disjunct mountain
ranges of New Guinea. At one extreme, forty-two species, including
seven birds of paradise, were confined to the largest and highest
range. Fifty-three additional species occurred on this range plus
one or both of the next two largest and highest ranges, but on no
smaller range. At the opposite extreme, 10 species, such as the
rail Rallicula [forbesi] and the warbler Sericornis arfakianus,
occurred on all seven ranges.

12. Johnson (1975) summarised distributions of 92 montane
bird super-species on 31 mountain ranges of western North America.
Fourteen species occurred on 28 or more ranges, three of these
species (Hairy woodpecker, Mountain chickadee, Cassin's finch) on
all 31 ranges. About 25 species were on the largest range(s) but
were absent from all ranges too small to hold more than 40
breeding species.

   13. Brown (1971) reported distributions of 13 small, flight-
less, high-altitude mammals on 17 mountain ranges rising out of the
Great Basin desert of western North America. One species, Lepus
townsendi, was confined to the range with the richest fauna, while
at the opposite extreme the chipmunk Eutamias umbrinus occurred on
14 of the 17 ranges. In general, herbivores, small species, and
habitat generalists occurred on more and smaller ranges than carnivores, large species, and habitat specialists.

                    WHY DO AREA REQUIREMENTS EXIST?

As illustrated by the studies just summarised, the incidences of
most species increase with area, for at least five different
reasons. We begin by discussing the types of reasons that will
first occur to most people, then we discuss less often considered
but similarly important effects.

   1. Habitat. Each species has its characteristic habitat
requirements. The likelihood that any given habitat type will be
encountered on a census plot increases with the plot's area. In
addition, certain habitats cannot occur at all until an island
exceeds a certain area - e.g., high elevations and large rivers can
exist only on large islands. To cite a few of the numerous examples
of such effects in the 13 case studies just considered, some of the
New Guinea montane bird species confined to the largest mountain
ranges are species of alpine habitats above 10,000 ft., habitats
that exist only on the largest and highest mountains; and the
kingfisher is confined as a breeding species to British islands
large enough to have streams.   Species that simultaneously require
two habitats for different purposes may also require large areas -
e.g., the Red-winged blackbird and Yellow-headed blackbird of
North America, which build their nests in marshes but do much foraging on the ground. Conversely, Rock pipit and oystercatcher require
only rocky shores and can therefore live on even the smallest
British islands.

   2. Territory size. Given appropriate habitat, a species
requires an area of this habitat at least equal to the territory
size of one pair in order to maintain even the smallest possible
breeding population. For mobile species that disperse efficiently
in the interval between each breeding season and can flood potential
breeding sites with colonists each year, incidence rises rapidly
from 0 to 1 with increasing area as area approaches the territory
size. Thus, Moore and Hooper (1975, p. 247) noted such a correspondence for some bird species in their studies of British woods. If
a species can utilise adjacent habitats in addition to the habitat
patch studied, it can then be found in a study patch smaller than

one territory. This is why Whitcomb et al. (1976) found certain field-foraging species in the smallest North American woods, and why Moore and Hooper (1975) found blackbird and Wood pigeon in the smallest British woods and the land-feeding, water-nesting moorhen in the smallest ponds.

3. Temporal integration. Species that seek food in different places at different times must integrate resources over large areas. The simplest case involves regular seasonal movements each year, as of tropical fruit and flower-feeding bird and bat species that utilise different tree species in different months (cf. Table I of Crome 1976, showing that in a 23-month study in Australia, 24 different fruiting tree species contributed 10% or more of the cassowary's diet in at least one month, and that no tree species contributed over 10% of the diet in more than five of these 23 months). Ungulates whose seasonal movements are well-known include antelopes in the Serengeti of East Africa and caribou of the North American arctic. Ornithologists are becoming increasingly aware that population crashes of some European and North American long-distance migrant bird species are due to habitat destruction or pollution in their tropical winter quarters, so that area requirements encompass two continents - cf. the extinction or virtual extinction of Bachman's warbler due to habitat destruction on Cuba, the decline of Kirtland's warbler despite rigorous protection of its Michigan nesting ground but possibly due to habitat destruction in the Bahamas, the crash of the whitethroat in Britain due to the Sahel drought, and reproductive failure of peregrines on unpolluted Alaskan rivers due to pesticides on South American wintering grounds. Another case of large area requirements due to temporal integration, still within the life-span of an individual but on a time-scale of more than 1 year, involves some montane birds of paradise, of which young males up to several years of age live at altitudes up to thousands of feet below adult males (Diamond 1972a, p. 31). Finally, temporal integration by a whole population, on a time-scale longer than an individual's life-time, causes persistence of the population to be dependent on availability of "hot spots" of high local resource production, which enable the population to survive crashes in occasional bad years (Diamond 1975a, pp. 369-371).

4. Immigration-extinction equilibria. It is obvious why presence of a species requires an area equal at least to the territory size of one pair. For some species, such as the raven on British islands, incidence approaches 1 for islands or habitat patches little larger than one territory. For other species, however, incidence is negligible until one reaches islands or patches equal in area to thousands of territories. How can this be?

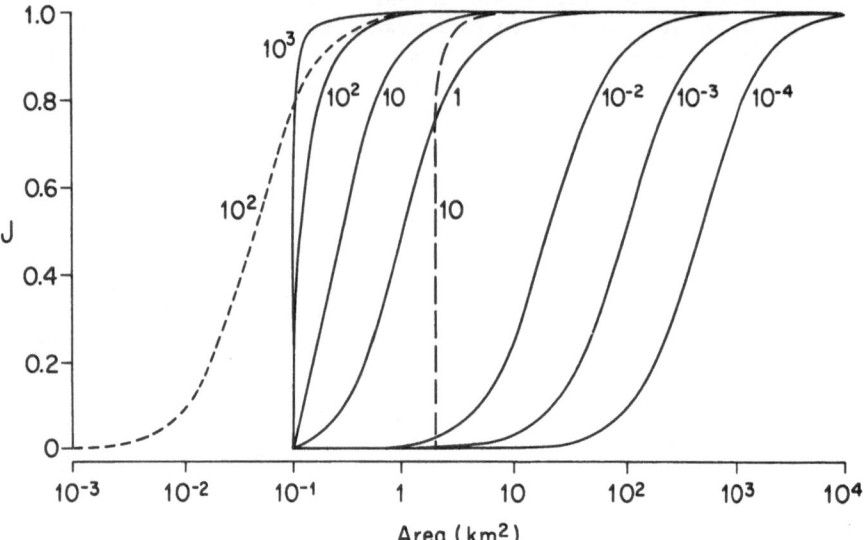

Figure 3

Individuals are not immortal, so that populations fluctuate in numbers. The smaller a closed population is, the higher is the probability per unit time that it will become extinct, due to deaths of all individuals of one sex within a short time. Extinctions can be reversed by immigrations from the outside. The higher the immigration rate is, the shorter will be the interval between extinction and recolonisation, and the higher will be the incidence for a given area or a given maximum population size. This is illustrated by Fig. 3, which plots the incidence-area relation predicted theoretically from a simple model of immigration-extinction equilibria, assuming extinction rates to vary inversely with area.

The practical message of Fig. 3 is that sedentary species require large areas to have any reasonable prospect of long-term survival. This is especially true for species of continental tropical rainforest, of which individual reserves should exceed 250 km^2 to have long-term conservation value for birds (Terborgh 1974, Diamond 1975b). But a surprising and important conclusion from the studies of Whitcomb et al. (1976) on eastern North American forest birds is that this message also applies in temperate-zone forests to many long-distance migrants, whose annual journeys of thousands of km between wintering and summering grounds may conceal a high degree of philopatry to their summer breeding site.

5. Extinction disequilibria. If immigration rates are zero, then it is only a question of time before even the largest population with the most extensive range fluctuates out of existence. At any given elapsed time after cessation of immigration, incidence will still increase with area, as in Fig. 3. However, the relation now represents an extinction disequilibrium instead of an immigration-extinction equilibrium, and the value of incidence for a given area decreases with time (Diamond 1972b; Terborgh 1974). The distributional patterns discussed above for small mammal species on North American mountains, for most mammal species on Bass Straits islands, and for some bird species on New Guinea satellite islands represent such disequilibria. These sedentary species colonised islands or mountaintops during Pleistocene periods of low sea-level or of expanded boreal habitat, but then were left stranded by rising sea-level or warmer climates without possibility of further immigration, and the stranded populations have gradually been succumbing to extinction. Extinction has been more rapid on smaller islands and for less abundant species. For instance, the wider distributions of herbivores, small species, and habitat generalists in Brown's study of North American mammals and in Hope's study of Bass Straits mammals reflect the greater abundance and lower extinction rates of such species.

The practical message of these disequilibria is as follows.
If, after a habitat has been fragmented into isolated reserves, one
finds a species still to have a large population and a large range,
this does not guarantee a secure future for the population unless
immigration is adequate to maintain the population in the face of
fluctuations. The extinctions that followed forest fragmentation
within a few decades in New Zealand reserves (Fleming 1975) and on
Panama's Barro Colorado Island (Willis 1974) warn us how rapidly
these disequilibria can collapse to their final solution.

REFERENCES

Brown, J. H. 1971. Mammals on mountaintops: nonequilibrium insular
    biogeography. Am. Nat., 105, 467-478

Crome, F. H. J. 1976. Some observations on the biology of the
    Cassowary in northern Queensland. Emu, 76, 6-14.

Diamond, J. M. 1972a. Avifauna of the Eastern Highlands of New
    Guinea. Nuttall Ornithological Club, Cambridge, Mass.

Diamond, J. M. 1972b. Biogeographic kinetics: estimation of
    relaxation times for avifaunas of southwest Pacific islands.
    Proc. nat. Acad. Sci. U.S.A. 69, 3199-3203.

Diamond, J. M. 1975a. Assembly of species communities. In: Ecology
    and Evolution of Communities, edited by Cody, M. L. and J. M.
    Diamond, 342-444. Cambridge, Mass: Harvard Univ. P.

Diamond, J. M. 1975b. The island dilemma: lessons of modern
    biogeographic studies for the design of natural preserves.
    Biol. Conserv., 7, 129-146.

Diamond, J. M. and Marshall A. G. 1977. Distributional ecology
    of New Hebridean birds: a species kalaidoscope. Submitted
    to Evolution.

Fleming, C. A. 1975. Scientific planning of reserves. Forest
    Bird no. 196, 15-18.

Ford, J. 1963. The reptilian fauna of the islands between Dongara
    and Lancelin, Western Australia. West. Aust. Nat. 8, 135-142.

Galli, A. B., Leck, C. B. and Forman, R. T. T. 1976. Avian dist-
    ribution patterns in forest islands of different sizes in
    central New Jersey. Auk 93, 356-364.

Goldstein, E. L. 1975. Island biogeography of ants. Evolution
    29, 750-762.

Johnson, N. K. 1975. Controls of number of bird species on montane
    islands in Great Basin. Evolution 29, 545-567.

Mayr, E. and Diamond, J. M. 1976. Speciation in the birds of
    Northern Melanesia. Bull. Mus. comp. Zool., in preparation.

Moore, N. W. and Hooper, M. D. 1975. On the number of bird species
    in British woods. Biol. Conserv. 8, 239-250.

Morse, D. H. 1971. The foraging of warblers isolated on small
    islands. Ecology 52, 216-228.

Terborgh, J. W. 1974. Preservation of natural diversity: the
    problem of extinction-prone species. BioScience 24, 715-722.

Whitcomb, R. F., Lynch, J. F., Opler, P. A. and Robbins, C. A,
    1976. Island biogeography and conservation: the limitations
    of small preserves. Science, N. Y., in press.

Willis, E. O. 1974. Populations and local extinctions of birds
    on Barro Colorado Island, Panama. Ecol. Monogr. 44, 153-169.

DISCUSSION: PAPER 2

R. W. WEIN	I have been concerned with establishing ecological reserves in Canada, and the work described appears highly relevant to the choice of such sites. How do critical areas vary with latitude? Does one need larger or smaller reserves in the tropics than in colder regions?
J. M. DIAMOND	Many factors influence the situation for any species. Tropical species are generally more sedentary: as a result their incidence (J) tends to be lower per unit area. If this holds generally, larger reserves would be needed in the tropics.
R. W. WEIN	In my view the opposite is the rule. If a tropical species is sedentary, a small unit is sufficient for protection. In subarctic and arctic areas, species are often migratory (e.g. caribou) and only the protection of the entire migratory route may protect the population in question.

A. D. BRADSHAW   I have heard it said that at Barro
Colorado the area requirements were not
necessarily directly those of the species that
were lost, but those of predators whose loss
modified the overall situation. Can this be
expanded on?

J. M. DIAMOND   The causes of extinction may be related
to the area requirements of a predator that
in turn affects the likelihood of extinction.
This is illustrated by what happened on Barro
Colorado Island. A disproportionate number of
the birds lost from Barro Colorado during
the first sixty years after it became an island
were ground-nesting species. Probably this
loss was due to predation by monkeys, which
increased when big vertebrate predators like
jaguars were lost because the island was too
small to support them. The hypothesis is thus
that the loss of the large mammals allowed an
increase in smaller mammal predators that in
turn affected the birds. It illustrates the
need to examine the whole system in such cases.

N. POLUNIN   Although much impressed by Dr Diamond's
work I am apprehensive at the lack of referen-
ces by zoogeographers, animal ecologists and
many others to plants on which whole ecological
systems after all depend. Surely vegetation
processes and plant population dynamics are
fundamental to all this work - for island
dilemmas are often plant (or plant habitat)
dilemmas.

J. M. DIAMOND   My paper does omit discussion of plants,
but the principles certainly should be applic-
able. The trouble is lack of good examples
from the plant kingdom. We need more data.

R. M. MAY   Phenomenological species-area studies do
exist for plants (see, for example May, 1976,
p. 164). It is the detailed information about
incidence functions and the like that is
lacking.

J. M. DIAMOND   A good deal of information would be needed
before the patterns could be analysed, and
the situation in plants may be more complex:
life cycles are very different and vegetative

spread introduces a new dimension. Animal biogeography has only expanded into this field recently, when biogeographers became prepared to look for overall patterns rather than intricate details of ecology. Botanists may also need to look at the general perspective of distribution - do more "dirty" science, instead of being preoccupied with detailed ecological analyses.

L. M. TALBOT

The area a species requires may vary even in one region, because of the interplay of many factors. In East Africa the Ngorongoro Crater, 300 square miles in extent, supports a wildebeeste population. The adjacent Serengeti Plains support a different population of the same species which ranges over 15,000 square miles. This shows how the same species can require vastly different areas if the habitat varies. The important factor is how well the area involved provides the basic resources for the species, and this is independent from the size of the area, as the Serengeti-Ngorongoro example demonstrates. An associated factor is the strategy which the species utilises to obtain its resources from a habitat - and this too is independent of the size of the area. The point is that this is not a linear relationship between species and area required. Each situation requires independent analysis.

M. W. HOLDGATE

The time scale is obviously important in these analyses. The area needed to provide a reasonably secure refuge for 50 years may differ from that needed for 1,000 years, over which period the amplitude of environmental fluctuation can be expected to be greater.

J. M. DIAMOND

Temporal fluctuations can be assessed by repeated surveys, from which we can calculate the fraction of a given period of years for which a species is present on a given island. The question you raise about survival for long periods can be assessed by examining islands that were connected to mainlands by land bridges at Pleistocene times of low sea-level up to 10,000 years ago - for example, Britain connected to Europe, Java to Asia, Trinidad to South America, etc. When the post glacial

rise in sea-level severed the land bridges,
populations of animals unable to cross water,
like the Javan rhinoceros and innumerable
other populations, were stranded in isolation,
and have been gradually fluctuating out of
existence at rates that vary inversely as
island area. These patterns suggest that an
area of several hundred km^2 is required to
save even a small fraction of a rich tropical
rain forest avifauna like that of New Guinea
or South America for thousands of years. To
save even just half of such an avifauna requires
an area of thousands of km^2.

Further Discussion of Critical Areas for Plants

In a discussion session on Monday evening, 5 July, Mrs. Balfour
and Professors Bradshaw, Diamond, Godron, and Margaris discussed
the concept of critical area as applied to plants. It was
concluded that plants and birds may tend to differ in two respects
with regard to critical areas:

1. Critical areas for some plant species may be smaller than
for some bird species, insofar as some plant species live at population densities (measured in individuals per hectare) far higher
than any bird species can attain. However, rainforest tree species
generally occur at low densities and will tend to have large area
requirements.

2. An individual tree generally lives much longer than an
individual bird, even under conditions where the tree cannot
reproduce. Thus, from occurrence of a tree species in forest tracts
of a certain size, one cannot conclude that the tract exceeds the
critical area unless one has determined that the tree is also
fruiting and producing seedlings in the tract.

3.   MODELS AND THE FORMULATION AND TESTING OF HYPOTHESES IN

GRAZING LAND ECOSYSTEM MANAGEMENT

G. M. VAN DYNE, L. A. JOYCE, B. K. WILLIAMS

Department of Range Science

Colorado State University, Fort Collins, Colorado USA

INTRODUCTION

The purpose of this paper is to describe different levels of hypotheses used in building and testing grazing land models, with
emphasis towards ecosystem managment. A large-scale, total-system
grassland simulation model is used as a point of focus. A brief
description is provided of the structure of the model and results
from example experimental runs testing management-oriented hypotheses are given.

SOME EXPERIENCES IN MODELLING GRAZING LANDS

Models play a particularly important role as a synthesis tool by
providing a mechanism for combining diverse experimental results
and integrating them with the literature. Another major value of
models is in research; if a model which simulates certain
phenomena can be sufficiently validated to provide confidence in
its output, it can be used for experimentation on those phenomena.
A rich testing ground is provided on which to examine hypotheses
at a different level from that of the hypotheses used in the construction of the model.

More than 100 simulation and optimisation models of grazing
land systems have been constructed and reported in a diverse
literature within the last two decades. Several independent but
complementary approaches to modelling and analysis of grazing land
and grassland systems have grown from International Biological
Programme studies throughout the world. Van Dyne and Abramsky (1975)

compared many of these and related models and specified whether they were simulation or optimisation models, whether they were based on difference or differential equations, what the time steps were for the models, what the general characteristics were for the driving variables and state variables, and related information.

In general, the magnitude of effort required for modelling has been underestimated in planning most such studies. For example, for one large-scale model of a shortgrass prairie system in North America, the development of a total-system, multiple-flow model involved inputs primarily from some 12 scientists. The modelling project occupied the majority of their time for several years, a total of more than 20 scientist years being devoted to the modelling work alone (Van Dyne and Anway, 1976). The documentation of that large model in terms of formal publication (Innis, 1977) and in internal documents (Cole et al. 1976) was a large and new experience in grazing land research. During the long development of the over- all model there were numerous changes in personnel and it is not surprising that full evaluation and use of that and other similarly developed models has not occurred.

Relatively little has been done in detailed review of the development, validation, or verification of models. The process of modelling grazing land and grassland systems has not been carefully documented. Particularly, the utilisation and testing of hypotheses related to ecosystem management has not been clearly specified.

HYPOTHESES AND MODELS

Hypotheses are formulated by inference from observation of data and nature. As such, hypotheses are expressions of natural principles. They are also used as tentative assumptions made in order to develop and test logical and empirical consequences. The formulation of hypotheses is a first step to the eventual development of a theory based on a set of azioms and a set of theorems or postulates, all of which would be mutually reinforcing.

Essentially there are three overlapping areas of hypothesis which may be constructed about ecosystems: (i) those concerning organs or species, (ii) those at the population or community level, and (iii) those at the total-system level. Hypotheses at the first level derive from ecophysiology studies and characterise the process functions in many grazing land ecosystem simulation models. Hypo- theses at the second level are the subject matter of studies on population dynamics and species interaction. Hypotheses at the third level result from comparative ecosystem studies. Each group includes some hypotheses with management implications and some which contribute to ecological theory.

An hypothesis can be rejected but not proved. Ideally, to evaluate an hypothesis one should use independent data or independent experience from that used in structuring the model. "Experience" is the overall information one has based on training and work. And, it may include both quantitative and qualitative information. In contrast, the term "data" is meant to represent published numerical or graphic relationships among variables.

A mathematical simulation model is in effect a "macrohypothesis" composed of many interconnected hypotheses. The hypotheses of which a model is constructed are generally of a low level of resolution, ie about individual processes and flows.

A model may be constructed from either data, experience, eco-logical principles, or a combination of these. A simulation model requires a set of driving variables and initial values for the state variables. The output of the model may be compared with separately collected data or separately obtained experience. If the model is to be tested against the experience of individuals independent of those who constructed the model, it is useful to structure the experience into hypotheses. Thus, individual scient-ists can be brought together to derive hypotheses based on their experience and on examination of the resource management-related literature rather than utilising the specific field and laboratory process data or literature from which the model is constructed. In some instances the only data available for testing a model are the data generated by experienced managers. This occurs in situations where no experimental control is possible over the system and thus it is impossible to collect a priori a defined set of time series data of the state variables of the system in response to management stresses. Thus, for example, models of the world, models of cities, and models of most systems involving humans as internal components must be treated in this manner.

The scientists constructing a model have essentially the same basic literature available to them as do the scientists and managers constructing hypotheses or collecting data. Therefore model con-struction and model evaluation cannot be completely independent exercises, and care must be taken to ensure as little circularity as possible in model building and evaluation. Van Dyne et al. (1976) have noted that essentially three separate groups of investigators are utilised in some large-scale modelling efforts. These groups are responsible for (i) model construction, (ii) collection of field data for model evaluation and (iii) conduction and interpretation of model experiments using the results from the previous two groups. Each group contains several members, with varied disciplinary backgrounds. The group's members interact intensively, but the kind and amount of interaction will depend upon the specific individuals composing the groups. Therefore, differences in original training, differences in

Table 1.   Example processes for hierarchical organization at the
           physiological, organismal, or population level of biotic
           components and equivalent abiotic levels of the ecosystem.

ABIOTIC PROCESSES
  heat transfer process

radiation	atmosphere	→ soil
conduction	atmosphere	→ soil
convection	atmosphere	→ soil
radiation	atmosphere	→ plant
conduction	atmosphere	→ plant
convection	atmosphere	→ plant
radiation	atmosphere	→ animals
conduction	atmosphere	→ animals
convection	atmosphere	→ animals
conduction	soil layer	→ soil layer
conduction	soil surface	→ animals
radiation	soil	→ plant

  water transfer processes

precipitation	atmospheric water	→ surface water
evaopocondensation	atmospheric water	→ surface water
evaporation	soil water	→ atmospheric water
interception	atmospheric water	→ plant surface water
snow melt	solid water	→ liquid water
infiltration	soil surface water	→ soil layer water
runoff	soil surface water	→ stream channel water

nutrient transfer processes

N_2 fixation	atmosphere	→ soil NH_4^+ and NO_3^-
mineralization	soil organic matter	→ soil NH_4^+
immobilization	soil NH_4^+	→ soil organic $\overset{P}{matter}$
exchange	soil mineral lattice	→ soil NH_4^+ food
oxidation	NH_4^+	→ NO_3^-
humification	dead $root_P^N$	→ soil organic $matter_P^N$
ammonification	dead $root_P^N$	→ soil NH_4^+
humification	litter N	→ soil organic matter
transformation	insoluble P	→ soluble P
CO_2 diffusion	soil layer	→ soil layer
mineral decomposition	CO_2 soil storage	→ soil solution CO_2
buffer transformation	CO_2 soil storage	→ soil solution CO_2

PRODUCER PROCESSES

shattering	standing dead	→ litter
physical decomposition	standing dead	→ litter
shattering	standing live	→ litter
expiration	standing live	→ standing dead
expiration	live root	→ dead root
CO_2 diffusion	external atmosphere	→ stomatal atmosphere
C fixation	stomatal atmosphere	→ organic compound
translocation	leaf	→ stem
translocation	stem	→ seed
translocation	stem	→ crown storage
translocation	crown	→ root
respiration	stem	→ atmosphere
respiration	root	→ soil CO_2
chemical transformation	labile	→ nonlabile compounds
absorption	soil solution nutrient	→ live root nutrient
exudation	live root nutrient	→ soil solution nutrient

Table 1 (cont'd)

<u>DECOMPOSER PROCESSES</u>

consumption	aboveground litter (mulch) ──────► microbiota
consumption	belowground litter. humic compounds. animal residues. dead roots ──────► microbiota
expiration	live active microbiota ──────► dead microbiota
state transformation	live active microbiota ──────► inactive microbiota
expiration	inactive microbiota ──────► dead microbiota
humification	dead microbiota ──────► humic compounds
extracellular decomposition	belowground litter, humic compounds, animal residues. dead roots ──────► degraded compounds

<u>CONSUMER PROCESSES</u>

emigration	in system ──────► out of system
immigration	out of system ──────► in system
diet selection	food available ──────► food handled
ingestion	food selected ──────► food consumed
"wastetation"	food selected ──────► food wasted
storage	food selected ──────► cache
transportation	food selected ──────► carried to young
digestion	food consumed ──────► food digested
defecation	food consumed ──────► feces
metabolization	food digested ──────► food metabolized
urination	food digested ──────► urine
eructation	food digested ──────► atmosphere
transformation	food metabolized ──────► basal metabolism
transformation	food metabolized ──────► heat maintenance
transformation	food metabolized ──────► reproductive tissue
transformation	food metabolized ──────► milk
transformation	food metabolized ──────► nonfat
transformation	food metabolized ──────► fat
transformation	food metabolized ──────► integument
transition	age/sex state$_i$ ──────► age/sex state$_j$

	state$_i$	state$_j$	
1	hibernating or dipausing	2-4. 8-10, 16-18	21
2	preproductive feeding	1. 3. 4. 5. 8. 9	21
3	preproductive nonfeeding	1. 2. 4. 5. 8.9	21
4	male sexually active	1. 5. 6	21
5	male sexually inactive	1. 4	21
6	male incubating	7	21
7	male nonincubating	1. 4-6	21
8	female sexually active	1. 9, 12. 17	21
9	female sexually inactive	1. 8	21
10	female lactating	11. 16	21
11	female nonlactating	1. 8. 9	21
12	female gravid	1. 10, 13. 14. 16-18	21
13	female nongravid	8. 9	21
14	female incubating	15. 18	21
15	female nonincubating	1. 8. 9	21
16	female sexually active lactating	8. 10	21
17	female sexually active gravid	16	21
18	female sexually active incubating	8, 14	21
19	female lactating gravid	10. 12	21
20	female gravid incubating	12. 13	21
21	dead		

the abilities and characteristics of the individuals, and different interactions produce groups with "semi-independent" views regarding the system being modelled. Then, more-or-less independent sets of knowledge and ideas break the circularity loop.

HIERARCHIES AND LEVELS OF HYPOTHESES

At least three kinds of hypotheses concerning ecosystems may be recognised:

(i)   hypotheses concerning the functional nature of individual abiotic and biotic processes

(ii) hypotheses concerning response of a given ecosytem to various stresses

(iii) hypotheses comparing general responses of a class of ecosystems.

Hypotheses of Subsystem Function

To make comparative analyses of the hypotheses used in constructing either empirical or mechanistic flows in a simulation model a list of processes is necessary. Such a list can be developed in the biotic portion of the ecosystem at the physiological, organismal, or population levels of organisational resolution. Similarly, processes may be structured at equivalent levels of resolution in the abiotic portion of the system. A preliminary list of such processes is provided in Table 1 (Van Dyne et al. 1976).

In order to make comparative analyses of the representation and utility of hypotheses describing processes in different grazing land models, it is necesssary to develop a hierarchial structure among such processes. Depending upon the purpose of a model, the modeller may form hypotheses about system structure and function at different levels of resolution. Hypotheses, and thus processes, at each lower level of resolution are designed so as to nest within the next higher level.

The mathematical representations of flow processes lend themselves to analysis by such questions as the following:

(a)   For a given process, what driving variable and state variable values are required in the calculation?

(b)   Over what range of values of the driving variables and state variables can a provess function given meaningful values?

(c)   How are the variables combined within the mathematical representation of the process hypothesis, i.e. what is the functional form?

(d)   Are the process functions nonlinear, do they include time lags, do they include discontinuities and thresholds, etc.?

To begin to answer such questions as above, it is useful to organise information about processes in simulation models in various ways. Construction of a "coupling matrix" is a useful step. Such a matrix has as rows and columns the state variables in the system model. An entry, x or 1, in a cell of the matrix indicates there is matter or energy flow between those two variables in the model. Another useful tabularisation of information is a "flow by variable" matrix in which the rows represent the different flows in the model and the columns represent the driving variables and state variables in the model. An entry, x or 1, in a cell of the matrix indicates there is an "information flow" from that driving variable or state variable used in calculating the flow function. Another, but more difficult step, is to classify the nature of the flow calculation, such as a constant flow, a constant proportion of the donor or receiver compartment or both, a "maxiumum-redution" calculation involving several variables, a "Liebig's Law of the Minimum" approach to calculating the flow, an optimisation algorithm calculation, and so forth.

In structuring hypotheses for grazing land model systems it is useful to have hypotheses at different levels of resolution. For example, if the primary emphasis in a grazing land ecosystem model is on animal production, it may not be necessary to have "mechanistic" flows in the plant production or decomposition segments. Alternatively, if emphasis is on plant production, the animal's dietary selectivity and nutrient metabolism segments may not require a great deal of "mechanism". Thus, it is valuable to have available a library of "empirical" flow functions. These functions perhaps should account for 75% or more of the variability in the flows. To our knowledge, detailed descriptions have not been published of grazing land system models decoupled into the individual flows or complexes of flows to account for process functions.   The flows thus derived would vary from "mechanistic" on the one hand to quite "empirical" on the other hand, and a documented library of hypotheses about these rate processes would be extremely useful in grazing land modelling. It would also be extremely useful to have a parallel "library of data" to evaluate individual flow process hypotheses. This is particularly important since it is extremely expensive to collect a full set of state variable and driving variable measurements. Such a complete set of information would be useful in model building and evaluation. However, it is possible to evaluate separately the individual hypotheses about the rate processes. These evaluations can be done singly before the individual processes are coupled

Table 2. A checklist of major ecological phenomena to be examined in testing system—stress related hypotheses. (See also Table 4.)

Ecological Phenomena	Grazing Intensity	Grazing Season	Grazing Animal Species	Insect Control	Herbicide Application	Water Manipulation	Fertilization	Predator Control	Fire Impacts	Air Pollution	Ionizing Irradiation	Pitting & Interseeding
Primary production	x	x		x	x	x	x	x	x	x		x
Secondary production	x	x	x	x		x	x	x	x	x	x	x
Dietary quantity & quality	x											
Nutrient cycling rates	x						x	x	x	x		
Biophage: saprophage shifts	x											
Large: small mammal completion	x							x				x
Grazer v. browser efficiencies			x									
System lag and pulse responses				x	x			x	x			
Temporal variability due to climate						x	x					x
Shoot: root ratios						x	x					
C_3 v. C_4 production										x		
Subsystem diversity								x				
Population dynamics feedbacks								x				
Water—use efficiency									x	x		
Phenophase progression											x	

together. Of course, the box and arrow diagram of the system is
an hypothesis itself of the major functional and structural inter-
relationships.

Ecosystem Management-Oriented Hypotheses

Ecosystem management-oriented hypotheses are specific to a
particular ecosystem, and they concern the system's response to
perturbations. These hypotheses are often generated by researchers
and managers who base them on personal experience or examination of
the scientific literature. This method of hypothesis formulation
obviates the circularity implicit in hypotheses developed from a
direct examination of field data from the ecosystem.

For example, scientists have met in several workshops to
structure hypotheses on grazing land responses to stress. Many
of the scientists involved in these exercises had conducted field
measurement studies in the shortgrass prairie which provided time
series of state variables under different treatments. Several of
these scientists made general, brief reviews of literature and
derived conclusions of the expected major response of the shortgrass prairie ecosystem to the main stresses noted. These management-oriented hypotheses are used to structure experiments to be
run on a large-scale simulation model of this grassland ecosystem.
A checklist of major ecological phenomena to be examined in such
hypotheses tests is provided in Table 2. Here an entry into the
matrix of the table indicates that a particular ecologicial
phenomenon will be examined in evaluating a particular type of
system stress. The system stresses vary from grazing, to pesticide
application, to agronomic treatments, and others.

The evaluation of these hypotheses will provide a test of the
utility of the model. But to test any given hypothesis will require
careful consideration of the number of experimental runs to be made
of the ecosystem model. The model experimental runs required to
test the hypotheses regarding grazing as listed in Annex I are shown
in Table 3. Thus, the 13 hypotheses regarding grazing will require
only 16 model runs for their evaluation. By judicious selection of
experimental designs, some treatments can serve more than one
hypothesis test.

For each set of experimental runs to test a given group of
hypotheses, certain model output values are required. These can be
selected from a large number of variable time traces or rate process
values available from the output of a model run. For a large-scale
model such as the one with which we are working there may be as many
as 1,500 such time traces. The model, described further below,
requires the following variables as driving variables: daily rain-

Table 3. Example hypotheses concerning management manipulations of
          shortgrass prairie for testing with advanced versions of
          the ELM prairie ecosystem model. Domestic livestock
          referred to herein are cattle and sheep. Time referred to
          herein is the Julian date.[1]

HYPOTHESES CONCERNING GRAZING INTENSITY

* Increasing grazing intensity by domestic livestock during the
  growing season on shortgrass prairie will at first increase
  levels of net primary production. but subsequently will decrease
  it.

* Increasing grazing intensity will decrease gains of individual
  animals, but not in a linear manner. having less impact at first
  than at higher intensities.

* Increasing intensity of grazing will at first increase, then
  decrease gains per unit area for domestic animals.

* Increasing grazing intensity will alter dietary selectivity and
  diets will become increasingly different from preferred diets.

* Increasing grazing intensity will slowly but continually decrease
  dietary quantity.

* Increasing grazing intensity will increase the nutrient cycling
  rate, on an annual basis, for both nitrogen and phosphorus.

* Increasing grazing intensity will decrease the flow of energy as
  respiration by decomposers as compared to lower intensities of
  grazing.

* Increasing grazing intensity of domestic animals will alter the
  competition relationship with small herbivores; at light levels
  of intensity grazing there will be no dietary competition, as
  reflected in domestic animal performance, when small herbivores
  are removed, but at higher grazing intensities, removal of small
  herbivores will increase secondary production of domestic
  animals.

Table 3 (cont'd)

HYPOTHESES CONCERNING SEASON OF GRAZING

* When the grazing season is increased in length, but centered about
   the center of the plant growing season (about day 200 based on
   the thermal growing season)and stocking rate is held constant
   (based on metabolic equivalents), both animal production and
   plant production are decreased.

* Under the above conditions, domestic animal production will be
   influenced to a greater extent than will be net primary
   production.

HYPOTHESES CONCERNING SPECIES OF GRAZING ANIMALS

* Grazers (cattle and bison) are more efficient large herbivores
   than are browsers (sheep and antelope) on the shortgrass prairie
   where efficiency is expressed in secondary production per unit
   of digestible organic matter consumed.

* Native herbivores (bison and antelope) are more efficient than
   domestic herbivores (cattle and sheep) on a yearlong basis, but
   less efficient during the normal grazing season for domestic
   animals.

* A mixture of large herbivores, where the relative numbers of
   animals of different species in the mix is related to the
   predicted available herbage in relationship to dietary
   preferences, are more efficient than is any one of the
   individual herbivores.

HYPOTHESES CONCERNING GRASSHOPPER POPULATIONS

* Complete control of normal grasshopper levels will not have a
   significant effect on the ecosystem as a whole in the year of
   control. But grasshopper control will have an effect the
   following year by reducing levels by 75% at midyear due to
   reduction of overwintered eggs.

* Insecticides will affect only the late stage nymphs and early
   adults and the result will be within one to two weeks following
   application with little residual influence.

Table 3 (cont'd)

HYPOTHESES CONCERNING HERBICIDE APPLICATIONS

* Herbicide application will have a long-term (two-year) affect on
  cactus but not on forbs, shrubs, and grasses.

* Herbicide control, influencing cactus in one growing season, will
  give grasses the competitive advantage.

HYPOTHESES CONCERNING IRRIGATION, NITROGEN, AND PHOSPHORUS
                        TREATMENTS

* Increasing additions of nitrogen will increase forage production
  and carrying capacity up to 50%. Beyond this level both will
  decrease and during dry years both may decrease below the control
  levels.

* The addition of phosphorus fertilizer even under conditions of
  high levels of water and nitrogen will have no effect on forage
  production or carrying capacity.

* Increasing the length of time during the growing season. during
  which the soil water potential is maintained at -0.8 bars, will
  increase forage production and carrying capacity in a linear
  manner.

* Irrigation and nitrogen fertilization of the shortgrass prairie
  will increase the shoot-root ratios and increase the amount of
  belowground biomass relative to the control.

* Nitrogen fertilization alone will increase total decomposition at
  low levels and decrease it at high levels. Increased diversity
  of plant biomass as levels of nitrogen increase, and increased
  ratio of cool-season plant biomass to warm-season plant biomass
  with increasing levels of nitrogen.

* Any treatments that increase net primary production will result
  in a proportionally smaller increase in secondary production.

* Irrigation of the shortgrass prairie will increase total decomposition, will not change diversity, will increase the
  biomass proportion of warm- season grasses as the length of
  irrigation increases.

* Water plus nitrogen will increase total decomposition, decrease
  diversity, and decrease the ratio of cool- season to warm-
  season plant biomass.

Table 3 (cont'd)

HYPOTHESES REGARDING PREDATION

* Prey species population levels, in terms of numbers and biomass,
  control predator species population levels.

* If precipitation is reduced by 25% for two years there will be
  lower jackrabbit and kangaroo rat population levels by 50% and
  a subsequent reduction of coyote levels by 25%. This subsequently will result in increase in grasshopper and other
  insect populations and therefore levels of prairie deer mice,
  13-lined ground squirrels, and grasshopper mice will remain
  relatively constant.

* If precipitation is increased by 25% for two years there will be
  higher levels of kangaroo rat population by 50% and jackrabbit
  population by 25% with subsequent increase of population levels
  of coyotes by 20%. The increased biomass of primary producers
  will subsequently increase the prairie deer mouse, 13-lined
  ground squirrel and grasshopper mouse populations by 20%.

* Predation levels control certain prairie species population
  levels over the short run.

* Eliminating all the predator species (coyote. grasshopper mouse,
  13-lined ground squirrel. and prairie deer mouse) from the
  system for one year will result in at least a 10% increase in
  total prey biomass with a subsequent return to nominal values.

* Doubling the nominal predator levels (of species noted above) will
  result in at least a 20% initial decrease in total prey biomass
  with a subsequent return to nominal values. This will result in
  an initial increase in primary producer standing crop by at
  least 5% with a subsequent return to normal.

* Dietary selection by predators shifts to take advantage of the
  most abundant species, thereby reducing impact on prey species
  populations which are low or declining and increasing impact on
  prey species populations which are high or increasing.

* Reducing lagomorph populations to 10% of normal levels will result
  in a reduction of the lagomorph in the coyote diet to 25% of
  normal levels.

* Doubling lagomorph populations will result in an increase in
  lagomorph proportions of the coyote diet to 110% normal levels.

* Predator species population levels are maintained at their normal
  level due to only partially overlapping diets.

Table 3 (cont'd)

* Removing the grasshopper mouse and the prairie deer mouse will
  cause a 25% increase in the 13-lined ground squirrels;
  removing the 13-lined ground squirrel and the grasshopper mouse
  will result in only a 10% increase in the prairie deer mouse;
  and removing the prairie deer mouse and the 13-lined squirrel
  will result in only a 25% increase in the grasshopper mouse.

* As plant cover increases, the vulnerability of specific prey
  decreases while the number of prey species significant in the
  diet will increase.

* Increase in plant cover by 50% should result in an increase of
  species diversity in the coyote diet by 100%; decrease in plant
  cover by 50% should decrease the species diversity in the coyote
  diet by 10%.

HYPOTHESES CONCERNING FIRE

* A complete removal of aboveground herbage by fire will result in:
  an average of 3°C increase at 2.5 cm in soil temperature; an
  increase in evaporation of 24 mm; an increase in decomposition
  rates of 15% and an advance of the date of regrowth by 15 days
  for phenophase 1; an advance in the development of grasshoppers
  by 30 days; a reduction in peak standing crops of blue grama
  grass by 50, 35, and 0% in succeeding years; and an increase in
  available nitrogen by about 10% due to death and decomposition of
  roots when the tops are burned.

HYPOTHESES CONCERNING AIR POLLUTION

* Air pollutants will decrease gross photosynthesis of all primary
  producer categories by 40%. This reduced gross photosynthesis
  will result in a decrease of translocation to crowns and roots;
  an increase in biomass of litter and of standing dead; a decrease in producer biomass and production; a decrease in
  consumer biomass, grasshopper production, and mammalian
  consumer production; an increase in microbial biomass and
  production; and an increase in soil water storage.

HYPOTHESES CONCERNING RAINFALL ENHANCEMENT

* Rainfall enhancement by increasing rainfall by 10% between 1 April
  and 1 October will increase range condition by one class and
  projected carrying capacity by 10% and will add a positive 10%

Table 3 (cont'd)

on the following variables: net primary production, net belowground production, net aboveground production, peak aboveground
live vegetation, microbial response, transpiration, evaporation,
total water loss, potential evapotranspiration, secondary
production, soil water levels in the 4-15 cm layer and small
mammal populations.

HYPOTHESES CONCERNING IONIZING RADIATION

* Application of 10r/hr radiation exposure will increase mammal
  mortality by 70%, decrease mammal fertility rate by 20%, decrease
  phenophase progression of warm-season grasses by 10%, cactus by
  20%, and result in no flowering; forbs by 40% and result in no
  flowering; and shrubs by 50%; increase plant mortality of the
  above groupings.

HYPOTHESES CONCERNING BOTANICAL AND MECHANICAL MANIPULATION

* Furrowing will enhance soil water availability for plant growth
  and thus increase primary productivity.

* Furrowing will result in soil temperature increases more rapidly
  in the spring because more bare surface is exposed to direct
  solar radiation and thus result in early initiation of growth.
  This earlier initiation of growth will result in earlier
  advancement through subsequent phenological stages and increase
  seed production by assuring completion of these phases—the
  reproductive processes—before normal summer drought adversely
  influence them.

* Although there will be somewhat greater amounts of soil water
  available for plant growth in the furrow community and in the
  control, the increase in primary production resultant will be
  proportionately greater.

* Interseeding the shortgrass prairie plant community with crested
  wheatgrass will make more use of early spring precipitation more
  efficiently than in either the furrowed or controlled communities.
  This will be measured in water-use efficiency (net primary
  production/precipitation) for the spring period of April
  through June.

* Decomposition rates within the soil will be increased in those
  treatments and experience higher spring soil temperature and
  soil water levels and this will result in more nutrients
  available for plant growth.

Table 3 (cont'd)

---

* Canopy temperatures in the furrowed treatment will rise more
   rapidly in the spring than in the control treatment because
   of re-radiation from the warmer soil and this will result in
   higher net primary productivity during the cool portions of
   the growing season.

* The biomass of native vegetation on furrowed treatments will have
   a higher water content than that on control. Due to this higher
   water content the palatability will be increased and domestic
   livestock intake of forage will be increased.

* In treatments including the interseeding of crested wheatgrass,
   the ratios of treatment-to-control annual net primary production
   will be greater in "cool-season" years as compared to warm-
   season years (respectively 1971 and 1972) and furthermore the
   ratios of live stock production will vary in a similar manner.

* Higher levels of seed-eating consumer will result from furrowing
   and interseeding as compared to the toll levels for these
   consumer groups.

---

[1]We acknowledge several program participants in the formulation of
these hypotheses including John Leetham, Jack Lloyd. Phil Sims,
Bill Lauenroth, Unab Bokhari, Dave Swift, Carl Marti, Bob Packard,
Len Paur. Dale Bartos, Joe Trlica, Jeorg-Henner Lotze, Freeman
Smith, Norm French, Jerry Dodd, and Herb Fisser.

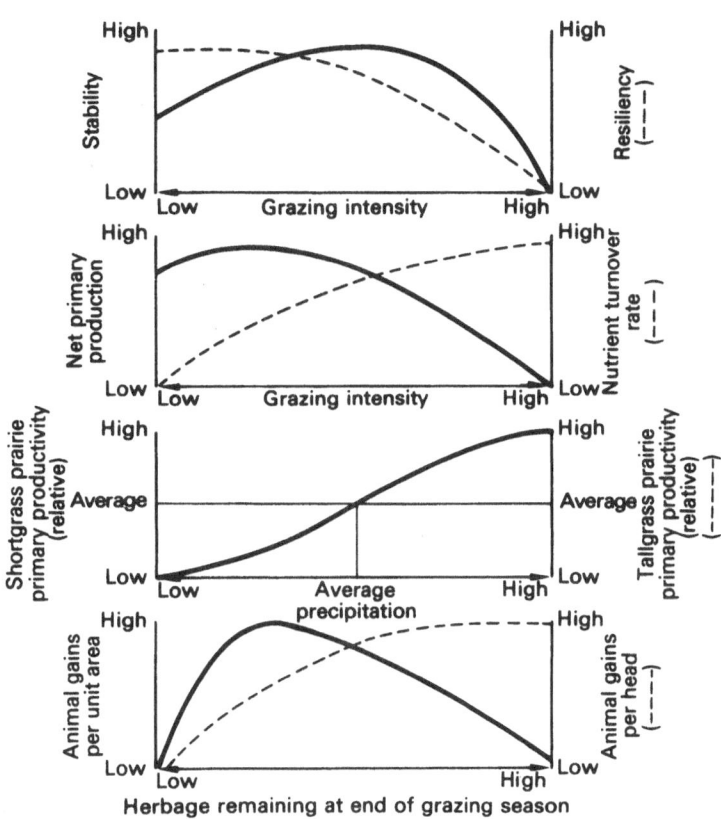

Fig. 1. Some hypotheses concerning across-grazing land comparisons.

fall (cm), incoming solar radiation not affected by clouds (ly/day),
minimum air temperature at 2m (^{o}C), maximum air temperature at 2m
(^{o}C), cloud cover (%), wind speed (mi/hr), average daily relative
humidity (%), and soil temperature at 2m (^{o}C). Table 4 lists the
state variables for the abiotic, consumer, producer, decomposer, and
nutrient submodels in the overall model. Additionally, there are
many flows whose values can also be used in model evaluation. The
state variables are expressed in various units: abiotic submodel
variables are in cm H_2O m^{-2} and ^{o}C; producer, consumer, and decomposer submodel variables are g C m^{-2}; phosphorus submodel
variables are g P m^{-2}; and nitrogen submodel variables are g N m^{-2}.
Flow values are appropriate units per time step, e.g., g C m^{-2} $(2d)^{-1}$.

The overall size of the model makes it important to select
a priori the group of time traces for variables and process rates
that one desires to utilize in evaluation of given hypotheses about
system response to experimental stress. For example, in an experiment related to application of irrigation water, nitrogen, and
phosphorus, the following outputs are to be plotted or printed or
both if marked with an asterisk:

* Root-shoot ratio at least every two weeks
* Aboveground biomass by functional groups very two weeks
* Belowground biomass by function groups and total every two
    weeks
* Nitrate and ammonia at all levels in the soil every two weeks
* Labile phosphorus at all levels in the soil every two weeks
  Total irrigation water added
  Date irrigation ends
  Total decomposition at two-week intervals
  Mean phenophase by functional groups at two-week intervals
* Nitrogen and phosphorus concentration in shoots, live roots,
    and aboveground and belowground litter very two weeks
* Aboveground and belowground litter biomass every two weeks
* Phosphorus concentration in plant groups every two weeks
  Runoff events as they occur

                    Theory-Oriented Hypotheses

In addition to the large number of hypotheses used in structuring
a grazing land model and the relatively large number of hypotheses
that can be structured about the within-grazing land responses to
stress, there is another level of hypothesis concerned with between-
grazing land comparisons. Examples of such hypotheses are illustrated in Figure 1.

This figure shows that qualitatively stated hypotheses can,
as a first step in quantification, be put into graphic form. This
shows the relative magnitude and type of relationships involved.

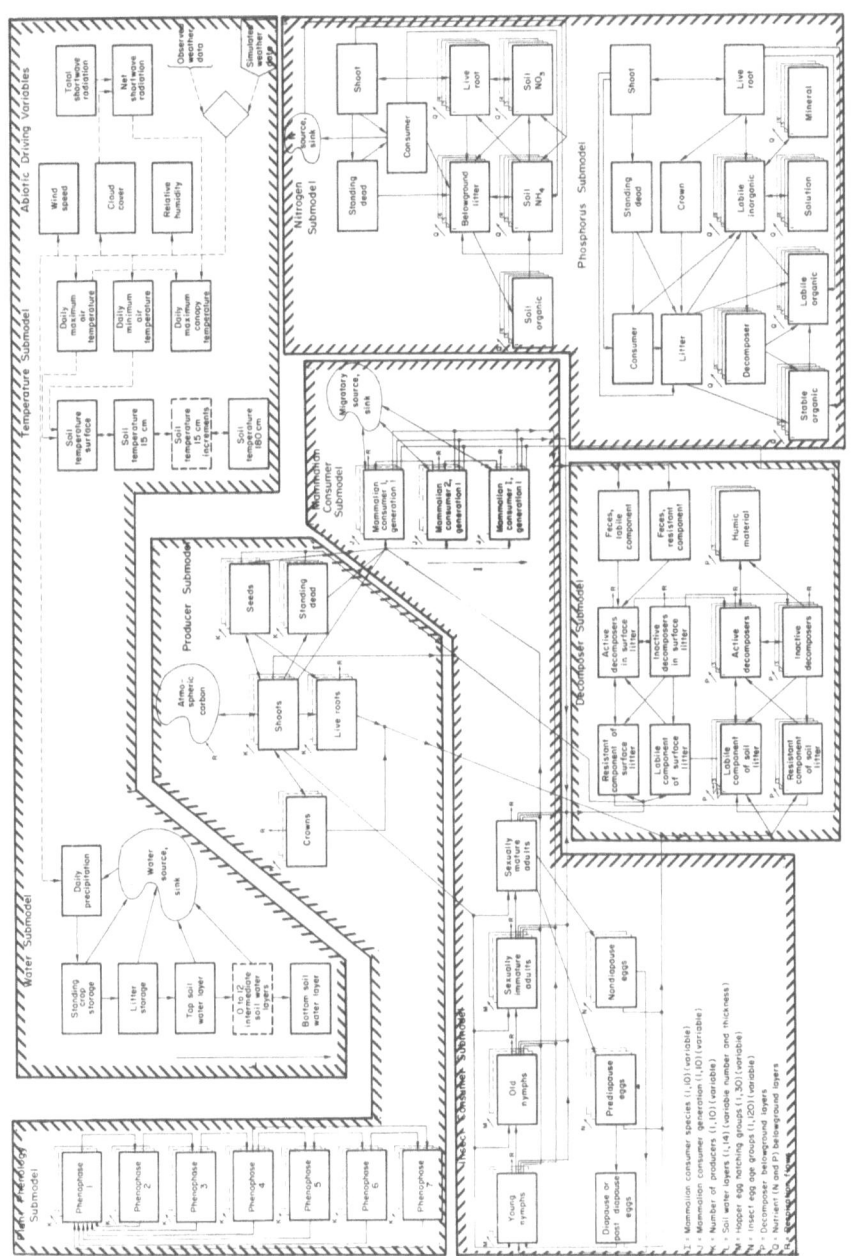

Fig. 2.   The grazing land system model is composed of submodels describing heat and water flow through the system, carbon and phenology flows in primary producers, carbon flow in decomposers and the movement of nitrogen and phosphorus in the system (adapted from Van Dyne et al. 1977).

Example hypotheses (H) are:

H: Increasing domestic animal grazing intensity will increase
      ecosystem stability to a point and then decrease stability.

H: Increased grazing intensity will tend to decrease resilience,
      i.e., the ability of the system to respond after a stress.

H: Nutrient turnover rates, or flow of nutrients through some
      point in the ecosystem will increase with grazing intensity.

H: Net primary production will increase to a point, then decrease
      with more intensive grazing pressure.

H: Water stress will have less impact on a normally semi-arid
      grassland (such as a shortgrass prairie) than on a less arid
      grassland (such as a tallgrass prairie).

H: As grazing intensity increases, gains per individual animal
      (of domestic animals) will continually decrease whereas gains
      per unit area will at first increase and then decrease.

H: As one changes the amount of herbage remaining at the end of
      the growing season there is a nonlinear response of gains per
      unit area and net primary production. Both of these curves
      reach peaks and drop off from either end. Maximum gain per
      unit area of the grazing animal will occur at a lower level of
      herbage left ungrazed than will maximum economic return per
      unit area. Maximum economic return per unit area will closely
      parallel maximum plant production.

                A GRAZING LAND ECOSYSTEM SIMULATION MODEL

The results of testing management-oriented hypotheses for a grazing
land system presented here are for a shortgrass prairie of northeastern Colorado, USA.

                          The Model System

The model that was utilised was an advanced version of a large-
scale, multiple-flow, total-system simulation model, specifically
the version ELM 1974a. Overall descriptions of a 1973 version from
the model sequence are given by Innis (1972), Innis (1975), and
Van Dyne and Anway (1976). Detailed descriptions of individual
submodels of ELM 1973 are provided by Innin (1977); code is
listed by Cole et al. (1976). The general structure of the model
is illustrated in Figure 2. Experiments with another version of

Table 4. Model experimental runs required to test hypotheses
regarding grazing. (See also Table 2.)

RUNS FOR HYPOTHESES CONCERNING GRAZING INTENSITY.

(i)   The control grazing treatment of light grazing during the
normal grazing season, days 120–300, about 50% of the year.

(ii)   Stocking densities at the same season but three times the
level of control, i.e., heavy grazing.

(iii)   Stocking densities at the same season but five times the
level of control, i.e., extra-heavy grazing.

(iv)   Control treatment, except that small herbivores are set to
zero.

(v)   Stocking at five times control level and the small
herbivores are set to zero.

ADDITIONAL RUNS FOR HYPOTHESES CONCERNING SEASON OF GRAZING.

(vi)   75% of the year, days 64–334.

(vii)   25% of the year, days 155–245.

ADDITIONAL RUNS FOR HYPOTHESES CONCERNING SPECIES OF GRAZING ANIMAL.

(viii)   Bison during the normal grazing season.

(ix)   Sheep during the normal grazing season.

(x)   Antelope during the normal grazing season.

(xi)   Cattle on a yearlong basis.

(xii)   Bison on a yearlong basis.

(xiii)   Sheep on a yearlong basis.

(xiv)   Antelope on a yearlong basis.

(xv)   A mixture of herbivores during the normal grazing season.

(xvi)   A mixture of herbivores on a yearlong basis.

this general model for a tallgrass prairie are reported by Parton
and Risser (1976). Other experimental results derived from use of
this model are reported by Van Dyne et al (1976).

The model has about 250 state variables (Table 4) and about
650 flow functions. The model is now parameterised for shortgrass
prairie and tallgrass prairie. Initial efforts have been made to
parameterise it for mixed prairie, desert grassland, shrub-steppe,
high mountain grassland and annual grassland. The model, even though
it appears complex, is still relatively simplified. It contains
several submodels, interacting to produce the total model. It
includes abiotic, producer, consumer, decomposer and nutrient
components.

Many of the model's flow functions are at the level of the
physiology of the organism or at equivalent levels in the abiotic
portion of the grazing land ecosystem. Most of the field-measured
state variables in the system model are at the "whole organism" or
the population level (Table 4). The resource manager frequently
deals with the community responses or even ecosystem responses. In
fact, many of the hypotheses formulated in Annex I are oriented to
that level. Van Dyne et al. (1976) have listed a large series of
testable hypotheses concerning structure, function, and utilisation
of grassland ecosystems. Their hypotheses are organised according
to groups related to biomass structure, energy flow, nutrient
cycling, ecosystem controls and ecosystem responses.

One should interpret model responses cautiously. Models do
not represent reality; rather they represent the modellers'
perception of reality. Model responses are therefore condtioned
on the modellers' perceptions as well as the time, effort and
computer hardware available.

Models are never perfect and they frequently evolve. Thus,
responses to experimental treatments may vary from version to version
of the same general model. Van Dyne et al. (1976) reported on
experiments with a 1973 version of ELM as compared to the ELM 1974a
model. They found the pattern of responses was similar for such
key variables as intercepted insolation, gross photosynthesis,
cattle production, decomposer CO_2 evolution, total ecosystem
respiration, and production/respiration ratios. When the two
different models with the same set of driving variables and initial
conditions were run they gave similar output; when variables
differed, the values generally were less than 15% apart.

A full simulation run with the grassland ecosystem model ELM
1974a produces about 1,500 "variables" which may be plotted against
time or each other. These variables include the system state
variables, driving variables, rate process values, and intermediate

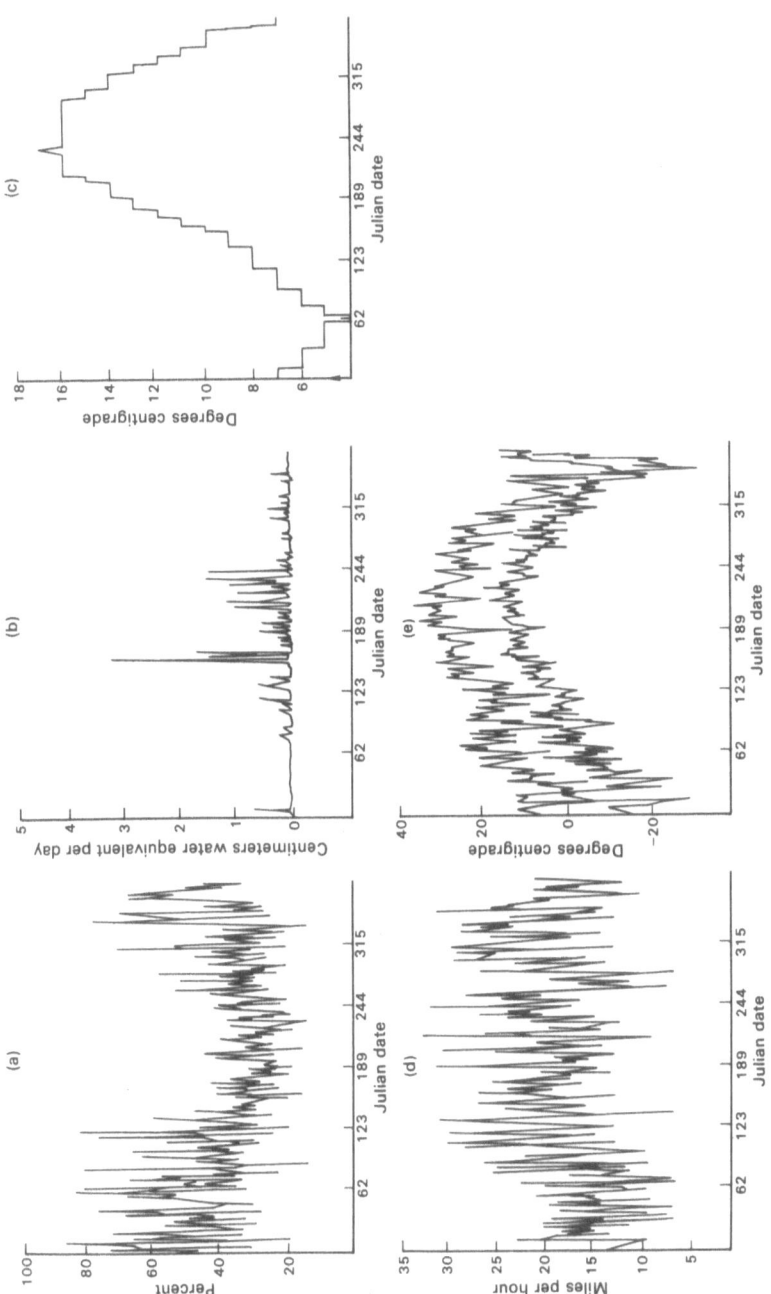

Fig. 3. Driving variable records for 1972 used in running the grassland ecosystem model in experiments reported herein: (a) relative humidity, (b) precipitation, (c) soil temperature at 72 cm below surface, (d) wind speed, (e) maximum (upper line) and minimum (lower line) temperature at 2 m. Relative humidity and cloud cover data are averages of eight measurements per day at the nearby weather station in Cheyenne, Wyoming; other data are from the US IBP Grassland Biome Pawnee Site.

Table 5.   State variables of the ELM model.

Abiotic Submodel

x(109) - x(117) — Soil moisture in each of 10 soil layers
x(118) - x(131) — Soil temperature in each of 13 soil layers

Producer Submodel

x(200) - x(204)   — Live shoots for individual categories, i.e. warm—season grass, cool—season
                              grass, forbs, shrubs, succulents
x(210) - x(214)   — Storage components for individual plant categories
x(220) - x(224)   — Standing dead of individual plant categories
x(230) - x(234)   — Seeds of individual plant categories
x(240) - x(244)   — Roots of individual plant categories
x(280)   — Resistant component of surface litter
x(281)   — Labile component of surface litter
x(290), x(292), x(294)   — Resistant component of belowground litter—upper, middle, lower layers
x(291), x(293), x(295)   — Labile component of belowground litter—upper, middle, lower layers

Consumer Submodel

x(300) — cattle   x(305) — ground squirrel
x(301) — coyote   x(306) — kangaroo rat
x(302) — rabbit   x(307) — bison
x(303) — grasshopper   x(308) — antelope
x(304) — deer mouse   x(309) — sheep

Table 5 (cont'd)

Decomposer Submodel

x(610) - x(612) — Humic material in belowground litter—upper, middle, lower layers
x(651) — Active decomposers in surface litter and faeces
x(652) - x(654) — Active decomposers in belowground litter—upper, middle, lower layers
x(661) — Inactive decomposers in surface litter and faeces
x(662) - x(664) — Inactive decomposers in belowground litter—upper, middle, lower layers

Nitrogen Submodel

x(801) — Standing dead
x(811) - 814) — Soil nitrate N in nutrient soil layers
x(821) - 824) — Soil ammonium N in nutrient soil layers
x(841) - 844) — Live root N in nutrient soil layers
x(846) — Consumer N
x(851) - 854) — Belowground litter N in nutrient soil layers
x(860) — Live shoot N
x(891) - 894) — Soil organic N in nutrient soil layers

Phosphorus Submodel

x(901), x(911), x(921), x(931) — Inorganic solution P in nutrient soil layers
x(920), x(912), x(922), x(932) — Labile inorganic P in nutrient soil layers
x(903), x(913), x(923), x(933) — Precipitated P and primary minerals in nutrient soil layers
x(904) — Live root P
x(906) — Live shoot P
x(907), x(917), x(927), x(937) — Decomposer P in nutrient soil layers
x(908), x(918), x(928), x(938) — Stable organic P in nutrient soil layers
x(909), x(919), x(929), x(939) — Labile organic P in nutrient soil layers
x(916) — Crown P
x(926) — Standing dead P
x(932) — Surface litter P
x(946) — Consumer P

calculations. In recent experiments we have conducted a series
of 1-year simulations with a 2-day time step. We have selected
for graphical presentation a few of the variables and rates in
order to illustrate differences in ecosystem dynamics due to
grazing treatments by cattle, sheep, bison, or Pronghorn antelope.

The simulation model was initiated by setting the run to
begin on 1 January 1970 or 1972. The 1972 driving variables,
plotted in Figure 3, are precipitation, maximum and minimum air
temperature, wind, relative humidity, and soil temperature at
six-foot depth. The records came from on or nearby the Pawnee
Site, a US IBP Grassland Biome study site located on the Central
Great Plains Experimental Range operated by the Agricultural Research
Service and portions of the Pawnee National Grassland managed by the
Forest Service, both USDA.

A series of five experiments were performed with the model for
the cliamtic regimes of 1970 and 1972. These included experiments
on grazing intensity and differential grazing effects of different
large herbivores. Some results are tabulated in Table 5. The
discussions which follow focus on Experiment 5, in which four large
herbivores were stocked; one species was included in each run of
the model, all at a light grazing intensity. The stocking level
for each species was chosen to yield equivalent metabolic weights.

Dynamics of Driving Variables

The curve of solar radiation input to the system (Figure 4) shows
the typical sinusoidal shape that one would expect near 40° north
latitude. There is, of course, considerable day-to-day variability
in the solar radiation input. Average maximum values in the interval
of Julian dates 160 to 240 were in the order of 450 Ly m^{-2}. Minimum
values in mid-winter were in the order of 150 Ly m^{-2}.

Air temperature values (Figure 4) followed approximately the
solar radiation input. The values plotted are daily maximum and
daily minimum. There was much more variability in air temperature
during the winter period than there was for solar radiation input.
Also, there was somewhat of a lag of air temperature values as
compared to solar radiation values; the former peaked later in
the summer. Minimum values during the year were about $-30^\circ C$.
Maximum values were about $33^\circ C$. The range between daily maximum
and daily minimum was in the order of $15^\circ C$ at many times during the
year. The same driving variable data apply to all four grazing
treatments.

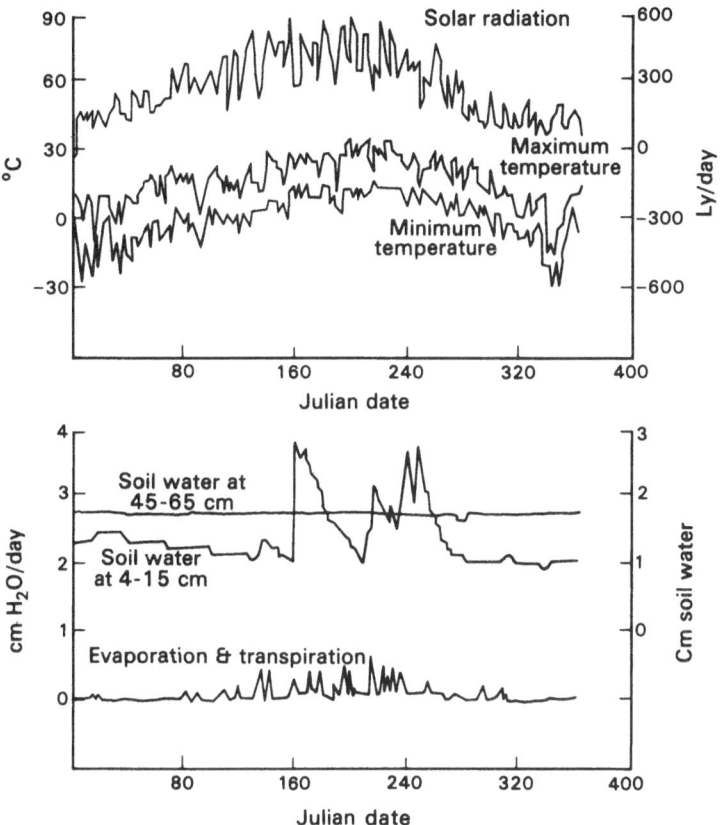

Fig. 4.   The dynamics of solar radiation, maximum and minimum air
          temperature for 1972 are plotted in the upper graph.   The
          lower graph depicts the soil water dynamics at the 4 to
          15 cm depth and the 45 to 65 cm depth.   The combined sum
          of evaporation and transpiration is also shown in the lower
          graph.

Abiotic Dynamics

Dynamics of two abiotic state variables and one abiotic process rate are plotted in Figure 4. Soil water at the lower depth (45 to 65 cm) in this soil profile varied little throughout the year and remained in the order of 2.7 cm water for that depth interval. The only variability in soil water content at this depth among the runs for the four herbivore grazing treatments was that with antelope and cattle there was a slight depression in soil water at about day 280. The reasons for this depression under antelope and cattle grazing treatments are unclear. Integrated over the year, under all treatments the water balance was similar: bare soil evaporation, 12 cm; transpiration, 13 cm; and precipitation, 27 cm (Table 5).

For all practical purposes, the dynamics of soil water in the upper depth layer plotted (4 to 15 cm) was identical for the cattle, antelope, and sheep grazing treatments. There was a difference, however, between bison grazing and the other treatments from about day 280 through day 330. There was slightly more water in the soil at this depth under bison than under the other grazing treatements. This could be related to the use of cool-season grasses by bison (as will be discussed in a later section) which decreased water use by these plants.

Evaporation and transpiration are calculated separately in the model and the combined sum "evapotranspiration" is plotted in Figure 4. For all practical purposes, the dynamics of the evapotranspiration rates were the same under all four grazing treatments. Evapotranspiration was near zero during much of the non-growing season when the soil profile was dry and the plants were not transpiring. Evapotranspiration was characterised by peaks in the spring and in the fall. There was a rather constant low value during the summer upon which was superimposed large losses following rainfall events (Compare Figures 3 and 4). Integrated over the 365 day year, bare soil evaporation was almost equal to transpiration, respectively, 12 and 13 cm.

Total precipitation was about 27 cm and evapotranspiration totalled 25 cm leaving 2 cm to be accounted for in runoff, deep percolation, and increase in soil water storage. Deep percolation is negligible, runoff occurs only when rainfall is very heavy and is small under light grazing treatments; therefore, soil water storage should have increased slightly although this cannot be ascertained from the plots on the scale given in Figure 4.

Producer Dynamics

Phenology codes, having a range of values from 0 to 7, are illustrated in Figure 5. The seven phenophases are as follows (Cole,
1976):

(1) winter quiescence or first visible growth,

(2) first leaves fully extended,

(3) middle leaves fully extended,

(4) late leaves fully extended and first floral buds,

(5) flowering (floral buds, open flowers, and ripening fruits),

(6) fruiting (buds, flowers, green and ripe fruit, and dispersing
     seeds), and

(7) dispersing seeds and senescence.

     The values 1 through 7 represent the number of phenophase, and
although ordered, they do not necessarily represent equally spaced
points on a continuous scale.

     There was no significant influence on phenological dynamics
of warm-season grasses, cool-season grasses, or forbs due to the
differential grazing of cattle, bison, sheep, and antelope. In
fact, at the resolution plotted in Figure 5 there is relatively
little difference in the dynamics of the various plants mentioned.
However, note that forbs changed to phenophase 4 at about day 120,
ahead of both warm-season and cool-season grasses. At about day
180 cool-season grasses advanced to phenophase 6. Warm-season
grasses continued their phenological development later into the
year not reaching phenophases 6 or 7 until after days 240 and 280,
respectively. At about day 320 there was a transfer of much of the
live vegetation categories into standing dead with limited material
remaining in the vegetative stage.

     The nitrogen content of the roots is plotted in Figure 5 for
two different depths in the profile. At the 7.2 to 12.7 cm depth
there appeared to be no significant difference in the nitrogen
dynamids due to the grazing treatment. There was about 3 g N m^{-2}
until the summer growing season when the nitrogen level increased
to 0.4, and by late summer and through the fall it fell to 0.5 before
decreasing.

     The dynamics of nitrogen at the shallower depth (2.54 to 7.62 cm)
was similar under all grazing treatments until about day 240. At
that time there was a slight divergence, with the bison and sheep

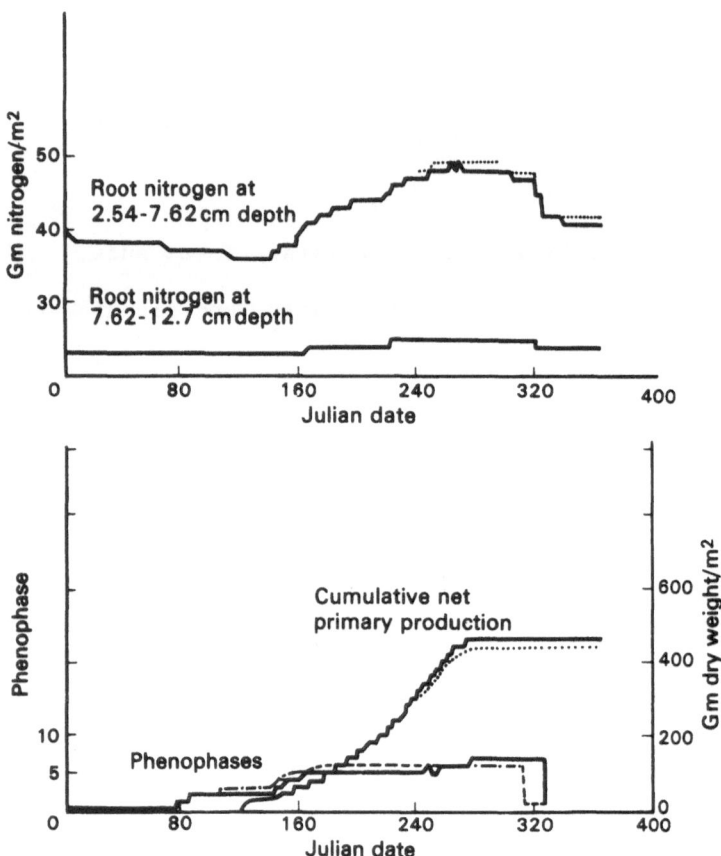

Fig. 5.   The upper graph displays the nitrogen content of the roots
          at the 2.54 to 7.62 cm depth and the 7.62 to 12.7 cm depth.
          The solid line represents values obtained during light
          grazing with cattle and the dotted line, values obtained
          under light grazing by bison. The lower graph shows net
          primary production under cattle grazing (solid line) and
          under bison (dotted line) and phenophases for warm-season
          grasses (————), cool-season grasses (——————) and
          forbs (—·—·—).

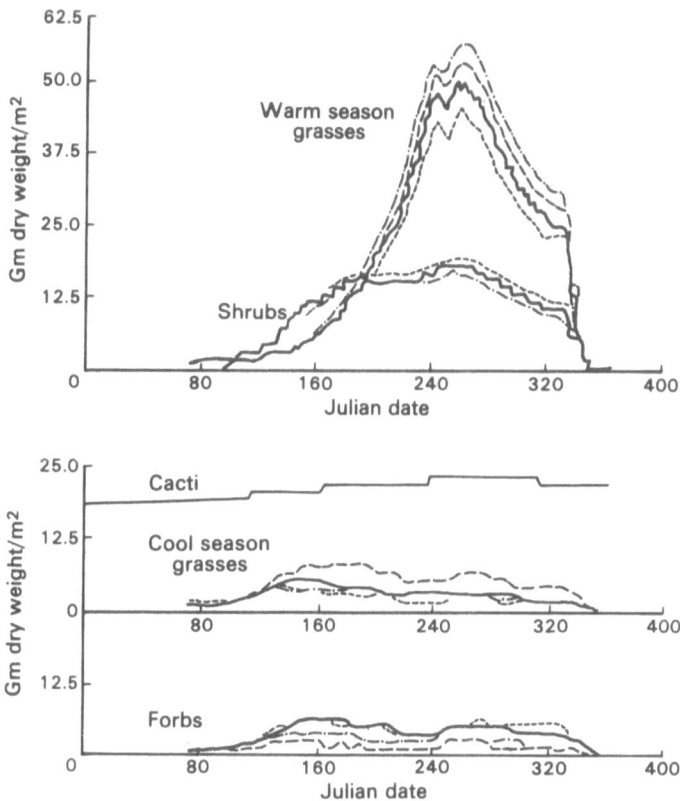

Fig. 6. Seasonal dynamics of aboveground herbage biomass for all
plant catagories are depicted for the grazing treatments
of cattle (————), bison (······), sheep (–·–·––·–) and
antelope (––––––). The response of cacti was similar
under all treatments, hence only values for the cattle
grazing treatment are shown.

treatments causing soil nitrogen to be somewhat higher than under
cattle and antelope treatments (Figure 5). The maximum difference,
however, was only in the order of 3%. Soil nitrogen dynamics at
this layer were similar to those at the lower depth, but the
magnitude of change was much greater. Low values of soil nitrogen
at the shallower layers were found up to day 160 when there was an
increase to just beyond day 240. This closely parallels, but is
not as rapid an increase rate as cumulative net primary production.
After day 320 there was a rapid drop of soil nitrogen in the shallow
soil layers under all grazing treatments.

Cumulative net primary production and photosynthesis varied
considerably among the grazing treatments (Figure 5). Total
integrated gross photosynthesis for the year was in the order of
820 g dw m^{-2} for the bison grazing treatment in contrast to 860
for the pronghorn grazing treatments; for cattle and sheep the
values were identical at 850. Net photosynthesis patterns were
similar for grazing with bison, cattle, sheep and pronghorn; the
values were respectively 49, 51, 52, 52 g dw m^{-2} (Table 6).
Cumulative net photosynthesis under the cattle grazing treatment
and the bison grazing treatment are plotted in Figure 5. For
all practical purposes differences in net photosynthesis among
the grazing treatments came about from day 280 through to the end
of the growing season.

The different species of grazing herbivores have a considerable
impact on the dynamics of the warm-season grasses, the cool-season
grasses, and the forbs (Figure 6). Although the warm-season grasses
were affected more in absolute magnitude, in relative magnitude
there were tremendous impacts on cool-season grasses and forbs.
The different grazing treatments did not affect the major timing
of seasonal events, but simply varied the magnitude of standing
crop biomass for the species.

The relative impact on warm-season grasses was greatest for
bison and least for sheep (Figure 6 top graph, top curves). The
differences in standing crop of warm-season grasses begins to appear
soon after day 160. The differences in standing crop of this plant
group were greatest by about day 265. Bison had a much higher
preference for grass than did sheep or antelope. Bison particularly
consumed cool-season grass (Figure 6 lower graph), but their
preference for forbs and shrubs was sufficiently low that the bulk
of their diet came from warm-season grasses. Thus, although the
treatment was light grazing, there were significant impacts of
herbivory on the growth and standing crop of the vegetation groups.

Sheep provide another extreme. They, along with antelope, showed a strong preference for forbs and thus when grazed by these two herbivore species the standing crops for forbs were low during much of the year (Figure 6, lower graph, lower curves). The impact of sheep and antelope on forb standing crops began as early as day 120 and continued until the end of the grazing season. The animals were put onto the range at about day 120 and removed at about 300 but there was still considerable difference in the standing crop of the vegetation after day 300 until driving snow and rains in the late fall transferred standing live to either standing dead or to litter$\frac{1}{2}$

There was an unexplained residual impact of both sheep and bison on the standing crop of cactus after the animals were removed from the system. This was the only difference in amount of biomass or dynamics of cactus as affected by the herbivore species. The late-season impact was relatively small. Possibly it was due to a residual effect on soil water levels which would affect the relative growth of cactus. (Compare the anomalous behaviour of soil water at the lower depth late in the year under antelope and cattle grazing treatments noted above). On the scale of the plot in Figure 4, however, there was no significant difference in soil water levels at the end of the year among the four treatments.

Consumer Dynamics

For many consumer variables in the model, there was no major impact due to differential grazing by cattle, bison, sheep and pronghorn (Table 6). However, there was a significant impact on the gain of large herbivores due to treatment. Respectively, values were 8, 7, 4 and 2 g Cb m^{-2} for cattle, bison, sheep and pronghorn. These animals were all stocked at the same rate, relative to animal unit equivalents. (An animal unit equivalent is the weight of the animal to the 0.75 power relative to the weight of a 454 kg animal to the 0.75 power).

Calculating stocking rate involved comparing the different animals to each other on the basis of metabolic size. For example, the calculation given here is for bison stocked 180 days. Stocking rates are determined so that the animal units are the same as for long-term light grazing field studies of cattle. Cattle graze 180 days at 8.23 x 10^{-6} animals m^{-2}. The average weight per animals for cattle is obtained from simulation runs of light cattle grazing for 180 days. Animal unit months (AUM) are determined accordingly:

Table 6.　Data from a series of experiments with ELM, a large-scale simulation model of the shortgrass prairie.　Data from Experiment 3 are from Van Dyne et al. (1976) and are

SIMULATED YEAR: GRAZING INTENSITY: LARGE HERBIVORE SPECIES: OTHER SPECIFICATIONS:	Units all y^{-1}	Multiplier power of 10	Experiment 5				Experiment 4		
			1972 Light Cattle	1972 Light Bison	1972 Light Sheep	1972 Light Pronghorn	1972 Light Cattle	1972 Heavy Cattle	1972 Extra-Heavy Cattle
FLOW DEFINITION			-ELM 1974a model, new l.e.. new d.v.-				-ELM 1974a, new l.e., and d.v.-		
ABIOTIC SUBSYSTEMS									
Evaporation, bare soil	cm·cm^{-2}	0	12	12	12	12	12	14	14
Transpiration	cm cm^{-2}	0	13	13	13	13	13	7	6
Solar energy Input	cm cm^{-2}	4	11	11	11	11	11	11	11
Intercepted insolation	cal cm^{-2}	3	13	13	13	13	13	7	7
Precipitation	cm cm^{-2}	0	27	27	27	27	27	27	27
PRODUCER SUBSYSTEM									
Crown to/from shoot	g dw m^{-2}	0	34	32	35	35	34	16	15
Crown death	g dw m^{-2}	0	5.2	5.2	5.2	5.2	5.2	4.6	4.6
Shoot to/from crown + roots	g dw m^{-2}	1							
Crown respiration	g dw m^{-2}	0	12	12	12	12	12	12	12
Gross photosynthesis	g dw m^{-2}	1	85	82	85	86	85	49	45
Net photo., less crown + root resp.	g dw m^{-2}	1	47	45	49	48	47	23	22
Net photosynthesis	g dw m^{-2}	1	51	49	52	52	51	26	26
Root respiration	g dw m^{-2}	0	26	26	26	26	26	24	24
Seed production	g dw m^{-2}	-1	35	34	35	36	35	16	15
Shoot to standing dead	g dw m^{-2}	1	11	11	11	12	11	3.3	3.3
Seed germination	g dw m^{-2}	0	3	3	3	3	3	3	3
Root death	g dw m^{-2}	1	32	32	32	32	32	32	32
CONSUMER SUBSYSTEM									
Mammal production	g Cb m^{-2}	-2	7.7	6.7	4.6	2.4	7.2	-4.5	-8.6
Grasshopper production	g Cb m^{-2}	-3	150	150	150	140	150	170	170
Gains for large herbivores	g Cb m^{-2}	-2	7.6	6.6	4.5	2.3	7.0	-45	-86
Gains for coyotes	g Cb m^{-2}	-6	65	65	65	65	60	60	60
Gains for rabbits	g Cb m^{-2}	-4	11	11	11	11	12	13	13
Gains for grasshopper mice	g Cb m^{-2}	-5	160	160	160	160	2.4	2.5	2.5
Gains for prairie deermice	g Cb m^{-2}	-5	1	1	1	1	2.1	2	2
Gains for 13-lined ground squirrel	g Cb m^{-2}	-5	5	5	5	5	7	7	7
Gains for kangaroo rats	g Cb m^{-2}	-5	2	2	2	2	3	3	3
DECOMPOSER SUBSYSTEM									
Decomposer CO$_2$ evolution	g Cb m^{-2}	1	30	30	30	30	30	33	33
Mechanical litter to soil transfer	g Cb m^{-2}	-1	22	22	22	22	22	21	21
Decomposer production	g Cb m^{-2}	0	121	119	119	119	120	120	120
Active decomp. resp.: surface litter	g Cb m^{-2}	0	74	74	74	74		76	76
Active decomp. 0-4 cm layer	g Cb m^{-2}	0	122	122	122	122		128	128
Active decomp. 4-15 cm layer	g Cb m^{-2}	0	72	73	72	71		81	82
Active decomp. 15-75 cm layer	g Cb m^{-2}	0	30	32	30	29		43	46
Inactive decomp. resp.:surface litter	g Cb m^{-2}	-2	36	36	36	36	36	39	40
Inactive decomp. 0-4 cm layer	g Cb m^{-2}	-2	56	56	56	56	56	62	62
Inactive decomp. 4-15 cm layer	g Cb m^{-2}	-2	45	45	45	45	45	48	49
Inactive decomp. 15-75 cm layer	g Cb m^{-2}	-2	18	18	18	18	18	19	19
SYSTEM LEVEL INDICES									
Gross photosynthesis (P)	g Cb m^{-2}	1	34	33	34	33	34	20	18
System respiration (R)	g Cb m^{-2}	1	45	45	45	45	45	44	43
P/R ratio	none	-2	76	73	76	73	76	45	42
Gross photosynthesis/biomass	d^{-1}	-4	53	51	52	51	53	33	31
Gross photo./Intercepted radiation	none	-3	25	25	25	25	25	26	26
Herbage grazed	g dw m^{-2}	0	15	14	13	11	15	34	41

provided for comparison. Different model versions (1973, 1974, and 1974a) and different initial conditions and driving variables (old v. new sets) were used in these experiments.

	Experiment 3			Experiment 2			Experiment 1				
	1970 Light Cattle	1970 Heavy Cattle	1970 Extra-Heavy Cattle	1972 -	1972 -	1972 -	1970 Light Cattle	1970 Heavy Cattle	1970 Extra-Heavy Cattle	1970 Light Cattle No Small Herbivores	1970 Heavy Cattle No Small Herbivores
	ELM 1974a, old l.e., and d.v.--			--ELM 1974, old l.e., and d.v.,	No Large Herbivores	3-yr sequence--ELM 1973, old l.e., and D.V. (all)				Herbivores	Herbivores
	10	11	12	20	20	19	10	10	12	10	12
	13	12	9	8	8	7	14	13	10	14	11
	11	11	11	11	11	11	11	11	11	11	11
	18	15	8	17	20	19	20	17	11	20	11
	24	25	13		47	50	52	46	47	52	48
	25	25	24	5.5	8.3	11	27	27	26	27	26
							35	32	27	35	28
	24	24	26	11	11	12	25	25	27	27	27
	81	76	53	91	98	100	81	78	64	81	64
	41	42	25	54	60	64	49	45	35	58	36
	46	47	30	57	64	68	54	50	40	54	41
	25	26	25	25	25	26	23	23	22	23	22
	53	45	26	46	50	52	52	42	24	53	24
	12	9.5	4	14	16	18	12	9	3	12	4
	.05	.05	.05		4.6	5	5	4	0	6	0
	39	39	36	33	36	39	33	32	29	33	29
	8.4	20	-84	.14	6.9	2.8	8	22	-52	7	-51
	240	27	290	0	0	0	23	23	32	0	0
	7.3	19	86	0	0	0	7	21	-53	7	-51
	20	20	20		64	64	13	13	13	-10	-10
	43	43	43	64	6	1.9	62	62	62	0	0
	133	133	133	6	.8	.25	170	170	170	0	0
	65	65	65		.4	4.6	120	120	120	0	0
	320	320	320		3.6	2.1	320	320	320	0	0
	75	75	75		0	0	66	66	66	0	0
	25	26	33	27	18	19	21	22	28	20	28
	20	19	19		5	3.2	21	21	19	21	19
	93	93	85	92	49	50	86	88	82	86	82
	36	37	37	59	17	13	37	38	38	37	38
	76	80	96	145	93	99	58	62	78	58	77
	79	89	113	53	63	71	66	72	97	65	96
	50	57	81	9	7.6	8	44	47	71	44	69
	16	16	18	23	13	7.3	16	16	18	15	18
	43	44	50	49	55	61	38	40	46	38	46
	40	42	46	35	30	28	38	39	43	38	43
	15	16	17	15	13	12	15	15	17	15	17
	32	30	21	36	39	40	32	31	25	33	26
	40	40	44	42	33	34	33	35	40	33	40
	80	76	48	87	117	119	97	88	64	98	65
	46	43	32	55	56	54	42	41	35	43	35
	17	20	24	20	20	20	16	18	23	16	23
	16	45	58								

$$AUM = \frac{(\text{no. days range grazed})}{30} \, (\text{density}) \, (\frac{\text{average wt. (1b)}}{1,000 \text{ 1b.}})^{0.75}$$

Hence, AUM for cattle is

$$AUM = (180/30) \, (0.00000823) \, (29.89 \text{ kg Cb}/54.5 \text{ kg Cb})^{0.75}$$

$$= 3.16 \times 10^{-5}$$

Note that 1,000 1b. is 54.5 kg Cb. To determine the bison density:

$$AUM = (180/30) \, (X) \, (30.7/54,4)^{0.75}$$

Here X is the density of bison, and 30.7 kg Cb is the expected
weight of bison in mid-season based on a Brody-type curve. Letting
$AUM = 3.16 \times 10^{-5}$ and solving for X, the bison stocking density is
8.1×10^{-6} animals m^{-2}.

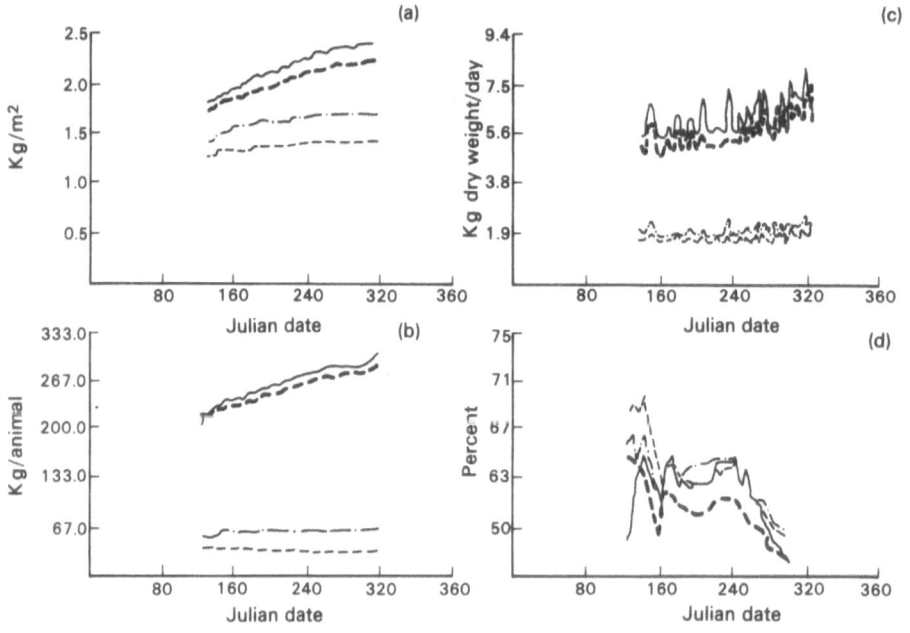

Fig. 7. Animal weight expressed per unit area (a), per animal (c),
         total intake per animal (b) and total forage digestibility
         (d) for the grazing treatments of cattle (————), bison
         (······), sheep (—·—·—·—) and antelope (——————).

There were differences among species of grazing herbivores both in the weight per individual and the weight per unit area (Figure 7). The gains of cattle and bison closely approximated to each other. Initial weights of bison were somewhat greater than for cattle, but cattle gained more rapidly early in the growing season. Toward the end of the season cattle exceeded bison in individual animal size by about 5 to 10%. Antelope made some gain although this was not discerned when plotted on the same weight scale as for the other individual animals (Figure 7, lower graph). Their gain per unit area was not as great as for the other three herbivore species (Figure 7, upper graph). This may be explained in part by their higher maintenance energy requirements and in part by the fact that the relative age of antelope when stocked is greater than the other large herbivores and thus the expected growth rate is lower. The end-of -the-season weights reached by the herbivores seem reasonable, particularly for cattle where more data are available for comparison. The weights match fairly well the projected Brody growth curves.

Total herbage intake, summed over all plant groups, is plotted for the large herbivores in Figure 7. As expected, the intake values increase over the grazing season as the animals gain body weight. There is considerable variability from day to day, largely associated with rainfall events and changes in plant growth and palatability. Relatively, the antelope are grazing more herbage than are the larger herbivores, cattle and bison.

Forage intake rate data by plant category are summarised in Figures 8 and 9. These data explain a great part of the dynamics shown for standing crop of these plant groups. (See discussion in the section on Producer Dynamics and also Figure 6).

Warm-season grasses are consumed heavily after day 160 by all four large herbivores. Surprisingly, consumption of warm-season grasses by antelope reaches levels as high as by cattle and sheep in late summer. This requires further study. The high intake of forbs by antelope early in the season is particularly noteworthy as is the high intake of forbs by bison at about day 160 to 200.

The digestibility of the herbage for each of the herbivores in general decreases over the summer grazing period (Figure 8). However, there were notable increases at various times at the diet shifted to plants higher in digestibility, in part in response to rainfall. Sheep selected the most digestible diet and cattle the least digestible diet. These fractions for digestibility are, however, a combination of the ability of the animals to digest food and of the selectivity of the diet. The general "digestive power" was expected to be higher in antelope than cattle, for example, because antelope grazed a diet higher in nitrogen content and

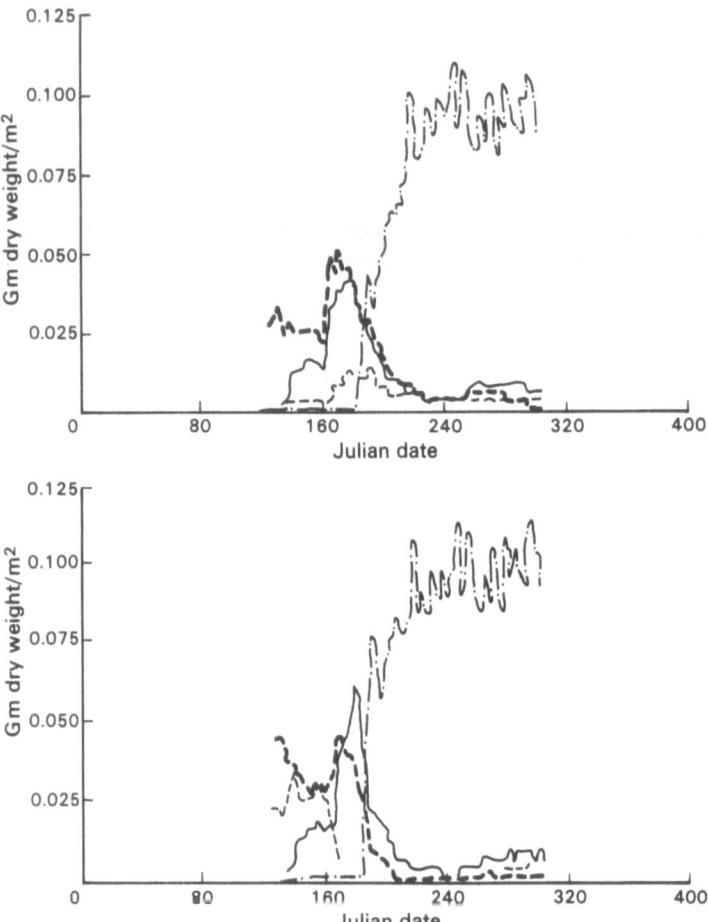

Fig. 8. Daily forage intake by cattle (upper graph) and bison
         (lower graph) of the plant groups: warm-season grass
         (−·−·−·−), cool-season grass (·····), forbs (———)
         and shrubs (------).

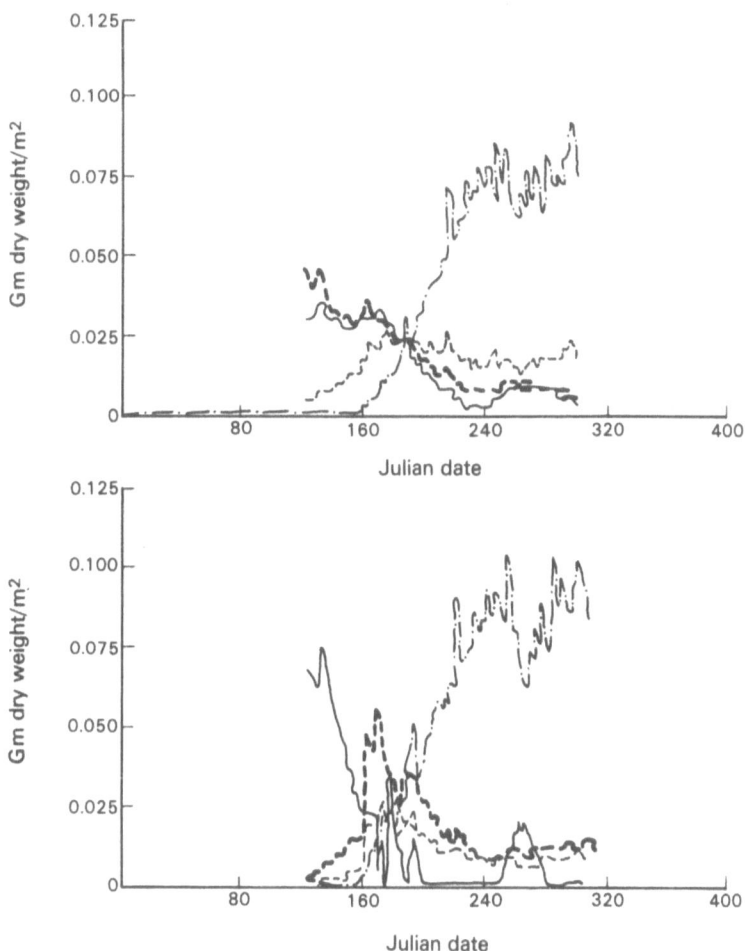

Fig. 9. Daily forage intake by sheep (upper graph) and antelope
(lower graph) of the plant groups: warm-season grass
(- .- .- .-), cool-season grass (·····), forbs (————)
and shrubs (-----).

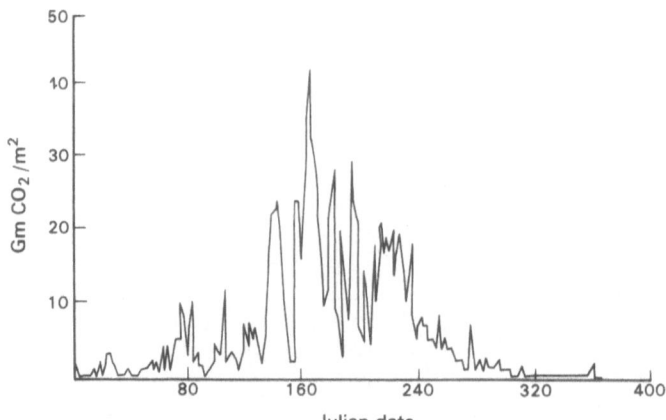

Fig. 10.   Evolution of carbon dioxide by decomposers during the
           simulation year. This graph is for the cattle grazing
           treatment, but there was no major difference in the re-
           sponse between treatments.

maintained a more favourable environment for rumen microbes.   Bison
do not graze as high a quality diet as do some of the other
herbivores, but they have a higher digestive efficiency, as has been
demonstrated in experimental studies on this rangeland (Peden, et
al. 1974).

Decomposer Dynamics

There was little evidence of differential impacts of the large
herbivores on various measures of decomposer activity (Table 6).
The only measurable difference was the biomass estimate for active
decomposers in the 15 to 75 cm layer of the soil. Here there was
a range in values of 29 for pronghorn vx. 32 g Cb m^{-2} for bison.
Reasons for this difference are not clear at present. As noted
in previous experiments, it requires very heavy grazing to make a
major impact on the decomposer system. Differences between light
and heavy grazing generally were not -large (Table 6). The seasonal
dynamics of decomposer CO_2 evolution from all depths in the profile
did not vary between herbivore species. Thus, only one set of data
are plotted in Figure 10. These data show, as expected, that in
times with good combinations of soil water levels and high temperatures that CO_2 evolution rate is high. It would drop off drastically,
however, as the soil dried. Thus, during the growing season there
were day to day variations in CO_2 evolution that were 2- to 3-fold.
Even during the non-grazing season, there were brief intervals,
particularly in spring, when CO_2 evolution was occurring.

SUMMARY

Mathematical modelling of ecosystems of real-life complexity, with
an aim toward developing information useful in generating and testing
ecological theory and providing a base for management decision-making,
is a large effort whose overall cost has been underestimated
previously. The consequence was that in large-scale programmes
including such modelling, relatively little such work has been done
on the utilisation or evaluation of the models. Almost all of the
effort allocated to modelling was spent on model construction. This
paper is an initial effort in showing the relationships between
hypotheses and simulation models, focusing on large-herbivore
grazing of a semi-arid rangeland.

    The relationship between simulation models and hypotheses is
discussed. A large-scale simulation model in fact is a "macrohypothesis" composed of many inner-connected "micro-hypotheses"
about how the system is structured and how it is functioning in its
various parts. But these micro-hypotheses about individual abiotic
and biotic processes are a different class of hypothesis to those
related to predicted responses of ecosystems to various kinds of
stresses. These two sets of classes of hypotheses are arrived at
from the same general base of information, but largely by different
groups of scientists. There is even a third set of hypotheses, those
that pertain to general responses of a class of ecosystems, commonly
generated by a third set of scientists. There is, of course, some
intersection of the sets of scientists and sets of hypotheses. This
parallels the intersection of the sets of data, theory, and models
which are available for use in the decision making process for ecosystem management.

    Initial ideas are outlined towards organising hypotheses on
abiotic and biotic processes in ecological systems into a hierarchical structure with separate consideration for the physiological,
organismal, and population levels.

    A procedure is outlined for setting up hypotheses a priori for
testing ecological simulation models. Example results from
modelling experimts are presented. Discussion concentrates upon
comparison of cattle, bison, sheep, and Pronghorn antelope grazing
on a shortgrass prairie ecosystem and their differential consequences. The model experiments are evaluated by means of
comparison of integrated values for selected flow rates and by
comparison of the dynamics of selected state variables within the
system. Various flows and variables are depicted for the abiotic,
producer, consumer and decomposer levels in the system.

ACKNOWLEDGEMENTS

This work has been supported in part by grants and contracts from
the National Science Foundation (DEB 76-11139) and the Council on
Environmental Quality (EQ-AC-005).

REFERENCES

Bear, G. D. 1973. Physiological studies. Game Research Report.
    Colorado Department of Game and Fish and Parks. PR W-40-R-13.

Blaxter, K. L., Wainman, F. W., Wilson, R. W. 1961. The regulation
    of food intake by sheep. Anim. Prod., 3, 51-61.

Blaxter, K. L., Clapperton, J. L., Wainman, F. W. 1966a. The extent
    of differences between six British breeds of sheep in their
    metabolism, feed intake, and utilisation and resistance to
    climatic stress. Br. J. Nutr. 20, 283-294.

Blaxter, K. L., Wainman, F. W., Davidson, J. L. 1966. The voluntary
    intkae of food by sheep and cattle in relation to their energy
    requirements for maintenance. Anim. Prod. 8, 75-83.

*Cole, G. W. (Ed.). 1976. "ELM" Version 2.0. Range Science Depart-
    ment Science Series, No. 20. Fort Collins: Colorado State Univ.

Crampton, F. W. and Lloyd, L. E. 1959. Fundamentals of nutrition.
    San Francisco. W. H. Freeman and Co.

Esminger, M. R. 1970. Sheep and wool science. The interstate
    Printers and Publishers, Inc.

Hoover, R. G., Til, C. E. and Ogilvie, S. 1959. The antelope of
    Colorado. Colorado Department of Game and Fish.

*Innis, G, S, 1972. ELM: A grassland ecosystem model. Presented
    at 1972 Summer Computer Simulation Conf., 14-16 June, San
    Diegeo, California.

*Innis, G. S. 1975. Role of total systems models in the Grassland
    Biome study. In: Systems analysis in simulation in ecology.
    Vol. 3, edited by B. C. Patten, 13-47. New York: Academic Press.

*Innis, G. S. (Ed.). 1977. (Title to be provided at a later date).
    Springer-Verlag (in press).

Laycock, W. A., Buchanan, H., Krueger, W. C. 1972. Three methods
    of determining diet, utilisation, and trampling damage on sheep
    ranges. J. Range Mgmt., 25, 352-356.

Meagher, M. M. 1973. The bison of Yellowstone National Park.
    National Park Service (Scientific Monograph Series No. 1).

*Parton, W. J. and Risser, P. G. 1976. Osage site version of the
    ELM grassland model. U.S. IBP Grassland Biome study Preprint
    No. 193. (Presented at the 1976 Summer Computer Simulation
    Conf., 12-15 July, Washington, D.C.).

Peden, D. G. 1972. The trophic relations of Bison bison to the
    shortgrass plains. Ph. D. Thesis. Fort Collins: Colorado
    State Univ.

*Peden, D. G., Van Dyne, G. M., Rice, R. W. and Hansen, R. M. 1974.
    The trophic ecology of Bison bison L. on shortgrass plains.
    J. Appl. Ecol., 11, 489-497.

Schwartz, C. C. and Nagy, J. C. 1975. Pronghorn diets relative to
    forage available in Colorado. J. Wildl. Mgmt., 40, 469-478.

Van Dyne, G. M. and Heady, H. F. 1965. Botanical composition of
    sheep and cattle diets on a mature annual range. Hilgardia,
    36, 465-492.

Van Dune, G. M. 1974. The status of grazing experiments. October
    10. (Memorandum on file).

Van Dyne, G. M. and Rice, R. W. 1974. Computer modelling experiments
    regarding grazing: shortgrass prairie synthesis volume.
    August 14. (Memorandum on file).

*Van Dune, G. M. and Abramsky, Z. 1975. Agricultural systems
    models and modelling: an overview. In: Study of agricultural
    systems, edited by G. E. Dalton, 23-106. London: Applied
    Science.

*Van Dyne, G. M. and Anway, J. C. 1976. A research programme for
    and the process of building and testing grassland ecosystem
    models. J. Range Mgmt., 29, 114-122.

*Van Dyne, G. M. et al. 1976. Evolving conceptualisation of
    ecological interrelationships on the C-b tract, p. 441-512
    (Chapter VII). In: C-b Shale Oil Project, Ashland Oil, Inc.
    and Shell Oil Co. Oil shale tract C-b: First year environ-
    mental baseline programme - annual summary and trends reports -
    November 1974 through October 1975. Published by C-b Shale
    Oil Project (Shell Oil Co., operator) 1700 Broadway, Denver,
    Colorado, U.S.A.

Van Dyne, G. M. et al. 1977. Analyses and syntheses of grassland
    ecosystem dynamics. BioEcos (in press).

Wallace, J. D. 1969. Nutritive value of forage selected by cattle
    on sandhill range. Ph. D. Thesis. Fort Collins: Colorado
    State Univ.

Wesley, D. E. 1971. Energy and water flux in Pronghorn. Ph. D.
    Thesis. Fort Collins: Colorado State Univ.

Yoakum, J. 1967. Literature of the American Pronghorn. Bull. Dep.
    Inter. (U.S.) (Unnumbered).

*   Literature cited in text

4. GENERAL PRINCIPLES FOR ECOSYSTEM DEFINITION AND MODELLING

J. N. R. Jeffers

Institute of Terrestrial Ecology

Grange-over-Sands, England

INTRODUCTION

As increasing attention comes to be focussed on environmental problems, the special characteristics of these problems begin to be appreciated. In part, the attention arises from the recognition of the damage done to the environment, not only by industrial and urban development, mining and recreation, but also by changes in agricultural and forestry practices. In part, our concern stems from the realisation that the non-renewable resources of the world are being used at an increasing rate, and that many of these resources will be depleted, even at present rates of use, within our lifetimes. Of special concern to this conference is that class of problems represented by the need to rehabilitate ecosystems which have already been damaged by our search for an increasingly higher standard of living.

The characteristics which distinguish environmental problems from all those other problems for which human institutions have been developed and adapted in our forms and patterns of central and local government, in the management and administration of public and private corporations, companies and industries, and in the making of personal choices and decisions may be summarised broadly under three headings:-

1. Complexity

    Environmental problems differ from almost all other problems in their complexity. Not only are we concerned with the geology of the area and the soils derived from the rocks, or

laid over the rocks by wind, rivers and glaciers, but we have
to predict the response of one or more ecological systems to
changes, both natural and imposed. These systems will, in
turn, generate a response in the form of human activity as
part of a feedback loop which may itself be of considerable
complexity. Imposed upon all of these factors, short and
long-term changes in climate will impose additional instabil-
ities or effects. Simple generalisations, adopted as forms
of response or management of such systems are, therefore,
almost certain to fail. As Beer (1975) has emphasised, a
problem only needs two feedback loops and five independent
variables to generate optimum solutions which are counter-
intuitive, and almost all environmental problems are more
complex than this.

2.   Timescale

Environmental problems frequently span timescales which are
considerably larger than the working life of one manager,
administrator or politician, and certainly longer than the
time spent by any one of these in one job. It is not, therefore, sufficient that the responsible individual should have
a clear understanding of the problem and its solution; it must
be possible to communicate both the problem and the solution
to successive managers. Otherwise, environmental systems are
necessarily subjected to continual perturbations as "generations" of managers, administrators, and (particularly)
politicians search for new definitions of the problems and
new solutions to redefined problems. Furthermore, as many of
the effects imposed on environmental systems operate over
long timescales, these effects themselves become difficult to
identify, and are usually confounded with many other time-
dependent effects.

3.   Multidisciplinary approach

Almost all of our traditional patterns of formal education
have concentrated on the creation of specialists and experts
in each of a range of disciplines, and it is tacitly understood that economics should be left to economists, accountancy
to accountants, chemistry to chemists, etc. Environmental
problems in contrast, require understanding and expertise
from a wide range of formal disciplines, including most of
the physical and biological sciences, economics, and sociology.
The advance in scientific knowledge is now so great that it is
unlikely that any one person can acquire extensive expertise
in more than a few disciplines, and most will be really expert

in only one. The solution of environmental problems, therefore, requires what has come to be called the "orchestration"
of a multidisciplinary team. Indeed, the difficulty of
combining the necessary disciplines in a concerted attack has
been identified as one of the major impediments to the improvement of our ability to deal with such problems (Mar, 1974).

That the three characteristics of complexity, long timescale, and
need for a multidisciplinary approach have almost always defeated
the traditional methods of decision-making for environmental
problems does not require any elaboration. The conference would
hardly have gathered in Reykjavik to discuss the rehabilitation of
damaged ecosystems if our traditional methods of managing those
ecosystems were even adequate. Many variants of these methods have
been tried from earliest historical times, ranging from simple rules
of thumb, through dependence on intuitive judgements, to more or
less refined experimental approaches which regard the environment
and ecosystems as direct analogues of physical processes. None of
these methods has prevented serious damage to environmental and
ecological systems resulting from man's mismanagement of natural
resources. In desperation, attempts have also been made to use
complex mathematical analogues of physical processes as the basis
for the generation of effective solutions to environmental problems,
perhaps in the belief that complexity of the solution is sufficient
to ensure some kind of match with the complexity of the problem.
It has not taken long to demonstrate that mathematical elegance
and formal complexity are neither necessary nor sufficient for the
management of our environment: indeed, they are usually positive
disadvantages. Our best strategy would appear to lie in the
application of systems analysis, although the term "systems analysis"
is here being used in the special sense defined by Raiffa (1973).
This paper is intended to give a brief description of applied systems analysis, with special emphasis on the definition and bounding
of ecosystems, and the effects of such definition on hypothesis
formulation and validation. Examples of the application of systems
analysis to environmental problems will be drawn from the current
work of the Institute of Terrestrial Ecology(ITE).

SYSTEMS ANALYSIS

Systems analysis is not a mathematical technique nor a group of
such techniques. It is essentially a broad framework of thought
designed to help those who have to make decisions to choose a
desirable (or, sometimes, "best" - when "best" can be defined)
course of action. In some senses, it is a compromise between the
traditionally intuitive solution of complex problems and the formal
solution of such problems through the deductive logic of mathematics.

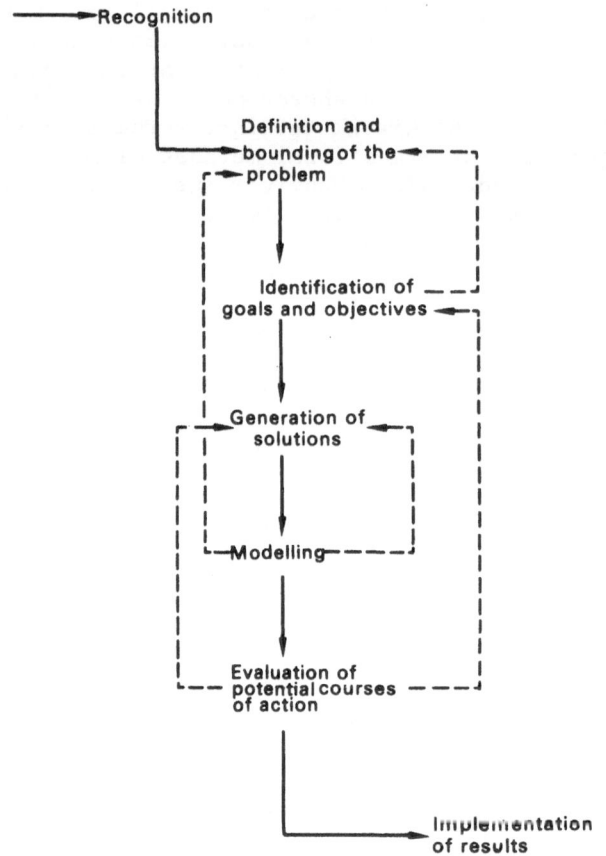

Figure 1. Main phases of applied systems analysis.

It aspires to promote good decision-making and is intended to
focus, and to force, hard thinking about large and complex
problems.

The main phases of applied systems analysis are summarised in
Figure 1. Beginning with the recognition of the problem to be
solved, the analysis proceeds through the definition and bounding
of the problem, the identification of goals and objectives, the
generation of alternative solutions, the formal modelling of
those solutions, and the evaluation of potential courses of
action, to the final implementation of the results. Systems analysis is, however, an iterative process and it may be necessary to
return to redefinition and bounding of the problem after preliminary
attempts have been made to identify goals and objectives, and
even after several attempts have been made to model the alternative
solutions. Similarly, it may be necessary to return to the generation of alternative solutions, or even to the identification of
goals and objectives, after the initial evaluation of potential
courses of action. The close inter-relationship between the
generation of alternative solutions and the modelling of those
solutions provides an "inner loop" of the method of systems analysis, i.e. a loop which will be traversed many times. All of these
recursive pathways are indicated by the dashed lines of Figure 1.

It is perhaps inevitable that much of the emphasis in the
published literature of systems analysis should be concentrated on
the alternative formulations that can be used to model the solutions
generated by the procedure. After all, most interest in systems
analysis occurs amongst mathematicians and mathematically-orientated
ecologists, and this interest is, therefore, most likely to be
expressed in terms of mathematics and computing. Within this group
of scientists, the major concern will rest with the choice of the
modelling strategy in terms of its realism, its degree of abstraction, the existence of optimum solutions, and the extent to which
the model can be regarded as a "black box" phenomenon.

Nevertheless, the primary focus of attention of ecological
research at all times should be on the definition and bounding of
the problem. Only in this way can we ensure that the research
resources are correctly allocated, and that the problem solved is
relevant to the original universe of discourse. Although there
is a strong tradition for scientists to insist that their work
should be unfettered by demands for relevance and practicality,
systems analysis does not belong to that tradition. Yorque (1975),
for example, has listed eight common myths about the conduct and
content of environmental impact studies, and these myths properly
emphasise the need to define and bound problems.

The main thrust of this paper, therefore, is on the definition and bounding of the problem as the central phase of the logical framework of systems analysis. This phase is directly analogous to the formulation of hypotheses in the scientific method, in that unless the hypothesis is carefully formulated most of the subsequent work will be wasted. Indeed, as scientists, our definition and bounding of the problem should take the form of the hypothesis, or hypotheses, to be tested. Only in this way can our subsequent models be capable of verification, and hence of rejection or temporary acceptance.

Before listing the desirable properties of the definition, it is perhaps worth mentioning three caricatures of hypotheses which have inhibited the practical application of systems analysis to ecological problems:-

1. The Red Queen hypothesis - any hypothesis which modifies itself by additional constraints or conditions so as to remain at the centre of the investigation falls into this category. Such an hypothesis is, of course, impossible to reject, and further progress with the research is delayed because an inappropriate model continues to be modified and adapted instead of being replaced.

2. The "hail of bullets" hypothesis - usually a compositive hypothesis with so many options or permutations that one, at least, of the options is incapable of being rejected. Such hypotheses again inhibit the development of research by preventing the abandonment of inappropriate and ineffective models.

3. The hypothesis of essential triviality - usually a generalised and weak hypothesis which does not lead to any useful logical deduction. In contrast to the other two types, such hypotheses inhibit management action and decision-making rather than the research itself.

ECOSYSTEM DEFINITION

As Yorque (1975) suggests, "systems analysts have been especially and properly fond of telling decision-makers about the need to carefully define and bound problems". Jeffers (1973) has stressed the importance of a "word model" as a preliminary to any application of systems analysis and modelling. The concept of the word model has been criticised by Mellanby (1976) who points out that a "word model" is nothing more than the most precise description that can

be given of the ecosystem and of the impacts upon that system, and
that there is, therefore, no need to invest the perfectly ordinary
process of describing something with a complicated name! Nevertheless, a good description will go a long way to help in selecting
an appropriate family of mathematical model and will, at the very
least, highlight the areas where knowledge of the system is
inadequate, and there seems little harm in distinguishing between
this description and the later mathematical expression of the
description.

Descriptions of ecological systems will usually need to be
bounded by the constraints of space, time and subsystems for
which decisions have to be made. Statisticians have always stressed the necessity of defining the population about which inferences
are to be made as a preliminary to any form of experimentation or
sampling, and the argument is readily extended to the modelling of
ecological systems. Our model is intended to facilitate inferences
about some population, and our initial definition and bounding of
the problem should be sufficiently explicit to identify that
population.

An interesting, if localised, problem in the United Kingdom
has been the treatment of areas which have been covered by
pulverised fuel ash, a waste product from coal-burning power stations. Pulverised fuel ash (pfa) is an almost inert material
containing little or no organic matter and varying quantities of
elements which are harmful to plant and animal life. Having
covered productive ecosystems with this material to a depth of
several metres, our concern is to develop new systems which are
capable of surviving and developing on pfa. Experiments on various
forms of treatment and management will necessarily be concentrated
on the earliest areas covered by pfa, but the relevance for later
areas of the results derived from such research will be doubtful
if, as seems likely, the industrial process or its raw materials
change with time. The ecosystem, in this case, has been defined
and bounded within space and subsystems, but not in time. The
systems analysis for the rehabilitation of such an ecosystem by
the application of fertilisers and organic material, either as a
surface dressing or incorporated in the surface layers of the
material, and the introduction of earthworms and other inverte-
brates (Satchell, 1972) will, therefore, require a sequential
component to test the applicability of the results of earlier
research to later material. The possible effects of climatic and
weather cycles on rehabilitation treatments of damaged ecosystems
impose similar constraints on the design of the investigations.
Sequential techniques have been available for many years (see,
for example, Bross (1965)) but are seldom used.

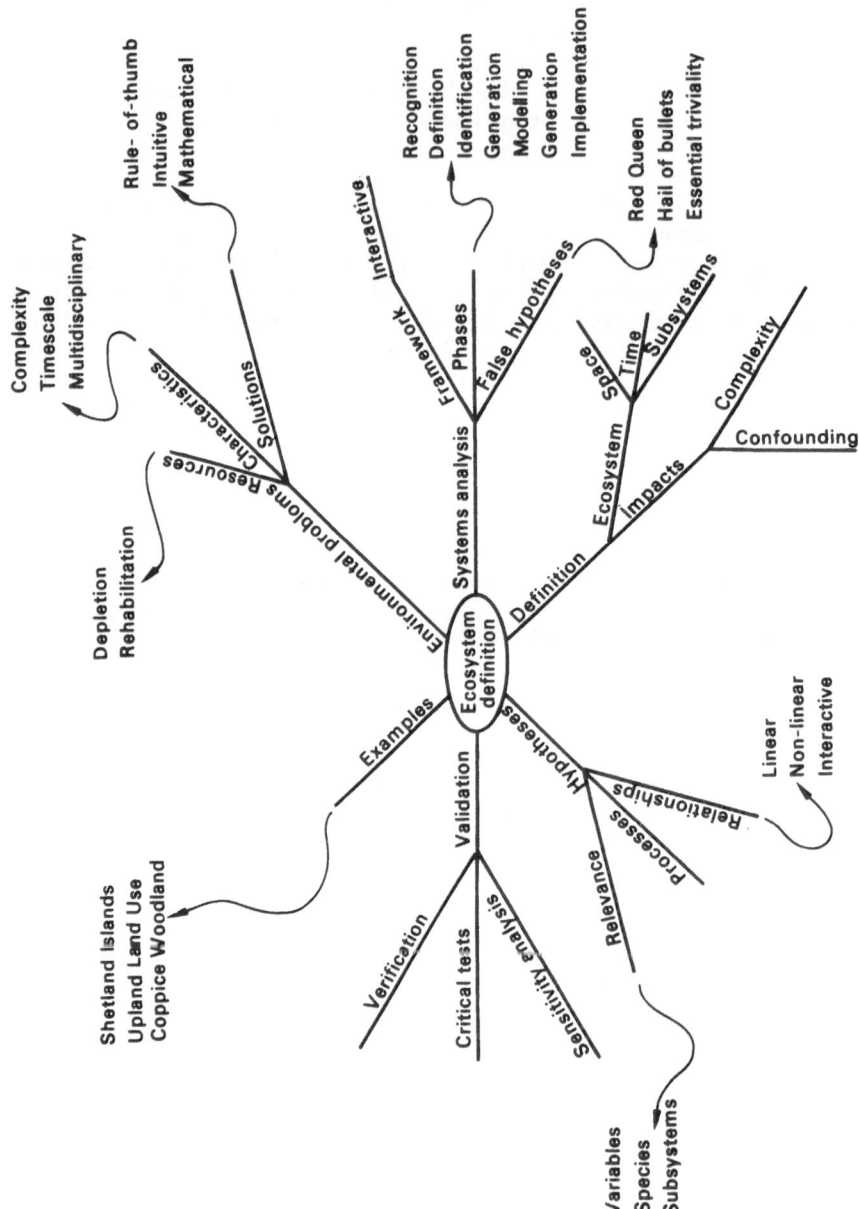

Figure 2. Ecosystem definition.

The bounding of the problem in space and time, however, is usually easier, and consequently usually more explicit than the bounding of the ecological subsystems to be incorporated in the models. Many of the projects of the International Biological Programme (IBP) assumed that it was necessary to model the whole ecosystem and that it was, therefore, unnecessary to define the subsystems of the ecosystem. When the final synthesis was attempted, many IBP projects found that there were major gaps in the system which could not be filled by any of the experimental or survey results, and these gaps were frequently emphasised by the absence of any preliminary synthesis. The experience of IBP had led many ecologists to question the need for studies of whole ecosystems (Heal, in press), and to focus attention upon carefully defined sets of subsystems. In the synthesis of the tundra ecosystems, for example, most attention was concentrated on the decomposer and nutrient cycles as a basis for the prediction of the effects of environmental impacts on the tundra.

The application of systems analysis to ecology is relatively new, and few guides are therefore available for the construction of ecological models. As a result, untested hypotheses are frequently incorporated into model development, and the optimum number of subsystems to be included in the model is difficult to predetermine for a defined acceptable level of accuracy. It can be argued that a more complicated model should be able to account more accurately for complexities in the real system, but, while this argument may appear to be correct intuitively, there are some additional factors to be considered. For example, the hypothesis that greater complexity leads to greater accuracy has been tested by analysing the total uncertainty accompanying model predictions. In general, systematic bias, resulting from abstracting the system into a few subsystems, is inversely related to complexity, but there is a concomitant increase in uncertainty due to errors of measurement of individual parameters within the model. As increasing numbers of parameters are added to the model, these parameters have to be qualified in field and laboratory experiments, and the estimates of the parameter values are never error-free. If these errors of measurement are carried through into a simulation, they contribute to the uncertainty of the predictions derived from the model (see, for example, O'Neill, 1971).

Goodall (1974), however, has emphasised the importance of niche structure in the dynamics of ecosystems, and suggests that an ecosystem model which ignores special differences runs the risk of neglecting important elements in its dynamics. Unless circumstances permit direct comparison of a simplified model with the observed behaviour of a representative range of ecosystems, he recommends that the acceptance of the simplified model should be based on a demonstration that deviations from the behaviour of an alternative model which takes biological diversity fully into account are negligible for the purpose in question.

Perhaps even more important than the need to define the levels
of complexity of the subsystems is the need to define the impacts
to be made upon the system. One of the myths identified by
Yorque (1975) reads "Environmental impact assessment should consider
all possible impacts of the proposed development". No model or
research investigation can possibly foresee all (or even most) of
the impacts likely to be made upon the system, and any investigation will need to be qualified by a series of hypotheses about the
relevance of managerial treatments and impacts. Ideally, the basic
structure of the investigation will enable the interaction of the
various factors to be investigated. Where experiments are to be
conducted, the design of the experiments can incorporate the
factorial structure of the impacts in such a way that the effects
of the impacts are not confounded. The many devices for the control
of factorial structures in experiments enable such investigations
to be carried through with economy and precision. Even where
direct experimentation is not possible, however, it is still necessary to enumerate the relevant impacts and to sample the defined
system so that the effects of these impacts are, if at all possible,
unconfounded. There may, indeed, be little point in continuing
an investigation for which it is not possible to separate the effects
of two or more impacts.

A particularly pertinent example of this difficulty can be
observed in the current attempts to model the relationships
between the variability and concentration of acid rainfall on tree
growth. With some difficulty, it is possible to measure very
small fluctuations in the growth of individual trees, but, so far,
only a relatively small proportion of the total variability of
growth in a period of, say, one hour can be accounted for by climatic variables, including temperature, moisture, wind speeds,
evapotranspiration, etc. When the possible lags between the
variations in climate and the growth response of trees is considered, however, it is not surprising that simple growth models are not
markedly successful. If, however, we add the fluctuations of
concentrations of acid rainfall to this model, we face the further
difficulty that the acid rainfall is itself closely correlated
with the same climatic variables used to characterise the growth
response of the trees. It follows that recording of growth and
climate at a limited number of sites is unlikely to provide any
useful information on the complex interaction between climate,
acid rainfall and tree growth unless some way can be found of
separating the confounded effects of climate and acid rainfall.

HYPOTHESES

In essence, any investigation of the rehabilitation of devastated
ecosystems which purports to be scientific and which attempts to

use systems analysis as the method of investigation requires the
definition and bounding of the problem to be framed as hypotheses
which can be tested formally, even if that test can only be
conducted after a chain of deductive reasoning from one or more
hypotheses which are incapable of direct verification.   Three
basic classes of hypotheses may be distinguished:-

1.   Hypotheses of relevance, identifying and defining the variables,
     species and subsystems which are relevant to the problem.

2.   Hypotheses of processes, linking the subsystems within the
     problem, and defining the impacts imposed upon the system.

3.   Hypotheses of relationships, and of the formal representations
     of those relationships by linear, non-linear and interactive
     mathematical expressions.

These three classes of hypotheses may well be linked within a formal chain, leading to processes which can be summarised by a
decision table enumerating all the hypotheses, and combinations of
hypotheses, that must be specified in order to solve a particular
problem.   It also specifies, for each combination of hypotheses,
the decisions or actions that should be taken to ensure that the
problem is correctly solved (Lewis, 1970).   Because decision
tables provide a clear and concise format for specifying a complex
set of hypotheses and the various consequent courses of action,
they are ideal for describing the conditions for interaction between component parts of a model (Davies, 1974).   The extension of
these techniques to the enumeration of the necessary combinations
of hypotheses for particular courses of action where uncontrolled
events may intervene, so that we are unable to control or predict
with certainty, has been the main thrust of recent research into
decision analysis (Raiffa, 1968).

VALIDATION

The emphasis I have laid upon the necessity for formal hypotheses
defining and bounding the problem for investigation should be
sufficient to dispel any lingering impressions that systems analysis
is some form of higher magic through which the abstractions and
algorithms of mathematics will enable problems to be solved without
careful thought.   Indeed, the need to formulate hypotheses so that
they are capable of being tested - usually by the use of null
hypotheses - will itself focus the major research effort on logical

thought rather than on computation, mathematics and computers. If
that thought has insufficient logic, no amount of computation will
rescue the model from inevitable failure, no matter how enjoyable
the computational exercise.

The emphasis on hypothesis formulation also helps to clarify
the distinction between the verification and validation of systems
models (Mihram, 1972). Although the usage of these terms is not
consistent, verification may be regarded as the process of testing
whether the general behaviour of a model is a "reasonable"
representation of that part of the real-life system which is being
investigated, and whether the mechanisms incorporated in the model
co-incide with the known mechanisms of the system. Verification
is, therefore, a largely subjective assessment of the success of
the modelling rather than an explicit test of the hypotheses
underlying the model. Validation, in contrast, is the quantitative
expression of the extent to which the output of the model agrees
with the behaviour of the real-life system, and is the explicit and
objective test of the basic hypotheses, made by means of a deline-
ation of test procedures, primarily statistical, which are applicable to the determination of the adequacy of the model. In most
ecological applications of systems analysis, this process of
validation has hardly been attempted, mainly because of inadequate
definition and bounding of the problem.

In addition to establishing that a model is or is not valid
for a particular purpose, it is important to investigate whether
changes in the input variables produce large or small changes in
the performance of the model. This investigation is called
sensitivity analysis and, ideally, should begin as soon as any
modelling is attempted (Miller, 1976). Parameters to which the
model behaviour is sensitive can then be made the subject of close
scrutiny and subsequent modification, and it may then be necessary
to undertake further experimental work or data analysis to ensure
that those mechanisms are more precisely modelled. Sensitivity
analysis, particularly if carried out early in the research
project, may greatly aid decisions about the allocation of
resources to various parts of the research programme.

Uncertainties in model performance can also be investigated
by sensitivity analysis, and, because actual uncertainties in the
knowledge of each parameter can be estimated, experimental model
runs can be made with variations in the appropriate order of
magnitude deliberately introduced in each one. For large and
complex models, sensitivity analysis can be a long and expensive
process, but it is essential to discover how models behave within
the full range of variation of the basic parameters, and studying
the effects of change in one parameter at a time provides no
information about interactions. For an example of sensitivity
analysis see Lawless et al (1971).

PRACTICAL EXAMPLES

Examples of the application of systems analysis to practical
problems of the management of ecosystems in Britain include a
study of the effects of oil exploration and exploitation on the
Shetland Islands, a study of upland land use, and the rehabilitation of coppice woodland in southern England.

The natural environment of the Shetland Islands has been
described in a volume edited by Goodier (1974). In 1974, the
Institute of Terrestrial Ecology prepared a project plan for a
multidisciplinary survey of Shetland with the objectives of
providing a baseline for biological surveillance, developing methods
for environmental survey to be used in a wider United Kingdom
context, and furnishing the necessary background and data for the
mathematical modelling of a development area. The habitat and
vegetation surveys were duly completed in 1975 and analysis of
the data obtained from the surveys has defined the habitat and
vegetation types which can be used as a basis for various kinds of
mathematical models. Currently, the most promising approach is a
Markov model which estimates the probabilities of the transitions
from one habitat or vegetation to another corresponding to a
variety of environmental impacts resulting from oil exploration
and exploitation.

Although it is probably not reasonable to regard the Shetlands
as a devastated ecosystem, the monitoring and prediction of changes
in the Shetlands which will result from oil exploitation is important for four reasons:-

1.   Shetland is intrinsically interesting because of its isolation
     and geographical position, and the status of its biota.

2.   Early warning of trends induced by recent developments are
     necessary to protect the Shetland biota by regulatory or
     other methods.

3.   Trends in selected Shetland biota may indicate a potential
     hazard to the ecosystems of the islands, and some of these
     systems are of commercial importance to Shetland, e.g. the
     fishing industry.

4.   In view of the relatively low levels of pollution over much
     of Shetland, trends induced by oil pollution may be more
     easily interpreted in terms of ecosystem function than
     elsewhere.

For this investigation, there was little difficulty in the
bounding of the problem in space or time and the terrestrial,
freshwater, and marine subsystems were equally easy to define.
Initially, the impacts were confined to those of oil spillage,
but it quickly became apparent that other impacts might be of
even greater importance. For example, the limited quantities
of freshwater on the islands impose constraints on oil storage
and processing plants, and it is therefore necessary to include
large transition probabilities in the Markov model for freshwater
habitats.

About one-third of the total area of England, Scotland and
Wales consists of rough grazing, upland forest and moorland and is
the least intensively used area of land. In every sense of the
word, these uplands represent a devastated ecosystem. Much of
the area was originally woodland felled for charcoal, fuel and
timber or cleared to provide grazing for cattle and sheep.
Subsequent management has resulted in the progressive deterioration
of soil fertility, and, even where land is returned to forest, only
the least demanding species can now be grown, at least until a
forest ecosystem has been recreated. A variety of land uses are
concerned with upland Britain, including agriculture, forestry,
recreation, water, mineral extraction, wildlife conservation, etc. -
each represented by one or more national organisation. Multiple
use of the uplands is a natural consequence of the low intensity
of any particular use resulting in a greater interaction between
users than occurs in the lowlands.

The upland land use study was originally initiated to indicate
the relationships between the various ITE projects concerned with
the uplands, to place these projects in a broader context of the
present and future use of uplands, and to identify aspects of
importance for future research. Again, there is no difficulty in
bounding this in space. The bounding of the problem in time is
more difficult as we cannot be sure that future changes will be
comparable with those of the past because of the changing
technology, and the changing economic and social conditions. The
definition and bounding of the relevant subsystems and impacts
has proved to be even more difficult and a desk study has therefore
been commissioned to identify:-

1. the main land uses;
2. their relative importance in terms of area;
3. the associated management practices;
4. the economics of the various management practices;
5. the physiognomic and ecological consequences of the alternat-
      ives;
6. the rates of change.

The coppice system of woodland management is of great antiq-
uity and was the only form of foresty known to be systemcatically
practised by the early Greeks and Romans. In Britain, it has
been practised for some thousand years because many broadleaved
trees produce new shoots from the stump when felled and coppice
regrowth is more or less inevitable on cutting broadleaved
woodland, unless steps are taken to prevent it. Such self-
perpetuating woodland, therefore, provided an early and unsophist-
icated system of forest management which yielded small timbers,
conveniently requiring little or no working. Where a scattering
of trees was left uncut for a longer period, to provide larger
timbers, a coppice-with-standards system developed.

In southern Britain, the original broadleaved forest was
fragmented in Anglo-Saxon times, and the remaining woods were
maintained under a coppice-with-standards system. Although this
system of management continued for at least one thousand years,
changing demands and technology resulted in the gradual abandonment
of this traditional form of management, although the majority of
broadleaved woodlands in the south of England still show signs of
their former coppice management.

The woodland area of lowland Britain occupies some 7 per cent.
of the land surface. In 1965, 33 per cent. of this woodland area,
or approximately 220 thousand hectares, was coppice or scrub woodland which is regarded as having little or no useful end product.

CONCLUSION

The message of this paper is a simple one. Systems analysis is one
possible way of solving complex problems through the generation of
alternative mathematical models of the problem. If systems
analysis is itself to be regarded as a valid expression of the
scientific method, the heart of the method lies in the definition
and bounding of the problem, and this must be done in a way that
provides formal hypotheses that are capable of explicit testing
and rejection. If advocates of systems analysis wish to continue
to claim that validation of systems models is usually impossible,
they must also abandon their claim to be contributing to any part
of science. For systems analysis to become an effective method of
investigation, it is therefore necessary to suppress the natural
tendency for ecologists to become so interested in the ecology
that they lose sight of the problem, and for mathematicians to
become so interested in the mathematics that they transform the
problem into the solution.

Limited experience of the application of systems analysis to
problems of the rehabilitation of damaged ecosystems suggests that

it is usually relatively easy to bound the problem in space and time, although special care may be necessary to deal with changes which are linked with time, as in changes of technology, and of economic and social pressures. Greater difficulty has usually been experienced in identifying the relevant subsystems, and the definition of these subsystems and the formulation of the formal hypotheses necessarily depends upon past research and practical experience. Systems analysis does not itself provide any method for this difficult and essential phase of the investigation. However, even greater difficulty has usually been experienced in identifying the relevant impacts on the ecosystem and of invest-igating these impacts so that their effects and interactions can be determined. The avoidance of the confounding of effects requires an application of basic principles which has frequently been lacking in published examples of the use of systems analysis.

REFERENCES

Beer, S. 1975. Platform for change. London; New York: Wiley.

Bross, I. D. J. 1965. Design for decision. New York: The Free Press.

Davies, N. R. 1974. Decision tables in discrete-system simulation. Simulation 22, 39-44.

Goodall, D. W. 1974. Problems of scale and detail in ecological modelling. J. Environ. Manage, 2, 149-157.

Goodier, R. (Editor) 1974. The natural environment of Shetland. Edinburgh: The Nature Conservancy Council.

Heal, O. W. and Perkins, D. F. (in press). IBP studies on montane grassland and moorlands. Phil. Trans. R. Soc. B.

Jeffers, J. N. R. 1973. Systems modelling and analysis in resource management. J. Environ. Manage., 1, 13-28.

Lawless, R. W., Williams, L. H. and Richie, C. G. 1971. A sensitivity analysis tool for simulation with application to disaster planning. Simulation, 17, 217-223.

Lewis, B. N. 1970. Decision logic tables for algorithms and logical trees. CAS Occas, Pap. 12, London: H.M.S.O.

Mar, B. W. 1974. Problems encountered in multidisciplinary
    resources and environmental simulation models development.
    J. Environ. Manage. 2, 83-100.

Mellanby, K. 1976. Mistaken models. Nature Lond., 259, 523.

Mihram, G. A. 1972. Some practical aspects of the verification and
    validation of simulation models. Opl. Res. Q., 23, 17-29.

Miller, D. R., Butler, Gail and Bramall, Lise (In press). Valid-
    ation of ecological system models. J. Environ. Manage. 4.

O'Neill, O. V. 1971. Error analysis of ecological models. Oak
    Ridge National Laboratory, Tennessee, 24 pp.

Raiffa, H. 1968. Decision analysis - introductory lectures on
    choices under uncertainty. Reading, Mass: Addison-Wesley.

Raiffa, H. 1973. A provisional research strategy. International
    Institute of Applied Systems Analysis.

Satchell, J. E. and Stone, D. A. 1972. Earthworm activity in
    pulverised fuel ash sites restored to agriculture. Report to
    Central Electricity Generating Board.

Yorque, T. (Editor) 1975. Ecological and resilience indicators
    for management. Progress Report, Institute of Resource
    Ecology, Vancouver, B. C. First Workshop.

DISCUSSION: PAPERS 3 and 4

H. REGIER   Jeffers emphasised at the start of his
                            paper the need to define and bound the problem
                            and identify goals and objectives. How waa this
                            done in constructing the model described by Van
                            Dyne?

G. VAN DYNE   Our aim was to predict the dynamics of
                            the main variables in the system under normal
                            extremes of weather and stocking density. We
                            sought to match the measured state variables
                            within 10% of the mean 80% of the time. We had
                            a priori a list of variables we wanted to predict.

                                   May I turn a question back to Jeffers?
                            He said that some people were disillusioned
                            with total system models i.e. those including
                            producers, consumers and decomposers. We did
                            not try to measure every variable in the system,
                            but only those variables for which we could get
                            field data. One of the values of doing a total
                            system model is that you do not ask or answer
                            a question about a sub-system in isolation.
                            It is not possible to predict the dynamics of
                            the animals in the system without information
                            about the plants - they are constrained within
                            the total system. The degree of resolution
                            required for various sub-systems varies, but
                            major components cannot be omitted from the
                            model.

J. N. R. JEFFERS   My criticism is that people call some-
                            thing a total system model when it is not, and
                            that much time and effort have been wasted in
                            attempts at defining so-called total system
                            models. The I.B.P. experience showed this very
                            clearly. As work proceeded gaps were uncovered
                            and work was done to plug them in an ad hoc
                            way that made the total synthesis unconvincing.
                            My suggestion is that it is only possible to
                            address carefully bounded problems, if we wish
                            to work within the framework of the scientific
                            method.

M. W. HOLDGATE   The inference is that only a sequential
                            approach, from defined problem to defined
                            problem, possibly involving many steps, will
                            work.

J. N. R. JEFFERS   Yes, this sequential approach must be
                          stressed - but it is often unsatisfactory to
                          funding agencies, who like to buy a complete
                          package in one step.

N. S. MARGARIS   As a delegate from a small country with
                          limited resources - what is the cost of building
                          a model of the Van Dyne type?

M. W. HOLDGATE   And - can it be extrapolated to other sites
                          of the same generic type to the one for which
                          it was built?

G. VAN DYNE   Eight or ten types of grassland were
                          examined in constructing our models, so as to
                          obtain a range of parameters and help in
                          transference - for example to the kinds of
                          grassland found in Macedonia. The cost of
                          "systems analysis" including several simulation
                          and optimisation modelling efforts, was about
                          12% of about $12 million spent in grassland
                          research over an 8 year period. The cost of
                          transferring models to other sites would be
                          quite small.

5. A SYSTEMS APPROACH TO THE ROLE OF NUTRIENTS IN CONTROLLING REHABILITATION OF TERRESTRIAL ECOSYSTEMS

B. Ulrich

Institute of Soil Science and Forest Nutrition

Goettingen University, Federal Republic of Germany

INTRODUCTION

Damage to ecosystems can be classified in a sequence of increasing severity with respect to biogeochemical cycles as follows:

a) Partial or complete destruction of the vegetation (by man, e.g. by harvesting, by fire etc.)

b) Loss of surface soil horizons, by surface erosion

c) Burying of the ecosystem by sedimentation

d) Loss of soil mantle by gully erosion

In case a) a recovery of the ecosystem is possible if, as in forestry, measures are taken to control the development of new vegetation. A lot of scientific knowledge and practical experience of ecosystem management has been collected in forestry and agriculture in relation to the rehabilitation of ecosystems damaged by harvesting.

The second category - surface erosion - has played and still plays a big role in primitive agriculture. Usually the damage done to the ecosystem is severe enough to prevent recovery of the previous vegetation. If we define an ecosystem as a particular type of vegetation, interacting with a particular climate and type of soil, a new ecosystem will develop which is more or less related to the previous one.

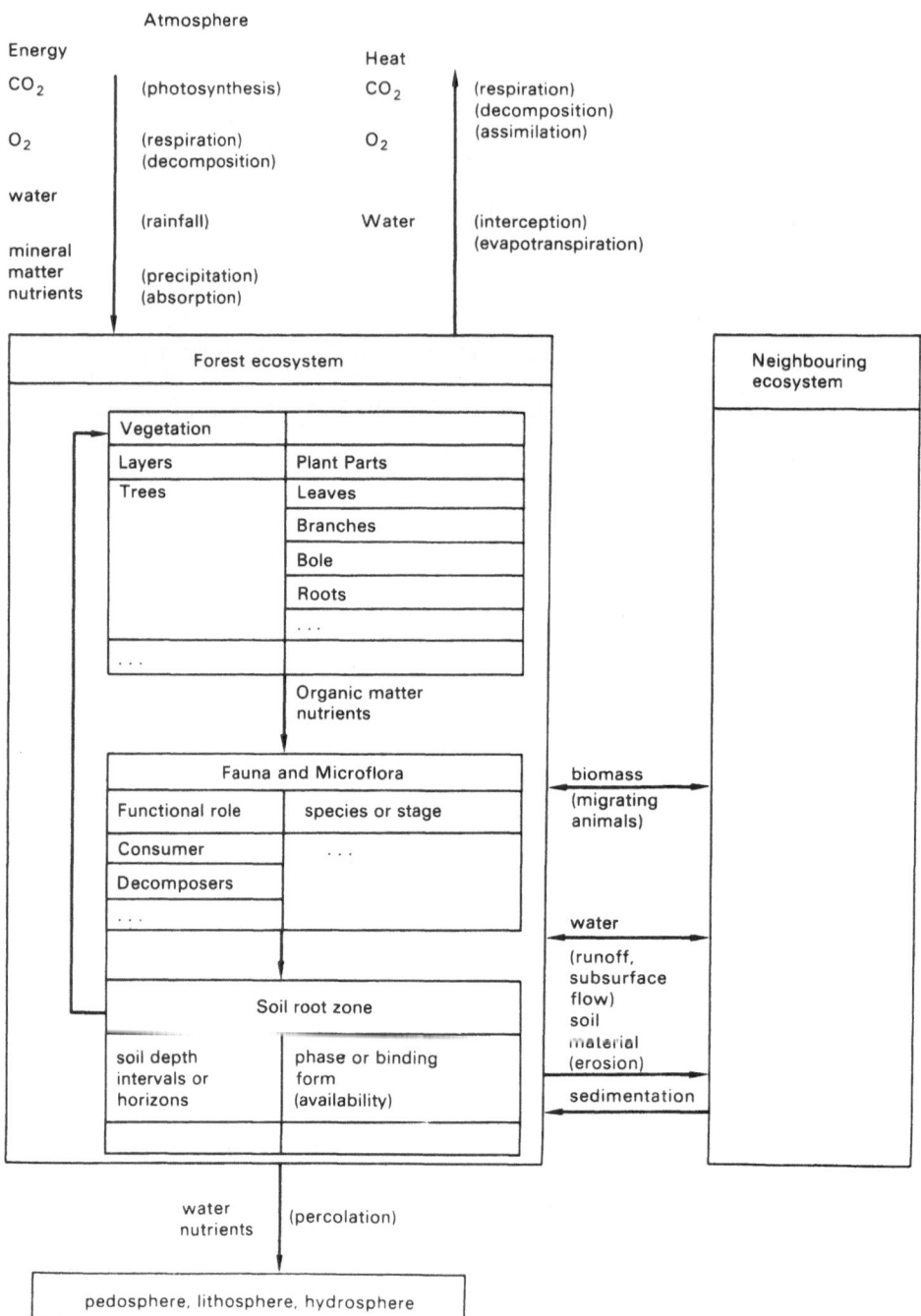

Fig. 1. Compartment model of forest ecosystem.

The same is true in situations (c) and (d) with the restriction that usually no relation exists any more between the previous
and the newly developing ecosystem.

In any of these cases the changes in the growth factors carried
by the soil, that is water, nutrients and oxygen for root respiration, play an important role in controlling the rehabilitation of
the ecosystem. In this paper an attempt is made to give a description of the hydrology, and especially the biogeochemical cycle,
in quantitative terms. This description should indicate the points
that ultimately determine the kind of rehabilitation. It should
further indicate the data base necessary for the development of
mathematical models simulating the rehabilitation, as far as biogeochemical cycles and soil processes (as regulating mechanisms)
are concerned.

THE BIOGEOCHEMICAL CYCLE IN AN ECOSYSTEM

Fig. 1 represents the basic features of the biogeochemical cycle
in an ecosystem in the form of a compartment model. The model sub-
divides the ecosystem in three compartments (vegetation, fauna and
microflora, soil) which can each be further subdivided following
ecological reasoning, into two further hierarchical levels. The
model shows in addition the inputs, out-puts and internal fluxes.
The fluxes of chemical elements represent transport processes
occurring as passive transport (mass flow), with water (rain water,
soil water, xylem sap) and organisms (litter fall, animals) as
transport media.

If the biomass of fauna and microflora is very small compared
with the biomass of vegetation or the mass of organic matter in
soil, the compartment representing fauna and microflora stores
only a small quantity of chemical substances and may be omitted
Consequently the action of these ecosystem components appears in
a systems model as processes.

The matter exchange with neighbouring ecosystems is very often,
especially on flat sites, negligible.

Making use of these simplifications the compartment model shown
in Fig. 2 results. Here the compartments represent the stores (in
kg. ha^{-1}), that means the amount of chemical elements present in
the respective ecosystem components and the arrows represent the
fluxes (in kg. ha^{-1}. yr^{-1}) connecting the compartments with each
other and the environment. Each flux represents one or several
processes. Each process depends upon environmental parameters
(driving forces) or ecosystem parameters (e.g. growth rates of
vegetation or quantity/intensity relations in soil) and may be
described mathematically as functions of the parameters.

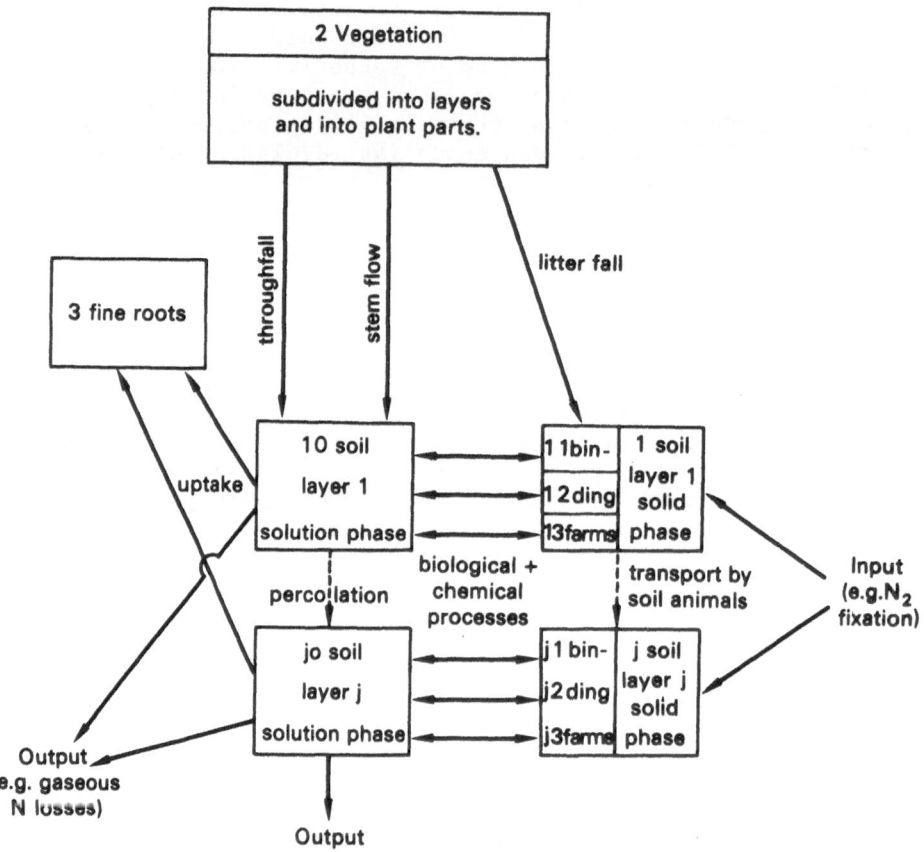

<u>Fig. 2.</u>   Compartment model of mineral cycling in forest.

STORES OF NUTRIENTS

Figures for the nutrient stores in the vegetation are obtained
from measurements of the biomass and the bioelement content of
that biomass. To get data useful for judging the effects of human
impact the biomass should be subdivided into parts affected to
different degrees. For many ecosystems there are enough data on
biomass and nutrient content in literature or other sources for
rough estimation of the nutrient stores in the vegetation without
experimental work.

In respect of soil stores the problem is much more complicated
for several reasons. Soils may contain high amounts of nutrients
in completely unavailable form. To make the model sensitive to
changes in nutrient stores, a separation into binding fractions of
different ecological significance is necessary. There is, however,
still no common agreement between soil scientists concerning such
a fractionation. We have found the following fractions useful:

Cations like K. Ca, Mg: exchangeable fraction. Any other binding
                          form is not considered as part of the
                          system: a possible contribution e.g. by
                          weathering is treated as an input.

Nitrogen: There is agreement about the necessity of
                          differentiating between NH_4^+, NO_3 and
                          organic bound N. A subdivision of the
                          latter according to rate of mineralisation
                          and formation would be useful, but is dif-
                          ficult to achieve due to lack of informa-
                          tion.

Phosphorus: We separate into Calcium-, Aluminium-,
                          Iron-, occluded and organic phospates.

In many countries soil mapping is in an advanced stage.
Usually soil types are mapped and during mapping descriptions of
soil profiles are prepared. But there is no way to derive from a
soil map quantitative data about nutrient stores. If the modelling
is to be more than an academic exercise on a special site, soil
scientists have to develop ways to derive quantitative data for
nutrient stores from qualitative soil descriptions. Research to
achieve this objective is in progress (Shrivastava 1976).

An example of the kind of nutrient inventories which can be
derived from common forest descriptions, knowing the yield class
of the forest and the soil type, is given in Table 1 (Ulrich et
al., 1975).

Table 1. Site types, forest types, management and stores of N, P, and K in the Pleistocene plain of Lower Saxony, Federal Republic in Germany.

		1	2	3	4	5	6
SITE TYPE		POOR SANDS	MEDIUM SANDS	RICH SANDS	SAND LOESS	GLACIAL LOAM	BOULDER CLAY
FOREST TYPE		Pinus silv.	Douglas fir	Quercus sessil.	Picea abies	Fagus silv.	Fagus Carpinus Alnus Acer
stand age at cutting (U)(yrs)		140	80	240	90	140	140
increment $(m^3 yr)$		4.4	10.5	4.8	8.4	6.0	8.1
Thinning: number of interferences		7	5	14	3	6	9
yield (m^3)		320	365	625	260	425	785
Nutrient stores in kg ha^{-1} (stand at age U, sum of bole, bark, branches and leaves; soil up to 50 cm depth)							
(t) = total; (m) = P_t-$P_{occluded}$; (a) = exchangeable							
N	(t) stand	260	690	370	710	500	436
	(t) thinning	430	790	1060	530	610	1030
	(t) humus layer	1700	2100	1100	2400	1000	700
	(t) mineral soil	2300	2500	10500	2900	4600	10500
P	(t) stand	33	100	20	110	70	60
	(t) thining	50	110	60	80	90	150
	(t) humus layer	70	90	60	130	50	40
	(m) mineral soil	170	420	970	630	1180	1660
K	(t) stand	165	380	200	532	280	250
	(t) thinning	270	420	590	390	330	580
	(t) humus layer	150	150	70	150	100	50
	(a) mineral soil	110	260	860	250	380	1890
Ca	(t) stand	210	730	400	450	380	330
	(t) thinning	310	850	1100	340	450	770
	(t) humus layer	150	150	75	150	100	50
	(a) mineral soil	250	650	4900	410	480	10500

FLUXES

The fluxes only have a common meaning if the ecosystem is in
steady state. A natural (unmanaged) ecosystem will be in steady
state if the mean annual outputs are equal to the mean annual
inputs: the mean annual fluxes as well as the mean annual stores
will be constant. For a managed (harvested) ecosystem like a
forest a steady state may be assumed if the inputs are balanced
by the sum of increments and outputs.

In a forest the following fluxes can be determined in the
field by measuring the flow rate of each transport media and its
nutrient concentration (cf. Fig. 2):

- input by precipitation
- flow by precipitation inside stand (sum of canopy drip and
  stem flow)
- flow by litter fall
- flow by percolation inside soil (suction lysimeter plates
  or candles) including output
- increment

From these fluxes the following can be calculated:

- input through filter action of the canopy and bark (see
  Mayer et al., 1974)
- uptake (by flux balance)
- changes in soil store (by input-output balance, assuming
  steady state): function of soil as source (weathering) or
  as sink (filter).

In case of N the quantitative estimation of fixation and
denitrification may be difficult.

Table 2 gives as an example the fluxes in an even aged (120 yrs)
beech forest on acid soil (podzolic brown earth) in the Solling
district. In the following especially the total input (I), the
percolation output (O) and the total uptake (A) for N, P, K, Ca
and Mg will be used for drawing conclusions about the effect of
human impact on the nutrient cycle.

FLOW REGULATING PROCESSES IN SOIL

The fate of nutrients in an ecosystem is governed by the interaction
of several regulatory processes which are located either in the

Table 2.   Flux Balance of Beech on podsolic brown earth

	H	Na	K	Ca	Mg	Fe	Mn	Al	Cl	S	P	N
						$(Kg\ ha^{-1}\ yr^{-1})$						
IP Rain Input	0.90	8.2	3.8	13.9	2.2	0.9	0.6	1.2	17.9	24.5	0.71	23.7
IF Filtering	-	3.8	7.3	8.6	1.3	0.5	1.5	0.5	12.5	14.9	0.09	4.8
IT Total Input = IP + IF	-	12	11.1	22.5	3.5	1.4	2.1	1.7	30.4	39.4	0.8	28.5
F2.10 Precipitation at soil surface	1.21	12.5	22.9	26	3.7	1.2	3.0	1.5	31.6	44.2	0.55	24.7
F2.1 Litter fall	-	0.7	16	16	1.6	1.8	5.1	0.5	0.8	3.2	4	49
Stand Increment	-	0.1	6.6	7.7	1.7	0.7	3.4	0.1	0.1	0.5	2.1	13
O Percolation Output	0.30	10.6	5.9	16.6	3.4	0.1	5.6	12.7	33.2	27.4	0.05	6
Internal Cycling by Washout	-	4.3	11.8	3.5	0.2	0.2	0.9	-0.2	1.2	4.8	-0.25	-3.8
US Uptake from Soil	-	5.1	34.4	27.2	3.5	2.3	7.6	0.4	2.1	8.5	5.85	58.2
UI direct uptake from Input	-					0.2		0.2			0.25	3.8
UT total uptake	-	5.1	34.4	27.2	3.5	2.5	7.6	0.6	2.1	8.5	6.1	62
changes of soil store	+1.7	-1.3	-1.4	-1.8	-1.6	+0.9	-6.9	-11.1	-2.9	+11.5	-1.35	+9.4
changes of soil store in kval/ha	+1.7	-.06	-.03	-.09	-.13	+.03	-.26	-1.23	-.08	+.72	-.13	+.67

environment (driving forces), the vegetation or the soil. To judge
the effect of impact on the vegetation in a constant environment
soil processes are most important.

Percolation Rate

The flow rate of percolating soil water determines the velocity of
the downward transport of dissolved nutrients. In water saturated
soils, the transport occurs at very high rates according to the
saturated permeability of the soil. Usually, the upper part of the
soil, however, is not saturated with water and the transport of
nutrients occurs according to the unsaturated permeability, which
varies with the moisture content of the soil. The amount of water
percolating per annum, F_W, can be estimated by the water balance
equation assuming that no surface flow occurs:

$$F_W = P - (I + E + T)$$

where P = precipitation, I = interception, E = evaporation, T =
actual transpiration. For loess soils in Central Europe $P \sim$ 700 mm,
$I+E+T \sim$ 500 mmm, so $F_W \sim$ 200 mm per annum. This volume of annually
percolating water can be compared with the amount of water stored
in the soil at field capacity (i.e. condition after drainage by
gravity of a saturated soil). Assuming a field capacity of 40%,
then the amount of water stored in a soil layer with a thickness
of 50 cm, is in balance with the annual amount of percolation. It
may be assumed then that the downward movement of nutrients dissolved
in the soil solution in that case is also in the order of 50 cm
per annum.

Damage to the vegetation reduces interception and transpiration
without increasing evaporation accordingly. Thus, the depth of water
infiltration per year is increased and may easily be doubled. Since
in many soils the depth of rooting does not exceed 50 to 100 cm,
this means that dissolved nutrients may be transported out of the
rooting zone within one year after the injury, before restoration
of the vegetation could occur. Since the capability of soils to
accumulate soluble salts is very limited, there is a tendency for
the percolation output to increase according to the reduction in
plant uptake. If this situation holds for some years, appreciable
losses from exchangeable K, Ca and Mg stores are to be expected in
acid sandy soils (cf. Table 1 and 2: ratio of soil store to uptake
in the site type "Poor Sands" for K = 7.5; Ca = 9.2; Mg = 24). As
a further consequence soil acidification will increase.

Physico-chemical Sink and Source Relations in Soil

As already stated, a soil cannot store a soluble salt to any appreciable extent. Cation storage takes place mainly at the exchange sites present on clay minerals and organic matter. For a soil with a definite clay and humus content, the effective cation exchange capacity (CEC_e) is a constant soil property only if the pH remains constant. The CEC_e decreases with decreasing pH and approaches very low values (< 10% of CEC_e at pH 7) at pH 3, especially in mineral soils. Soil acidification means therefore a decrease in the storage capacity for exchangeable cations. The extent of this reduction becomes evident by comparing the Ca stores in the different site types listed in Table 1.

The relations between the cations are governed by cation exchange equilibria, which again are pH dependent, due to changes in the surface properties of soil exchangers. The exchange equilibria determine the change in storage if the fraction of a cation in the solution is changed. The change in the relative composition of the solution reaching the soil surface caused by damage to the vegetation becomes evident if the composition of rain input and of precipitation below the vegetation at soil surface is compared. From the data in Table 2 the following ratios have been calculated:

	H	Na	K	Ca	Mg	Al
$\dfrac{F2,1o}{I\,P}$	1.3	1.5	6.0	1.9	1.7	1.3

If the vegetation is destroyed and the rain input reaches the soil surface directly, a tendency to a limited decrease in the exchangeable K store can be expected. Due to the easily weatherable K store present in many soils in illitic clay minerals this will be of no ecological significance. The phosphorus chemistry in soils is characterised by chemosorption prcesses approaching the behaviour of low soluble salts. This results in an almost constant output which is independent of changes in input.

Biological Processes

Biological processes play almost no role in the case of K, where even liberation during litter decomposition is predominantly a leaching process.

The other extreme is N, which exists almost entirely in organic binding (except fixed NH_4^+ in illitic clay minerals) and is completely governed by biological processes: by mineralisation ($N_{org} \rightarrow NH_4^+$), nitrification ($NH_4^+ \rightarrow NO_3^+$), denitrification ($NH_4^+$, $NO_3^- \rightarrow$

N_2O, NO, N_2) and assimilation ($N_2 \rightarrow N_{org}$).

As a consequence of damage to the vegetation layer, the radiation reaching soil surface may be increased. Since in many ecosystems of the northern hemisphere the temperature otpimum for the mineralisation process is not achieved, an increase of radiation at the soil surface will stimulate the mineralisation, thus leading to a pouring out of NH_4^+ and NO_3^-. In extreme cases (complete destruction of the vegetation and favourable microclimatic conditions) the organic matter accumulated in the humus layer on top of the mineral soil will be completely decomposed within a few years. According to Table 1 more than 2,000 kg N/ha, approaching half of the total soil store in site types characterised by moder or raw humus, can be mineralised. Under such circumstances denitrification which is negligible in steady state conditions in many ecosystems will also take place at ecologically significant rates. If no plant cover is present, that is if uptake is zero, the N input from rain as well as the mineralised nitrogen will be lost from the ecosystem partly by leaching and partly by denitrification. N losses of 1,000 to 2,000 kg ha^{-1}, corresponding to 100% of the easily mineralisable store, will lead to a sharp decrease in available N, thus reducing N uptake as well as growth drastically. In site types character- ised by mull humus (No. 6 in Table 1) however no increase in mineral- isation may occur due to the existence of a soil vegetation layer with high biomass production.

Ecosystems with good biological soil conditions, relatively high biomass of soil fauna and a high number of different species are consequently much less sensitive to partial or complete destruction of vegetation. Ecosystems with bad biological soil conditions, as indicated by raw humus, may - on the other hand - suffer considerably.

Rehabilitation depends predominantly on the amount of N avail- able for plant uptake. Table 2 shows that at present in Central Europe as a consequence of industrialisation the N input by rain amounts to > 20 kg N/ha $^{-1}$ yr $^{-1}$, that is 1/3 to 1/4 of the annual uptake of an ecosystem with mean biomass production. If we assume that at the beginning of the rehabilitation phase half of the up- take is stored in the biomass increment and half cycled, then an annual uptake rate of 40 kg N/ha will be reached again after only 4 years, allowing a quite well developed vegetation cover. If re- habilitation begins with a rain input of 4 kg N/ha^{-1} yr^{-1}, as the natural background may be without human influence, the vegetation cover can only develop much more slowly.

An increase of the N uptake rate and the corresponding growth rate may be caused by participation in the plant cover of species assimilating N_2 by help of microorganisms.

CONSTRUCTION OF MATHEMATICAL MODELS

As a first approximation, a mathematical model of the biogeochemical cycle in an ecosystem may be derived from the data base of stores and fluxes only, if the system has a high degree of feed-back. A balanced input-output model is in its simplest form represented by

$$\boxed{V_i} \xrightarrow{\quad F_{ij} \quad} \blacktriangleright \boxed{V_j}$$

where V_i and V_j are the stores (kg . ha^{-1}) at the beginning of the simulation period and F_{ij} the annual flux from compartment 1 into compartment 2 (kg . ha^{-1} . yr^{-1}). It is often reasonable to assume that the flux is proportional to the content of the donor compartment:

$$F_{ij} = A_{ij} . V_i$$

The proportionality constant A_{ij} can be calculated from measurements of V_i and F_{ij}. For a further discussion of this approach see Patten (1971).

Where the boxes i and j represent soil solution and solid soil phase respectively, the model is a description of quantity – intensity relations existing in soil. These relations can be expressed mathematically for each element, either by statistical expressions (regression equations) or by equations with a physico-chemical foundation (see Boast, 1973). Even for biological processes like the transformations in the nitrogen cycle, reasonable approaches exist (Frissel, 1975). These relationships can be fitted into the model in order to simulate the regulating function of the soil.

Such a model is adapted only to simulate the effect of changes in stores or fluxes. In terms of the model the destruction of vegetation means changes in compartment stores and can therefore be simulated within limits by such a simple model. The calculations can be done with a computer programme like Continuous System Modelling Programme (CSMP) from IBM (Van der Ploeg et al., 1975).

As examples Figs. 3 and 4 show a simulation of P uptake rate (Fig. 3) and of P store in the humus layer (Fig. 4) for the forest type 1 (pine on poor sands) and 2 (Douglas fir on medium sands) of Table 1; D means thinning (from Ulrich et al., 1975). Whereas in pine, according to the result of simulation calculations, the uptake rate as well as the P store in the humus layer increases with increasing age, the opposite is true for Douglas fir. The difference between both ecosystems is caused by a positive input/output + increment balance for pine and a negative one for Douglas fir. The rapid decline in humus layer store after clear cutting arises in the extreme case when no vegetation is growing up.

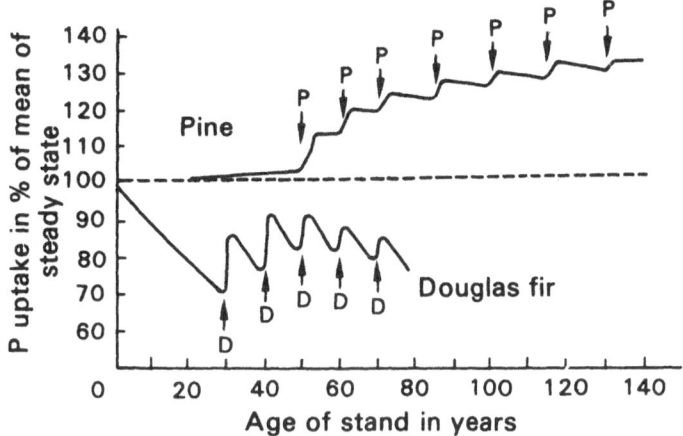

Fig. 3. Simulation of P uptake

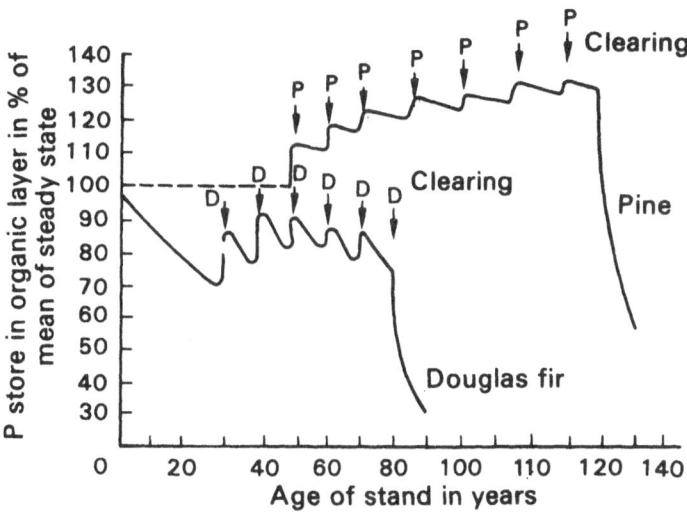

Fig. 4. Simulation of P store in soil organic layer

In this example we may consider nutrient uptake as the
controlling factor for growth rate. It shows how the recycling
of matter caused by thinning may influence growth. It shows further
that long term nutritional effects may exist, changing the vegetation
composition gradually in order to arrive at a balanced input/output
status. According to this result the fate of an ecosystem seems
to depend strongly on the input/output balance. Small shifts in
this balance will have great effects in the long run.

HISTORICAL REVIEW AND FUTURE OUTLOOK

Ecosystems are highly susceptible to changes in the input/output
balances if a time scale of decades is considered. This points to
the fact that destruction as well as rehabilitation of ecosystems
may occur at rates imperceptible by man.

Beginning roughly at 1,000 B.C. the devastation of natural
ecosystems by grazing, harvesting and following soil erosion
affected large parts of Central Europe to different degrees. Wit-
nesses of these events are the loamy sediments in the river valleys,
which are highly productive soils today, and in the river deltas.
In the Leine valley where Goettingen is located the loamy river
sediments were formed between 600 B.C. and 350 A.D. and again in
the Middle Ages (around 1,500 A.D.) (Wildhagen, 1972).

More recent evidence comes from maps of the 17th and 18th
century showing the distribution of forest and heath land. They
show that on the pleistocene plain north of Hanover large areas
of potential forest land has been occupied by heath. The forest
ecosystems were destroyed by over-exploitation of wood, sheep
grazing and by collecting the heather for cattle bedding down.
Over a few centuries the biomass production of the land fell to a
small percentage of that of the natural forest ecosystem. In the
mountain regions of Germany on large areas much the same has
happened as a consequence of collecting litter for cattle bedding
(Kreutzer, 1972).

Today this tremendous damage has more or less been cured, and
the forests are probably growing better than at any time. This
surprising change happened within the last 100 years for two reasons:
on the one hand the forest service rehabilitated forest ecosystems
by planting and re-establishing nutrient cycling (by keeping farmers
out of the forest) and on the other, industrialisation changed the
input/output balance. Nutrient gain, especially of nitrogen, from
the air, as shown in Table 2, should have resulted in a steady
increase in growth rate during recent decades in many unfertilised

ecosystems. The attentive observer finds in forests as well as in heath lands and peats many hints of a steady improvement in growth rate or - on heath lands and peats - the invasion of new plant species. Direct scientific assessment of the changes taken place during the last two to three decades is still missing, however.

In highly industrialised areas like Central Europe the rehabilitation of ecosystems therefore presents no problems at the moment, if the physical soil conditions are satisfactory and if the soil is not poisoned. As a proof for this statement, the rehabilitation of surface brown coal mining areas in the Ville district near Cologne may be cited.

Unfortunately there are reasons to assume (Ulrich, 1975) that the ecological situation will change again slowly but - in the long run - drastically. There are indications that the effectiveness of nitrogen input from rain in increasing growth rates of spruce on acid soils in Central Europe is already limited by a shortage of magnesium. The shortage of magnesium in soil will increase with time as a consequence of the acid rain caused by SO_2 emissions. The tendency to podzolisation in soils is increasing. Within the next decades the productivity of many poorer soils will probably decline if no fertilisers are applied.

REFERENCES

Boast, C.W.   1973 : Modelling the movement of chemicals in soils by water. Soil Sci.,115, 224-230.

Frissel, M. 1975. PUDOC Wageningen.

Kreutzer, K.   1972 : Uber den Einfluß der Streunutzung auf den Stickstoffhaushalt von Kiefernbestanden. Forstwiss. Zent Bl., 91, 263-270.

Mayer, R. and Ulrich, B. 1974 : Conclusion on the filtering action of forests from ecosystem analysis. Oecol. Plant., 9, 157-168.

Patten, B. C.   1971 : Systems Analysis and Simulation Ecology. New York: Academic Press.

Shrivastava, M.B. 1976 :   Quantifizierung der Beziehungen zwischen Standortsfaktoren und Oberhöhe am Beispiel der Fichte (Picea abies Karst.) in Hessen. Diss. Univ. Göttingen.

Van der Ploeg, R.R., Ulrich, B., Prenzel, J. and Benecke, P. 1975.
    Modelling the mass balance of forest ecosystems. Simulation
    Councils, Inc., P.O. Box 2228 La Jolla, California.

Ulrich, B. 1975 : Die Umweltbeeinflussung des Nährstoffhaushalts
    eines bodensauren Buchenwalds. Forstwiss. Zent Bl., 94, 280-287.

Ulrich, B., Mayer, R. and Sommer, U. 1975. Rückwirkungen der
    Wirtschaftsführung über den Nahrstoffhaushalt auf die Leist-
    ungsfähigkeit der Standorte. Forstarchiv 46, 5-8.

Wildhagen, H. and Meyer, B. 1972: Holozäne Bodenentwicklung,
    Sedimentbildung und Geomorphogenese im Flußauen-Bereich des
    Göttinger Leinetal-Grabens. Göttinger Bodenkundl. Ber. 21.
    1-158.

DISCUSSION: PAPER 5

H. P. BLUME	Professor Ulrich's last table showed a large output of nutrients, but what would the pattern have been under natural conditions? Our calculations showed an output of potassium of 2 kg ha^{-1} yr^{-1} in some woodland soils, under natural conditions. Can we interpret artificial changes until we understand the natural system?
B. ULRICH	Output depends on input. Even under natural conditions, rain contains some salts. The output from the soil follows because soils cannot accumulate salts as water moves down the profile. Output and input should balance. Today's potassium input is 4 kg ha^{-1} yr^{-1}: 2 kg of this comes from the sea and must have been balanced by a natural output.
H. P. BLUME	My figure of 2 kg was for weathering: if 2 kg also comes from the sea, the natural input and output must have been about 4 kg ha^{-1} yr^{-1}.
B. ULRICH	It is difficult to speak of natural inputs and outputs when the composition of all rain-fall over Western Europe is likely to reflect changes due to human activity.
N. POLUNIN	Would it not help if we stopped trying to discriminate between natural and modified conditions, since man is so universally

dominant? Even in the remotest places where
there are no records of his impact, his traces
can be found. Is not creative conservation
the thing to aim at, moving systems to the
conditions we believe to be most desirable.

D. R. SUKOPP   The main distinction may be between
integrated species, functioning as fully
established components of the ecosystem, and
non-integrated species which are still dependent
on man for their perpetuation, for example
in disturbed situations.

Part II: The Degradation of Land and Freshwater Ecosystems in Temperate Lands

Introduction

M. W. Holdgate

This part of the symposium was concerned with the history of man's impact on ecosystems in temperate areas, almost entirely in the Northern hemisphere. The first three papers review what has happened to terrestrial environments in three regions: North West Europe as a whole, Iceland, and the Mediterranean region, while the fourth (paper 9) describes the recent history of the Rhine as a classic illustration of degradation in a freshwater system. Wolff's paper is also illustrative of pollution by organic wastes and toxic substances as a factor in ecosystem breakdown and Wein's account (paper 10) analyses the role of another potent factor, fire.

AN APPROACH TO ECOSYSTEM DEGRADATION: OPENING REMARKS BY SESSION CHAIRMAN

H. Sukopp

Institut für Okologie
Technische Universitat Berlin
Federal Republic of Germany

In approaching this part of the volume, it may be helpful to focus attention on the question of how to compare the types of degradation caused by different impacts.

There are two main approaches:

1. The concept of retrogression (Whittaker and Woodwell, 1973) Chronic disturbances or stresses applied to natural communities by man can produce changes that are in some respects the reverse of succession, changes termed "retrogression". Reduction in any community characteristic that increases through succession might be used as a measure of retrogression, but many such measurements are ineffective. Reduction of particular sensitive species or other taxa may indicate the beginning of retrogression. For more general measure-

Table 1. Classification of Different Forms of Land Use Graded by Influence of Cultivation on Ecosystems in Central Europe

Hemeroby grades JALAS & SUKOPP	Ecosystems	Anthropogeneous Influence	Changes in Vegetation	Indicators Share of Neophytes in species of vasc
natural (ahemerobic)	Regions of rocks, moors and Tundra in some parts of Europe, in Central Europe only in parts of high mountains	Not existent	Not existent	0%
sub-natural (oligohemerobic)	Slightly thinned or grazed forests, growing dunes, growing flat or raised bogs	Minor wood withdrawal, pasture, air (e.g.SO_2) and water immissions (e.g.flooding of plain by eutrophic water)	Not stonger than to clearly show the original features of vegetation on the natural site	5%
semi-natural (mesohemerobic)	Forests of trees from alien habitats; heath, rough and dry grassland; landscape parks (extensive meadows & pastures)	Uprooting, seldom ploughing and/or felling, taking out of litter and heathsods, occasionally light fertilization	The natural forest is repressed in favour of heath and grassland	5-12%
cultivated (euhemerobic)	Forests of species introduced from other floristic regions. intensive pastures & lawns.	Fertilization, liming, mechanical weed-control, light drainage of trenches	Extension of intensely cultivated areas of high productivity in small distribution patterns	13-17%
β	Acres and fields	Levelling.steady ploughing.minor fertilization with minerals		
α	Special cultivations. e.g.fruit.wineyards. lawns.or field crop alternation with strongly selected weeds.	Deep ploughing. permanent and thorough drainage (and/or intensive irrigation). intensive fertilization and use of biocides.	Expanded areas of monocultures, settlements & waste grounds with biocoenosis of few competetive capacity,also many shortliving ruderal plant associations	18-22%
	Sewage farms	Adapting; strong irrigation with sewage	on sites which originate and disappear within short time and not periodically.	
artificial (polyhemerobic)	New refuse and rubble dumps. areas with ruin remains	One time extinction of the biocoenosis while biotope covered with alien material		>23%
	partially urbanized areas (e.g.paved lanes.macadamized installation of tracks)	Biocoenosis strongly reduced. biotope permanently and strongly changed		
devastated (metahemerobic)	poisoned ecosystems	biocoenosis extinct		—
	completely urbanized ecosystems (e.g. buildings. asphalt)	biocoenosis extinct	Extreme and onesided so that all life is extinguished (intentionally unintentionally)	

Share of Therophytes ular Plants	Influence on soil forming processes	Habitat changes and changes of edaphic qualities	Indicators Changes of Diagnostic Characteristics comp.w/natural soils
	not existent	not existent	not existent
<20%	Decomposition of litter, increasing acidity(by air borne acids)or alkalinity (by burdened river water)	Minor changes of nutrient supply	humusform; increasing Cl-, SO_4-contents in the soil solution
	different intensity of Decomposition and humification, partially increasing podolization(by litter of heath or conifers) or pseudogley formation (if transpiration was lowered)	Minor changes of nutrient supply, of water- or oxygen supply	humusform more dystrophic or more eutrophic
21-30%	Decomposition, humification & aggregation increased; acidification, podzolization & gley formation decreased	Increased nutrient supply with pH-changed availability of nutrient reserves; changed water- and oxygen supply	not O-Horizon pH-Increase
30-40%	as above; in addition shallow disturbance, erosion	as above; in addition shallow changes of the rooting process in the upper soil	formation of Ap-horizon pH-increase
	as above; profound disturbance, erosion, transposition	Strongly increased nutrient supply (and nutrient washout).simultaneous with decrease in nutrient disposal caused by redox changes; increasing rooting dept of the subsoil. increasing oxygen- and water supply	Formation of cultosols with humic,homogeneous topsoil>30-80 cm pH-increase
	Hydromorphologic processes humus accumulation. decomposition of soil	Strongly increased nutrient & water supply (& outwash) combined with decreasing	Rusty spots; of Na-increase saturation
>40%	(Part)fossilization if sediment layed on	Change of all habitat qualities; new combination or extreme concentration of factors	covered with anthropogeneous rocks
	Decrease in litter decomposition and bioterbation	Decrease of root penetration capability	O- and A_h-horizon not existent
—	Strong decrease of biogene processes (decomposition, humification,bioterbation)	Prevailing of obnoxious substances	CO_2 set free strongly diminshed down to not existent
		loss of space penetrable by roots	

ments three approaches seem more useful: (i) weighted averages, which may use "decreasers, increasers, and invaders" as ecological groups, (ii) reduction of species-diversity from that of a control or undisturbed sample, and (iii) coefficients of community for disturbed compared with undisturbed samples. The second and third of these make possible definition of a 50% change in the community with retrogression, and which may be compared with the responses of a given community to different stresses, or of different communities to a given stress.

2.   The concepts of "degrees of naturalness" or "hemerobiotic grades" (Jalas 1955, Sukopp 1972).
The effects of anthropogenous changes on ecosystems can be divided into hemerobiotic grades (Table 1). By hemerobiotic we mean the total of all effects on ecosystems when man voluntarily or involuntarily influences them. The grade of hemeroby of the ecosystem is deduced from the effects on the respective habitat with its organisms. It can be determined by either habitat research or by analysis of its biocoenosis which supplement each other. A special system is the well known saprobiotic system in limnology.

The hemerobiotic grade is an integrated expression for the influence of cultivation which does not replace a factor-analysis but is used where the analysis of single effective factors is not yet possible or where one cannot be sure whether all factors were registered. For the classification of the present state of an ecosystem in reference to human influence we register the following basic dimensions of hemeroby: 1. Intensity, 2. Duration, 3. Range of influence. The effect during the first period following an interference equals the product of intensity plus duration. A strong interference can have the same effect within a short period as an influence of little intensity over a long period of time. After a certain period of steady and regular influence a condition is reached where the same intensity causes only small changes. A balance, caused also by man, has been established.

Table 1 column 1 shows from top to bottom the increasing influence of cultivation expressed by hemeroby-grades; column 2 contains typical modified ecosystems; column 3 records human influence; column 4 to 6 changes in vegetation; column 7 contains crucial processes in soil formation, the succession of which was either influenced by man (e.g. stopping podzolisation on account of calcium) or caused by man (e.g. ploughing can be understood as a special form of disturbance); column 8

shows changes of soil qualities which are of direct importance for soil fauna; column 9, finally, lists easily detectable soil qualities which reflect the grade of anthropogeneous
changes of a habitat. Proper interpretation though, depends
on the fact that the original state can be reconstructed which
is not always easy.

The columns 5, 6 and 9 of the Table contain key-character-
istics enabling a fast determination of the hemeroby-grades
to be made; they are not to be mistaken for definitions.

In this table the grade of anthropogeneous changes in soil
always increases from top to bottom, often also increasingly
changing soil horizons in greater depth. Thus soil morphology
is changed only within the upper centimetres with oligo-hemer-
obic habitats; with βeu-hemerobic ones the change already
affects the entire upper layer, while with many αeu-hemerobic
habitats the changes reach a depth of 80 cm and more.

Basically it can be said that anthropogeneously changed
sites, left to themselves, go through the grades of hemeroby
from bottom to top (of the Table), in order to return more
and more closely to their original condition.

REFERENCES

Whittaker, R. H. and Woodwell, G. M. 1973.   Retrogression and
    Coenocline Distance (53-73). Handbook of Vegetation Science
    5. The Hague.

Jalas, J. (1955). Hemerobe und hemerochore Pflanzenarten. Ein
    terminologischer Reformversuch. Acta Soc. Fauna Flora Fenn.,
    72, (11) 1-15.

Sukopp, H. (1972). Wandel von Flora und Vegetation unter dem
    Einfluß des Menschen. Ber. Landw., 50, 112-139.

6. PREHISTORIC MAN'S IMPACT ON ENVIRONMENTS IN NORTH WEST EUROPE

G. W. Dimbleby

Institute of Archaeology

London, England

INTRODUCTION

History and archaeology concern themselves with people. People
enter an area, adopt some form of settlement and may eventually be
driven out by pressure from a different group of people. There is
a continual process of change, sometimes slow, sometimes very rapid,
in the peoples who move across the face of the land. But they not
only move across the face of the land; they use the land and they
inevitably disturb the ecosystems which were present there in the
first place. Today we are very concerned about the impact of mankind on the environment; many people think this is a recent
problem and that 50, 100 or perhaps 250 years ago no such troubles
existed.

Ever since agriculture was devised, man has altered the
environment for his own purposes; he has also altered it unwitting-
ly. Even before agriculture he had potent ecological influences
at his disposal, such as fire, or the manipulation of populations
of wild herbivores, and we are able to detect the local impact even
of Mesolithic man upon the ecosystems of north west Europe.
Agriculture was brought to this region by Neolithic peoples using
crops that had been developed in the very different environments
of south east Europe and south west Asia. Such crops were light-
demanding and could not be grown at these higher latitudes without
the removal of overhead forest canopy; that is, the ecological
dominants had to go.

The first land clearance (Figure 1) - called landnam by Iversen
(1941) - was surprisingly widespread considering the small populat-

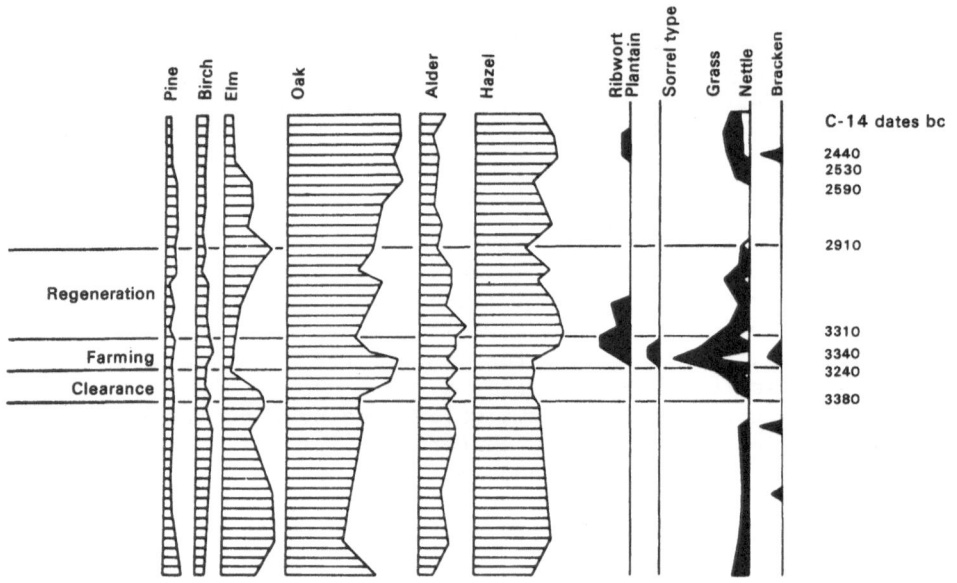

Fig. 1. Pollen diagram from Fallahogy, Ireland, showing a primary early neolithic episode of clearance for farming. (From J.G. Evans, <u>The Environment of Early Man in the British Isles</u>).

ions that were involved, though there were places, such as on
poor acid soils, that were apparently little influenced by the
Neolithic people; such places have not been left untouched, but
were subjected to clearance by later agriculturalists, particularly
of the Bronze Age. Not only could the effect of this prehistoric
people be detected over a wide area, but once agriculture was being
practised there was progressive deforestation, resulting in a
dramatic change of landscape. In Britain the uplands and the ligh-
ter soils of the lowlands were as devoid of forest by Roman times
as they are today, though the landscape differed in detail. Forests
on heavier low-lying ground persisted until post-Roman times, when
heavier ploughs and drainage techniques could be employed.

It can be shown then, that in many areas the condition of tree-
lessness that we see today was first brought about by prehistoric
man. However, his effect goes deeper than merely the removal of
trees; in places the fertility of the land was altered, usually
for the worse, and sometimes apparently irreversibly. Much of the
wet moorland of the north west European seaboard, or the extensive
heathland of the north European plain, is a man-made artifact dating
back in the main to prehistoric times; it arises as a consequence
of the destruction of the primary ecosystem, the deciduous forest,
and the land use which followed it.

Though the primary forest may have been destroyed, secondary
successions are always at work, tending to re-establish the primary
ecosystem or something very like it. Sometimes these regenerative
successions were able to proceed to a new climax; in other cases
the primary type of ecosystem was never restored. The situation
was complicated by the fact that the original clearance was not
the only impact of prehistoric man; there were often successive
waves of clearance and regrowth (Figure 2) before extensive clearance became permanent (Turner 1965).

As one would expect with living systems many different variants
exist, according to the environmental circumstances and the different land use practices. However, these variations usually fall
into one or other of the following categories:

(a)   After a clearance episode, the forest ecosystem is
      restored, virtually unchanged, by succession.

(b)   After a clearance episode, succession leads to a weaker
      representation of the forest ecosystem.

(c)   After clearance the forest does not return because of contin-
      ued use.

(d)   After clearance the forest does not return because of environ-
      mental deterioration.

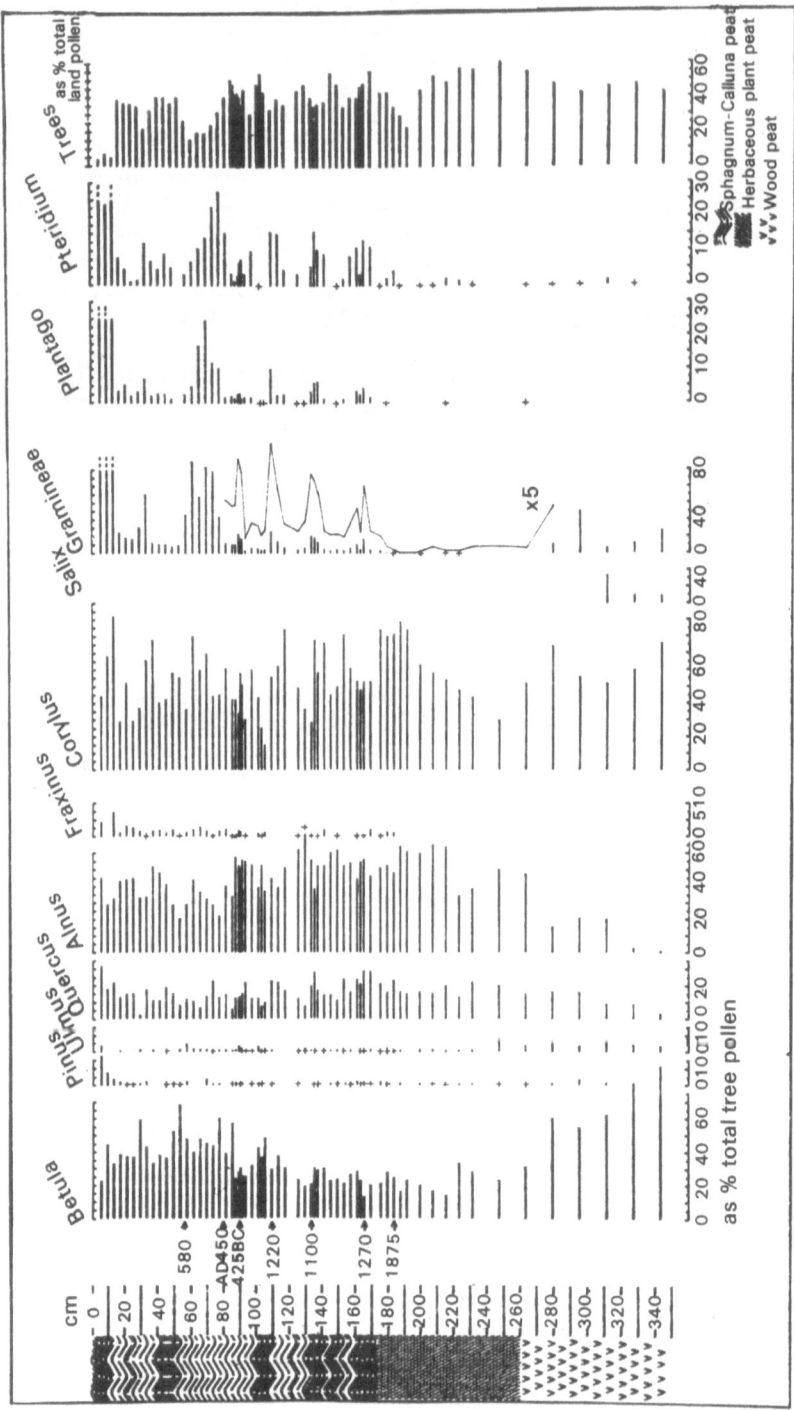

Fig. 2. Pollen diagram from Bloak Moss, Ayrshire, showing successive temporary clearances. Note curves for Gramineae, Plantago and Pteridium (From Turner, 1965).

RETURN OF ORIGINAL ECOSYSTEM

Pollen analyses of peat bogs can give evidence of clearance phases and the subsequent return of the forest. A good example of this pattern of forest restoration is shown by Smith (1975) for a lowland area of Northern Ireland. Here successive clearance phases, separated by centuries, are shown by the reduction of the tree pollen percentage and the increase of light-demanding grasses and herbs (Figure 3). The episodes of clearance give way to forest regrowth, and in each case the forest returns to its previous degree of dominance, as reflected by the non-arboreal pollen percentages.

It can also be shown by pollen analysis not only that the forest can return to its original dominance, but that the specific composition, upset by human interference, can restore itself. Table 1 (based on data from Dimbleby, 1962) shows the pollen analysis by tree genera of the forest on the watershed of the North York Moors,

a) before Mesolithic occupation,

(b) after Mesolithic influence (i.e. pre-agriculture),

(c) when the first clearance for agriculture took place, in the Bronze Age.

The percentages when the Bronze Age people attacked the forest were very similar to those before the Mesolithic people affected the forest, despite the fact that this Mesolithic occupation had led to a greater proportion of light-demanding species such as birch. The forest had clearly re-established its original composition once the temporary Mesolithic influence had passed away.

RETURN OF A WEAKENED FOREST COVER

This, too, is well illustrated by Smith (1975) from an upland area in Northern Ireland. Again the clearance phases are apparent, but following each one the forest does not return to its original dominance (Figure 4). Over a period of time there is a progressive replacement of forest by open conditions. There is no need to postulate any difference in the human influences in this situation as compared with the last; given the same pattern of clearance, the two situations would differ, due to the difference in the factors of the habitat which affect the regeneration succession - rainfall, exposure, soil fertility and depth etc. The pattern of such factors is of fundamental importance in determining the effect of prehistoric clearance. Jonassen (1950) came

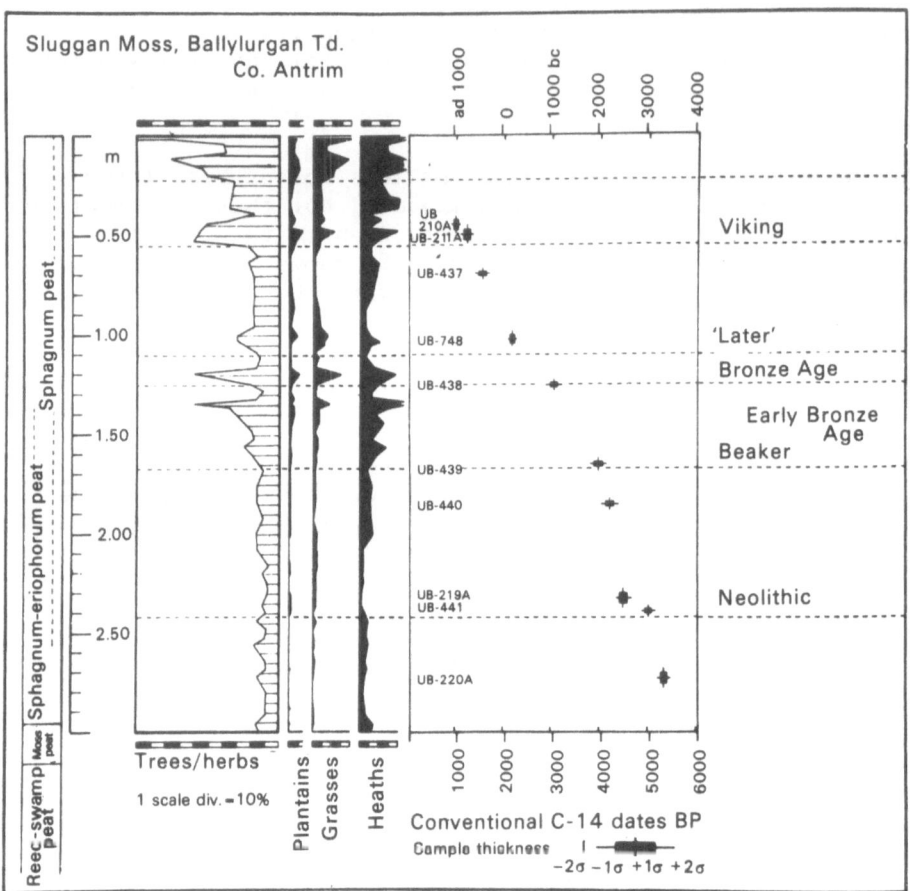

<u>Fig. 3.</u>   Dated pollen diagram showing phases of prehistoric clearance
         followed by full return of forest cover (From Smith, 1975).

Table 1. Percentages of pollen of tree species on North York Moors in Mesolithic and Bronze Age.

	BIRCH	PINE	ELM	LIME	ALDER	OAK	ASH	$\sum$ AP
			PERCENTAGES OF TREE POLLEN (AP)					
BEFORE MESOLITHIC	10	1	3	1	52	34	-	189
AFTER MESOLITHIC	15	1	3	2	44	34	1	515
BRONZE AGE (a)	9	2	2	1	56	26	3	331
BRONZE AGE (b)	9	1	2	1	57	27	5	132
BRONZE AGE (c)	5	3	-	+	55	36	2	66
MODERN	25	5	8	-	30	21	11	61

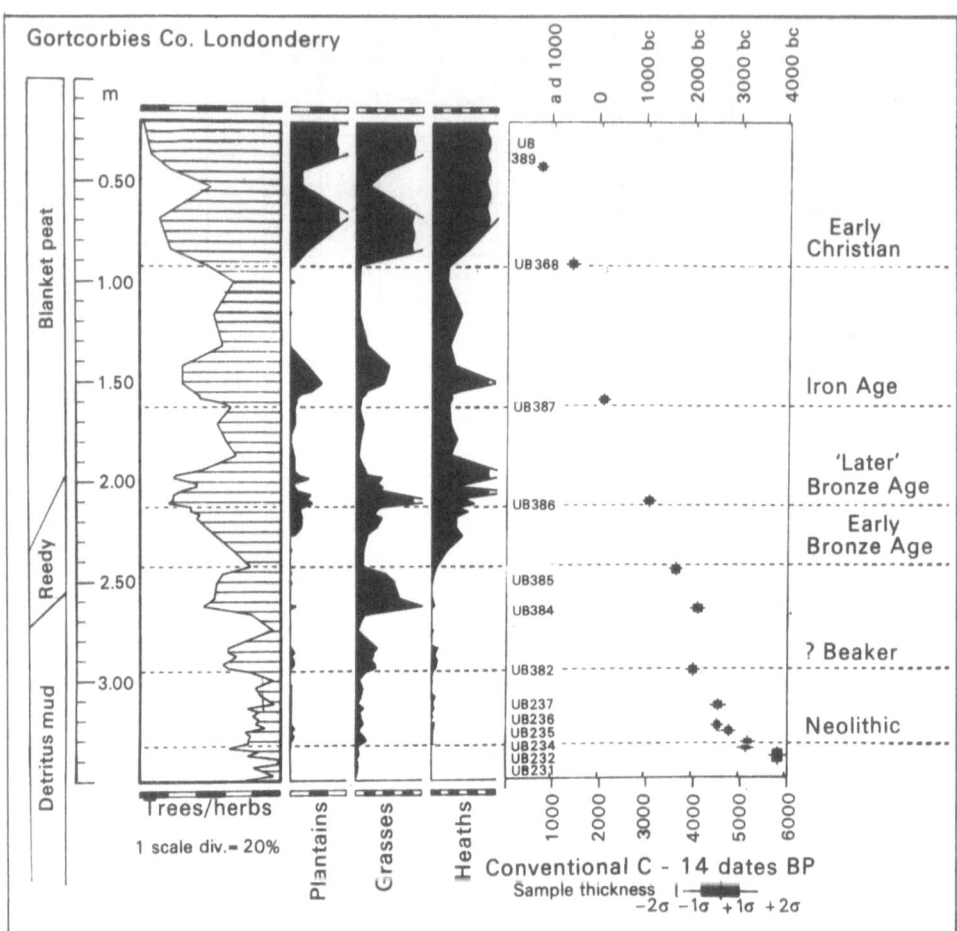

Fig. 4. Dated pollen diagram showing prehistoric clearance phases
           followed by progressively weaker regeneration of forest
           cover (From Smith, 1975).

to a similar conclusion on the Jutland heaths. Here on the better
soils Neolithic clearance was followed by vigorous regrowth of the
forest, whereas on the more base-poor soils the successive waves
of regrowth got progressively weaker until the forest ultimately
failed to return.

PREVENTED FROM RE-ESTABLISHMENT BY
CONTINUOUS LAND USE

This situation is found on the soils which were most favoured for
agriculture: soils which are fertile and easily worked. These are
often upland soils, such as those on chalk or limestone hills,
but may also be low-lying, as long as they are not too heavy for
cultivation with the ard. Calcareous river gravels may come in
this category - having been extensively exploited from the Neolithic onward.

Studies of soils buried beneath prehistoric earthworks on
the chalk of southern England have shown that clearance had
occurred even around early Neolithic sites (Figure 5); this is
indicated primarily by the land snails found in the buried soils,
and also by pollen analysis, though the preservation of pollen
is poor in base-rich soils. Occasionally there is a suggestion of
subsequent woodland regeneration, but more characteristically the
openness of the environment is perpetuated and extended into
later periods. The general picture is of widespread and permanent
clearance by the middle Bronze Age.

Nevertheless, these uplands continued to be exploited for
agriculture and even after intensified arable agriculture in the
Iron Age and subsequent times, which led to severe erosion off
steep slopes, the land continued to be used into medieval times.
When sheep-rearing replaced arable farming, these chalk soils were
still productive, and in times of special need, such as the Napol-
eonic wars, renewed agriculture was practised. It is this long
history of land use, and not loss of soil fertility, which has
prevented the regeneration of the forest cover, with possible
local exceptions.

FOREST REGENERATION PREVENTED BY ENVIRONMENTAL DEGRADATION

On base-poor soils, particularly if freely-draining or in a high
rainfall area, forest clearance created an imbalance by destroying
the deep rooting which maintained the fertility of the topsoil by
bringing up bases from the deeper horizons (Dimbleby, in press).
The leaching process, which is relatively more intense in acid
ecosystems, was therefore no longer compensated for by the action

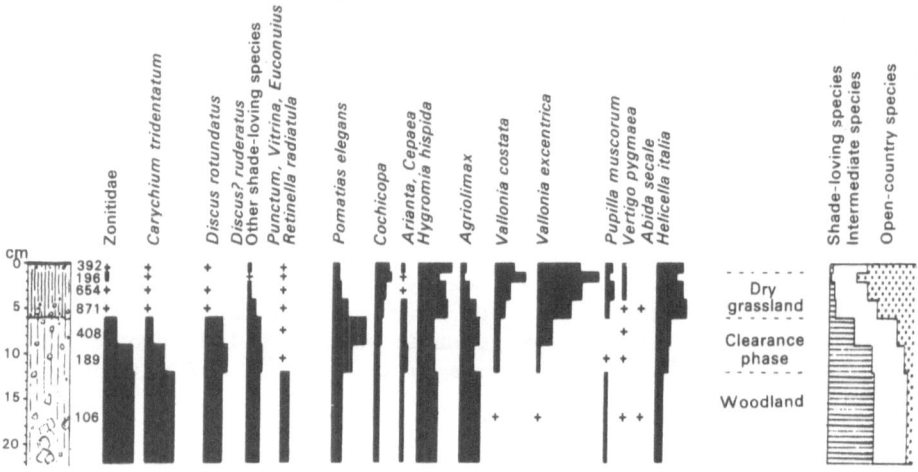

Fig. 5. Analysis of land snails in neolithic soil beneath the Avebury bank. A succession of species marks the change from woodland to grassland (From J.G. Evans, <u>Land Snails in Archaeology</u>)

of the deep roots, and the land became progressively more acid and
depleted of plant nutrients. Soils lost their structure, and various
forms of secondary soil condition developed - the thin iron-pan soil,
the podzol, or, in high rainfall areas, the build-up of blanket
peat. Such changes have occurred at many different times throughout the postglacial period and are clearly not primarily related
to climate; in all cases the hand of man can be detected. Recent
work by Moore (1975), has shown this to be true even of blanket
peat initiation, though obviously peat formation will only be
manifested in a wet climate.

Except in the case of blanket peat, which may be an irreversible change, the reversion to forest may have been possible by
secondary succession; it can be shown that such successions may be
correlated with a progressive change in the soil from the secondary podzol (or what ever form has been developed) back towards the
primary forest soil type, which in many cases appears to have been
an acid brown earth or similar type. Whilst this potentiality
existed, in fact it was seldom realised. It can be shown that
forest clearance of many of our moorland areas was followed by
use for agriculture, until, apparently, the agriculture failed,
presumably because of the reduced soil fertility. Even when this
happened, however, the land was still affected by man; the grass or
heath cover was liable to burn, and when burnt it could still be
used for rough grazing. So pressure against tree growth has
continued even though the land was no longer highly productive,
and the fact that the vigour of forest regeneration was so much
reduced on land which had been run down in this way no doubt
contributed to its vulnerability.

These man-induced trends are not to be confused with the similar but much slower process of 'retrogressive succession' which
Iversen (1969) believed was an inevitable process in climates such
as that of north west Europe. He maintained that progressive acidification on base-poor soils could, in the course of thousands of
years, lead to changes in the nutrient status of soils sufficiently
marked to bring about changes in the dominant tree species:
demanding species (e.g. ash, hazel, lime, elm) would be progressively replaced by less demanding species (e.g. birch). It can be
shown on archaeological evidence that far greater changes in
nutrient level could be brought about within a century or two through
prehistoric man's activity (Proudfoot, 1958).

Nor should it be assumed that the soils produced by the two
processes will be the same. There seems to be no natural counterpart of the thin iron-pan soil; and the same may be true of blanket
peat. Even where there are visual similarities, as with the podzols,
there seems to be marked differences in biology; for instance, the
forest podzols of Scandinavia have a very similar appearance to some

heathland podzols, but the latter are usually much more acid, and
C14 estimations on the humus B-horizon indicate that the rate of
turnover of organic matter is much slower in the heathland soils
(Perrin, et al., 1964) than in the forest podzols (Tamm and Holman
1967). The status of podzols in the deciduous forest belt needs
much more investigation, not an easy matter because so few areas
of old deciduous forest remain. But it seems clear that most of
the podzolized soils in this area today have arisen from more
fertile forest soils, many of which were brown soils, though
others may have been forest podzols of higher biological activity
than the present soils.

CONCLUSION

It is one of the attributes of man that he adjusts his use of the
land to what it can produce. If he is progressively degrading the
environment, he will therefore be progressively adjusting downwards.
It is commonly assumed that the soils we see today are the resultant of the interaction of climate and parent material; that these
soils are what the land provides and we have to make the best we
can of them. Archaeological investigations and paleo-ecological
studies show that major changes have been brought about in the
ecosystems of north west Europe; and on the poorer parent materials
these have resulted in a productivity far below the original
potential of the land. On the richer parent materials the changes
are less; indeed, nutrient levels may be higher under grassland than
under forest. To assess the present status of the land in relation
to its potential we have to realise that today we are using a
landscape which has been shaped by man, for better or for worse,
for 5,000 years or more. The most dramatic step-removal of the
primeval forest, was carried out by our prehistoric ancestors;
the environmental changes were most rapid at this stage and we have
to understand these changes if we are to construct an ecosystem
for our own purposes that has any inherent ecological equilibrium.

REFERENCES

Dimbleby, G. W. (1962). The Development of British heathlands
     and their soils. Oxf. For. Mem. No. 23.

Dimbleby, G. W. (in press). Changes in ecosystems through forest
     clearance. In: Proceedings of a conference on Man and his
     Environment, University of Birmingham, 1975, edited by J. Hawkes.

Iversen, J. (1941). Landnam i Danmarks Stenalder. En pollenanaly-
     tisk Undersøgelse over het første Landbrugs Indvirkning paa
     Vegetationsudviklingen. Danm. geol. Unders., RII Nr. 66.
     Nr. 66.

Iversen, J. 1969. Retrogressive development of a forest ecosystem
    demonstrated by pollen diagrams from fossil mor. Oikos,
    Suppl. 12, 35-49.

Jonassen, H. 1950. Recent pollen sedimentation and Jutland heath
    diagrams. Dansk bot. Ark., 13, 1-168.

Moore, P. D. 1975. Origin of blanket mires. Nature, Lond., 256,
    267-269.

Perrin, R. M. S., Willis, E. H. and Hodge, C. A. H. 1964. Dating
    of humus podzols by residual radiocarbon acticity. Nature,
    Lond., 202, 165-166.

Proudfoot, V. B. 1958. Problems of soil history. Podzol develop-
    ment at Goodland and Torr Townlands, Co. Antrim. J. Soil Sci.,
    9, 186-198.

Tamm, C. O. and Holman, H. 1967. Some remarks on soil organic
    matter turn-over in Swedish podzols. Meddr norske Skogsfors-
    Ves., Nr. 85, 33, 69-88.

Turner, H. 1965. A contribution to the history of forest clear-
    ance. Proc. R. Soc. B., 161, 343-354.

Smith, A. G. 1975. Neolithic and Bronze Age landscapes changes
    in Northern Ireland. In: The effect of man on the landscape:
    the Highland Zone, edited by J. G. Evans, S. Lumbrey and
    H. Cleere, 64-74. Council for British Archaeology Research
    Report No. 11.

DISCUSSION: PAPER 6

N. K. JACOBSEN   In Jutland I have seen sites like those
                  Professor Dimbleby described, in which there
                  were rapid soil changes, over periods of
                  100 or 200 years, near Iron Age settlements –
                  but climate had an influence as well as man,
                  since the wind removed the upper soil horizons
                  and made recovery impossible.

G. W. DIMBLEBY   Wind erosion can occur without climatic
                  change if man exposes the soil by destroying
                  the vegetation.

H. P. BLUME   There are thin iron pans in the Black
                  Forest, which biologists think were created by
                  human agency but which the pedologists, after

radiocarbon dating, believe antedate human
activity in the region. Man may have
increased the scale of their formation but
was not the sole cause.

A second point. Conversion of forest
areas to agriculture has been accompanied
by decreased soil fertility. Why? After
forest cover is removed there is lower
transpiration and therefore more water movement in the soil and a higher output of
nutrients into the ground. If this is the
case, soil fertility should decrease more
markedly in humid climates than in semi-
arid areas.

G. W. DIMBLEBY

In Ireland it has been shown that forest
clearance results in increased wetness of the
soil surface, apparently through reduced
evapo-transpiration combined with a worsened
soil structure. This can lead to peat
development.

I know no case of a thin iron pan soil
pre-dating man's impact on the forest; it is
difficult to see how such a soil could develop
under forest. It could conceivably have
occurred in an earlier tundra phase. Radiocarbon figures for the B_H horizons of British
heathland and Swedish forest podzols show much
higher values in the former, suggesting that
the rate of organic matter turnover is much
slower in the heath soils.

N. K. JACOBSEN

Surely the changes revealed by pollen
analysis - such as the cyclical oscillation
of grass:tree pollen ratios - could reflect
climatic change. On the Jutland salt marsh
coasts there is evidence of occupation from
100 BC to 200 AD and then a gap to 700 AD
and this may reflect climatic changes affecting the suitability of these coastlands for
occupation. Iron Age man faced catastrophe
when the climate changed sufficiently to make
an area no longer habitable.

G. W. DIMBLEBY   The Scottish analyses by Dr. Judith Turner went back to the Neolithic: the point they make is that man had already altered the environment long before the Iron Age. Climate may have aggravated these changes but did not initiate them. Sequences from forest to grassland and then to heath occur at all periods from the Mesolithic to modern times, but do not synchronise from site to site as they should if they were due to climate.

N. K. JACOBSEN   My data from 3,400 BC onwards show 7 or so changes which I believe do have climatic links.

B. ULRICH   In north Germany we have loamy river valley sediments deposited between 400 – 300 BC and around 600 AD and another layer from around 1,500 AD. The present problem is of soil acidification. Two thirds to three quarters of the acidity entering mineral soils today is of human industrial origin.

V. GEIST   As a large mammal scientist, I would stress the early date of major ecological changes. The point of departure for Professor Dimbleby's analysis was the Mesolithic, but we must remember that the large mega-fauna was eliminated in the northern temperate zone in immediate post glacial times. People adapted as best they could to settled life in northern regions after the disruption of this period. The Mesolithic, Professor Dimbleby's departure point, was not a "normal" period for humans, but the first great disaster. It saw severe reductions in populations in Europe, eruptions of cannibalism and warfare, technological and cultural decline, and poor phenotypic development of individuals. This can be related in part to the spread of forests in post-glacial times and to the extermination of the mega-fauna, a major source of sustenance between the Ice Age and the end of the Paleolithic.

G. W. DIMBLEBY   I took the Mesolithic as a starting point because this is as far back in time as I can get in my search for buried soils. Blown sand, possibly arising from damage to the soil, preserved old land surfaces. Such conditions have not yet turned up in the Palaeolithic. There is little doubt that locally Mesolithic man could have a major ecological impact.

7. THE DEGRADATION OF ICELANDIC ECOSYSTEMS

S. Fridriksson

Erdafraedinefnd Haskolans Islands

Reykjavik, Iceland

INTRODUCTION

For various reasons the restoration of a devastated ecosystem is
valuable, although it is realised tha tin its natural form, an
ecosystem is an unstable entity, flexible and everchanging, reacting
to the variable influences of the environment and possibly eventually meeting destruction. A static condition is stagnation and
lacks a successful response to sudden external changes, and even a
climax community is not stable. It is also evident that from an
aesthetic standpoint devastated areas often make magnificent
landscapes and scientifically such areas may be of great interest
as they provide habitats where one may follow a gradual development
of communities.

We wish, however, to save natural resources from being wasted,
to preserve habitats and communities, to maintain species and to
prevent rare genes from being lost.

A devastated ecosystem is of little use in organic productivity
and may be harmful and can accelerate the destruction in nearby
communities and is, therefore, undesirable from our point of view.

Due to the delicate links between ecosystems, a devastation of
one may threaten the existence of a significant member of a distant
ecosystem, as with migratory animals.

Restoration of an ecosystem is not an easy task, and may, in
many instances, be impossible because some of the numerous components may have been lost forever.

145

It is thus of great importance for the maintenance of species
and valuable for human welfare that detrimental influences to
plant and animal communities be recognised so that steps can be
taken in time to keep these at a minimum, and the ecological balance
may be re-established. It is easier to prevent the disaster than
to amend the damage done.

It may be pointed out that there is a major difference in our
reaction when the devastation is caused by natural forces, and when
it is partly or totally due to man. The latter type may be considered unnecessary, but is continuously becoming more serious as the
effects of man's activities become increasingly drastic.

The Icelandic environment has not been severely damaged by
mining or polluted by industrial activities. Just the same, there
are large devastated areas in the country. The destruction has
partly been caused by natural forces; by ice and fire, wind and
water, but with the arrival of man the balance of nature in the
virgin habitat of the island was drastically upset (Fridriksson,
1973).

DEVELOPMENT OF ICELANDIC ECOSYSTEMS

It is not known for certain whether any of the post-glacial biota in
Iceland survived the last ice-age, or whether all the present species colonised the country in one way or another following the
melting of the glacial dome ten to fifteen thousand years ago. It
is generally accepted, however, that a substantial proportion of
both plant and animal species are post-glacial arrivals (Löve and
Löve 1963).

One can thus consider the country at that time to have been
completely devastated by the ice-age glaciers. Its rehabilitation
was slow and the new communities were different from those that
had occupied it in the Tertiary, or in interglacial periods. As
Iceland was an island, the magnolia, sassifras and conifers of the
previous flora did not have a chance to retreat southward during
the Pleistocene glaciation as they did on the continents. The new
life forms that colonised the desolate post-glacial substrata
must have come from adjacent countries, mostly in northern Europe
and Asia.

It is possible to visualise the gradual dispersal of biota to
Iceland, its colonisation and succession by studying comparable
events, which are taking place today on fresh lava flows, on newly
released glacial moraines and on the recently formed volcanic island
of Surtsey (Fridriksson, 1975). The position of Iceland is such
that the distances from other countries are too great to allow for

an easy dispersal of many life forms. The ocean barrier has thus
affected greatly the quantity of species in the native biota, and
the climate being cool oceanic is selective for various sub-arctic
plants, lower animals and birds. Degradations of climate and
devastation by volcanic activities caused periodic set-backs, but
various species became well established and communities developed.

When Iceland was discovered by man there were probably 2-300
species of higher plants in the native flora, over 70 species of
nesting birds of which some 30 species were permanent residents,
with the others migrating to the south in winter. Of mammals the
Polar bear was an occasional visitor, but the Arctic fox gained
access on the pack ice and became established. Of fresh water fish,
salmon, Brook trout, char, eel and stickleback managed to find their
way to Icelandic rivers and lakes, but no reptiles got across. The
Icelandic ecosystems consisted of relatively few species, but the
vegetated habitats must have become rather rich in organic deposits
since there were few consumers, and the crop was nowhere removed to
any degree as no herbivorous mammals reached the country.

In the marshlands and dry grasslands organic matter fell to
the ground every year, accumulated and gradually decomposed. On
the forest floors the fertility was even greater. In some places
the birch brushwood was so thick as to be impassable for men, but
where the forests were less dense, there was a rich flora of herbs
and ferns. In the forest soil, minerals and organic matter accumulated, and the soil acquired a high moisture retention. The
vegetation advanced to the interior and covered approximately 40%
of the country. The climax community on the dry lowland was birch,
(Betula pubescens), which gradually occupied a great part of the
area.

MAN ENTERS THE ECOSYSTEM

In the ninth century A.D. most European countries, except Iceland,
were inhabited by man.

The first human settlers came to Iceland from the British
Isles.   They presumably brought with them various species which
up to then had been absent from the country's ecosystems. When
setting out on voyages of exploration they undoubtedly took
provisions to secure their sustenance after landing in a new home.
They were presumably accompanied by dogs, cats, goats, sheep, pigs,
hens, and ducks. .These animals are relatively easy to transport,
but they needed fodder during the journey, so that straw and hay
will have been collected for the herbivores and grain taken along
for the birds and for sowing. When the boats were laden, it is
possible that mice may have slipped on board with the cargo. Mice
were probably hitherto unknown mammals in Iceland's ecosystem.

It is believed that the first settlers lived in Iceland as
hermits and never became very numerous. When the Norse began to
arrive after 874, the effects of man for the first time became
considerable. The human population increased within a short time.
It is reckoned that the Norse settlers numbered about 20,000, but
in the 12th century the population probably reached 80,000. To
feed these people it was necessary to use the produce of domestic
animals, and additional sheep, horses, cattle, pigs, and goats were
imported by the Norse. Later, reindeer were brought to Iceland
from Norway in 1771-87; and unsuccessful experiments have been
made with the import of Musk oxen from Greenland.

Rats have been brought in with goods, and have multiplied in
densely populated areas. Hares, rabbits and even frogs have been
imported, but have not thrived in the wild. Lastly, amongst the
animals which have been deliberately introduced may be mentioned
the mink, which has now become widespread throughout the whole
country and has had a considerable effect on the bird and fish life.

Various lower animals followed man, his domestic animals and
cultivated crops. Amongst those may be mentioned various parasites
and scavengers, which live in proximity to man and domestic animals.
The import of goods has been accompanied by an increasing number of
species of invertebrate animals and new species are still being
added as a consequence of improved access to the country. Many of
these species have had important effects on the country's ecosystems.
The herbivores have caused the greatest disturbance to the original
vegetation.

INTRODUCTION OF PLANTS

Various species of plants which previously lacked the ability to
reach Iceland, or which could not thrive except in association with
man and his domestic animals, were introduced at or after the
settlement. When man began to clear forests or plough and spread
dung on the fields, conditions for colonisation by many annual and
nitrophile species were created. Some so-called weeds had perhaps
lived on beaches, outlying islands and bird-cliffs where they
benefited from bird manure and decaying marine life. These species
spread more widely when man provided them with improved growing
conditions in cultivated land, at the expense of the vegetation
which had been there before.

From analysis of pollen in horizons of peat dating from the
time of the settlement, it can be demonstrated that a sharp change
took place in the plant species during this period. There was an
increase in Stellaria media, Cerastium caespitosum, Rumex acetosa,
Rumex acetosella, Matricaria maritima, Achillea millefolium, and

several other herbs and grasses, but a decrease in birch
(Einarsson 1961). Carbonised remains, historical sources and the
present distribution of species also give evidence of introduction
of plants at various periods. The incidence of many species is
first and foremost restricted to harbours, cultivated areas or the
highroads, and it is possible to trace the import of some plants
to definite trading places or centres.

In this way, ever since the country was first inhabited,
individual plants have slipped in from time to time. Such immigr-
ants may have either increased the gene pool of native species or
in other instances, been individuals of species new to the flora
of the country. Over 180 new species of higher plants have been
recorded growing in the country since the turn of the last century,
and 26 of these species may be considered naturalised (Davidsson,
1967). Some botanists have estimated that up to a half of the
species of Icelandic flora which now contains 450 vascular plant
species were imported to the country by man (Steindorsson, 1962).

IMPACT ON ANIMALS AND PLANTS

When Iceland was discovered by man, there was an abundance of
birds and fish, and fishing and hunting were easy. With the arrival
of man, a new secondary consumer, who all but exhausted the population of some animals, was added to the ecosystem. The settler
used for his sustenance chiefly fish, larger birds, seals, and
whales. As man's settlement in the country progressed, the drive
to make use of the native fauna gradually increased. Geese,
ptarmigan, and cliff birds were caught and the eggs taken. Through
egg-collecting, hunting and the disturbance which he caused in
nesting grounds and seal-breeding grounds, man caused both a reduction in the numbers of certain animal species and a curtailment of
their areas of dominance.

Seals became restricted to remote shores, and the cliff birds
abandoned nesting places which proved easily accessible to egg
collectors. The numbers of fish in most rivers gradually diminished, in some even to zero. Attempts were also made to reduce the
number of foxes and birds of prey. Falcons were sold abroad and
attempts made to exterminate the eagle, whose numbers have steadily
dwindled up to the present day (Gudmundsson, 1969). The survival
of some species was endangered. Thus the Great auk (Pinguinus
impennis), became extinct when the last bird was killed in 1844
on the island of Eldey off the south western shore of Iceland.

There were few edible native herbs in the country, except
Angelica, which was dug up by man and heavily grazed by livestock:
it has become less common as a result. The "Iceland moss" was also

sometimes collected in such quantities that this lichen became
scarce in some areas. But man's effects on vegetation was greatest as a result of his exploitation of the forests (Finnsson, 1970).

THE EFFECTS OF LIVESTOCK

The entry of these new species into the ecosystem upset the prevail-
ing homeostasis. The chief factor in this disturbance was the
removal of the annual production of plants, which, with the arrival
of man and his domestic animals, was transported from the original
ecosystems, whereas before the nutrients had, for the most part,
been in constant circulation within a relatively closed system.

The transport of produce from the various communities home to
the human habitations resulted in a decrease in the fertility of
the soil which provided the harvest. Striking changes in the
amount of organic soil deposits took place. This may be seen in
the soil profiles, as layers deposited after the time of settlement are light in colour and contain less humus than earlier
layers. The eventual consequence of man's activities was that
various sensitive plant species could not survive the trampling
defoliation or loss of nutrients, so that their growth dwindled or
the species disappeared from the communities altogether.

At the time of the settlement, birch was a dominant species
in the plant communities of various regions, but now it was forced
to yield to grass species and various agricultural weeds.

DESTRUCTION OF FORESTS

Initially, the settlers made clearings for farmsteads and
hayfields, which were small openings of grassland in the vast birch
forest. But as the settlement became denser and the land had been
inhabited longer, the forest vegetation began to shrink. In some
districts, the forest was eliminated in early times, but in other
regions it survived better. As time passed, conditions changed so
much that the forested spots became like isolated oases in the
predominant grassland and desert. Today forest covers only 1%
of the country (Bjarnason, 1967).

Although forest was wasted by cutting, nevertheless, grazing
by livestock played the largest part in wiping out the birch
(Thorarinsson, 1961). Sheep which are grazed in forests defoliate
the trees and uproot seedlings along with other undergrowth.

When woodland is destroyed, it is not only the birch which
disappears from the habitat, but various forest plants go with it.
Thus, Geranium silvaticum, Filipendula ulmaria, Geum rivale and

various ferns disappear. Rare plant species are today sometimes only found in forests but without doubt many of them were more widespread before the time of the settlement, when their suitable habitat was more extensive. Destruction of woodland by man and domestic animals was followed by major wind and water erosion, especially in the volcanic zone extending across the country's centre from north to south.

The destruction of the woodland vegetation had consequent effects on the fauna. As the soil erodes, remnants of soil blocks are often left on the barren subsoil like islands in the wilderness. As the area of such islands diminishes the soil dries out and the number of species in these oases gradually decrease, although grasses and willows may hold out for some time. The lower fauna disappears and nesting birds such as the redwing (Turdus iliacus) evacuate. To some extent the biota of such oases may be compared with that of the smaller remnant volcanic islands from 1 - 50 hectares in extent in the Westmann archipelago off the southern coast of Iceland, which are gradually being eroded by the ocean. They have a varying number of species according to size of islands, with a loss of one vascular plant species for approximately every hectare decrease in area.

BURNING

In Iceland the custom of burning dead grass and sedges in the spring in order to improve grazing land in marshes and moors probably has a long history. Burning of wilted grass was done along with woodland clearing, or came in its wake, and was kept up in regions where there were extensive marshes and swamps.

The burning had a drastic effect on the vegetation. Birch, willows and berry bushes could not withstand the fire and quickly disappeared from the burned plant communities in favour of grasses and sedges. Burning was no doubt partly an indirect cause of the destruction of forests in Iceland, and it is still a questionable practice in many areas, since fire from the marshes can spread to heath and forest. Today, accidental fires also destroy heath and forest vegetation.

DRAINING OF MARSHES

Marshland in Iceland is reckoned to cover an area of about 10,000 km^2 or 40 - 50% of the land under vegetation (Johannesson, 1960). The ecosystem of the marshes has been very useful to man and his domestic animals. It supplied a larger part of the hay, and livestock grazed, and still graze, there at all times of the year.

   To increase the crop production of the marshes extensive
draining has been performed. Ditching was negligible up to 1920,
and draining did not really get under way until during the 2nd
World war. Today, drainage-ditches have probably been dug in one
tenth of all the marshlands, and bogs drained by numerous ditches
flanked by the excavated peat are now characteristic of large parts
of the Icelandic lowland.

   Draining changes the wet habitat of the marsh in such a way
that grasses increase markedly, whereas sedges and cotton grass
gradually disappear, as the marsh dries up. A drained bog becomes
a better pasture, but all plant and animal life is altered as the
wetland ecosystem is gradually changed into one of dry land.

   With drainage, the soil becomes dry, and the air has
increased access to organic matter in the soil which decomposes
rapidly. As a result, various minerals, which were previously
fixed in half-decomposed organic residues are released. Some of
them are put to use by living organisms, and the minerals once more
enter the cycle in the chain of life. The surface of the marsh
subsides with draining. Rainwater has an easier passage than before
from the surface down into the ditches and from there into streams.
Thus, there is an increased danger of leaching out of nutrients
from the top soil, and the soil loses its sifting capacity. The
drainage water undercuts banks of ditches. The top soil is washed
away and gradually fills the ditches or is carried with the drainage
water out into streams and lakes where it is deposited as silt.
When fertiliser is applied to a cultivated bog some of the nitrogen
and phosphate is washed away. As a result, vegetation also increases
in ditches, rivers and lakes, which in turn disturbs the oxygen
balance and changes the food chains in streams and ponds. As the
water table is lowered by extensive ditching, not only the marsh
dries up, but also neighbouring moorland and hills, and along with
this goes increased danger of erosion.

   In earlier days, the flooding of marshes and swamps provided
favourable conditions for various waders and web-footed birds,
which found food in the shallow pools. With draining, a complete
change has taken place in these habitats. The micro-fauna of the
bog was replaced by dryland fauna, and this in turn became food
for moorland birds instead of the godwit, ducks and other marshland birds which had occupied the area. The extensive draining
of bogs, which were breeding grounds for these migratory birds,
thus affected the population of various European bird species.

                            CONCLUSIONS

   Today, after man has lived in Iceland for 1,100 years, it is
obvious that there have been major changes in the native ecosystems.

It is estimated that half of the vegetation cover has been lost, its area decreasing from 40,000 to 20,000 km^2. Most of the birch wood is gone, and some of the more sensitive species have disappeared. The rate of devastation has been faster in Iceland than in other European countries due to the volcanic origin of the soils and the slow growth of plants.

It is fully recognised that man could not have survived in Iceland without upsetting the natural balance that existed. New species were introduced and extensive agriculture has altered various plant and animal communities. The conditions as they were at the time of the settlement will not be re-established, but an effort will be made to restore the devastated areas, whether they have been destroyed by over-exploitation, misuse or by natural catastrophic forces.

REFERENCES

Bjarnason, H., 1967. Skogsaken i Island og dens utvikling. Meddr norske SkogsforsVes, Vollebekk. 75-101.

Davidsson, I., 1967. The immigration and naturalisation of flowering plants in Iceland since 1900. Societas Scientiarum Islandica, Greinar IV 3 Reykjavik 1-37.

Einarsson, Th., 1961. Pollenanalytische Untersuchungen zur spat - und postglazialen Klimageschichte Islands. Sonderveröff. geol. Inst. Köln, 1-52.

Finnsson, H., 1970. Manfækkun af hallærum. Reykjavik. Almenna bokafelagid. 209p.

Fridriksson, S., 1973. Lif og land, Reykjavik, Vardi, 263p.

Fridriksson, S., 1975. Surtsey. London: Butterworths, 198p.

Gudmundsson, F., 1969. Bird life in Iceland. 65 Degrees, Reykjavik 24-27.

Johannesson, B., 1960. Soils of Iceland. Agricultural Research Institute. Reykjavik, Series B 12. 140 p.

Löve, A. and Löve, D., 1963. North Atlantic biota and their history. Oxford: Pergamon Press. 430p.

Steindorsson, S., 1962. On the age and immigration of the Icelandic Flora. Societas Scientiarum Islandica, Rit 35. Reykjavik 157.

Thorarinsson, S., 1961. Uppblastur a Islandi i ljosi oskulagaran-
    nsokna. Arsrit Skograektarfelags Islands, _Arsr. Skograektarf._
    Isl., 17-54.

DISCUSSION: PAPER 7

A. D. BRADSHAW   May I ask Professor Ulrich a question
                  arising from Dr. Fridriksson's paper? When
                  we are considering the degradation of the eco-
                  system as woodland is converted to grassland,
                  we should compare the nutrient position in
                  the two systems, for soil nutrient loss is
                  surely the key to degradation. My question to
                  Ulrich is what would have happened to the
                  nutrients when birch cover was replaced by
                  grassland?

B. ULRICH   I would not expect any great change in
                  nutrient storage or cycles on these young
                  soils. Nitrogen would alter most. Birch
                  forms open forest, and a ground vegetation
                  would already have been present. The problems
                  begin when over-grazing leads to topsoil erosion,
                  with loss of stored nitrogen. The problem
                  today in Iceland is probably one of restoring
                  nitrogen. In southern countries phosphorus
                  was also lost, but in these young Icelandic
                  soils, probably this is less of a problem. The
                  input of nitrogen in rain is very low - and a
                  vegetation is needed to collect it: the input,
                  however, can be too low to allow a vegetation
                  to form. A very long period is thus needed
                  for the system to recover naturally.

N. K. JACOBSEN   What about physical factors? Once soil
                  is lost, a long time is needed for more to
                  accumulate.

A. D. BRADSHAW   Studying soil and nutrients is important
                  because this helps to define the key stages
                  in degradation and hence also in restoration.

S. FRIDRIKSSON   In Iceland the critical step is the loss
                  of phosphorus and nitrogen when the topsoil
                  is blown away and the grass destroyed.
                  Soils are quick to form, but leaching is rapid
                  and nitrogen and phosphorus are washed out.
                  After loss of the sward and topsoil there is
                  a deficiency of available phosphate, hampering
                  recolonisation.

B. ULRICH   Dr. Fridriksson has just said that
                                Icelandic soils are low in available phosp-
                                horus. If the pH in the top layer is low,
                                for example because a lichen cover is adding
                                acid to the soil, phosphate may be precipitated
                                as iron phosphate, thereby blocking transport
                                from seed to shoot and causing germination of
                                seeds to be abortive.

M. W. HOLDGATE   I was interested in Dr. Fridriksson's
                                point that herbivores were deficient in the
                                original Iceland ecosystem. The same applies
                                in the Subantarctic and indeed in oceanic
                                islands generally. There are no native verte-
                                brates and maybe only 50% of the invertebrate
                                herbivore, microbivore and saprovore systems
                                predicted by models like that of Heal and
                                MacLean. This general island phenomenon may
                                help explain the vulnerability of island
                                systems to degradation when herbivores are
                                imported by man to systems that previously
                                functioned in a different way.

V. GEIST   The role of herbivores should not be
                                overemphasised. Many northern plants lack
                                the defences against herbivores found in
                                other regions - but are only exposed to
                                herbivory for a part of the year due, for
                                example, to protective winter snow cover and
                                the relatively low abundance of herbivores in
                                summer.

                                May I add a general comment about large
                                vertebrates? There has been little mention
                                of large herbivores or carnivores in our sess-
                                ion. I am accustomed to this omission in
                                discussions of ecological reserves in Northern
                                America. Today's landscapes are altered
                                because many large vertebrate consumers have
                                been absent for many thousands of years.
                                Large mammals, and especially the highly
                                social species, are a very vulnerable component
                                of an ecosystem, not only because they are
                                susceptible to direct impact but because their
                                traditional behaviour can be upset. They are
                                K-selected organisers whose traditions often
                                include migration patterns: their strategy
                                involves being in the right place at the right
                                time. Man can teach them new habits, restrict-
                                ing them to certain areas. This has an
                                implication for ecosystem restoration: these

animals cannot simply be introduced and left
to fend for themselves, but should be put in
a setting where they can be managed in a way
that implants knowledge of how to use the
landscape.

Time scales may be more important than
we realise. If we have a new glaciation, as
is likely, violent climatic contrasts may
result. These are likely to result in the
formation of extensive grasslands. The major
consumers that flourished in glacial times are
gone: can those that remain handle the
production that would be available? In North
America people often lament the extinction of
two thirds of the land mammals that were
native there - and the surviving relatives in
other parts of the world seem out of place in
an American setting. Yet camels have twice
escaped and survived, browsing shrubs not
exploited by other herbivores: horses and
asses were introduced and have run wild most
successfully. Maybe the spectrum of large
herbivores could be restored more easily than
might appear.

N. POLUNIN

May we hear of events on Surtsey, which
Dr. Fridriksson has monitored so thoroughly,
aided by the careful protection of this
unique study area from outside interference?
Has it not proved to be a unique microcosm of
the biotic invasion saga of Iceland?

S. FRIDRIKSSON

In my book Surtsey published by Butterworths 1975, I described the research on the
volcanic island of Surtsey and indicated how
it can throw light on steps in immigration of
the Icelandic biota as well as of many other
islands.

I explained that some of the smaller
islands among the Westmann Islands have very
few vascular plants, although they are quite
old. Surtsey, now 10 years old, has 15
species of vascular plants - but is bigger and
has a different substratum. It is obvious that
in the well developed communities of the
smaller islands there has been an extinction
of species after climax communities were
reached.

8. THE DEGRADATION OF BIOGEOCENOSES IN THE MEDITERRANEAN REGION

M. Godron

Centre Nationale de la Recherche Scientifique

34033 Montpellier-Cedex, France

INTRODUCTION

We are all supposed to think that "degradation" is specially important in the Mediterranean region, since a special paper has been
asked for on this problem. But if this is true, I am sure that it
is necessary, to begin with, to define the meaning of degradation,
in this particular case where the influence of man has radically
disturbed biogeocenoses. We cannot measure the degree of degradation of an ecosystem in relation to the climax for two reasons:
first, we often know only a para-climax and not the eu-climax, and
second, men may have optimized artificially the functioning of
'natural' biocenoses, for example by irrigation. Then we must ask
if a crop of wheat is more or less "degraded" than the surrounding
garrigue or maquis? So we are obliged to begin by characterising
each step of "artificialisation" (Dumont, 1959; Godron et al.,
1968).

The first and second steps of "artificialisation" are climax
and pene-climax. They are not always forests, for Mediterranean
climate may be very arid or very cold (O. de Bolos, 1954), but,
nevertheless, are not degraded. In the Mediterranean region, climax
vegetation is particularly rare, and it is not useful to dwell on
these stages.

At the third step of artificialisation, men use intensively
the "natural" (or, rather, "spontaneous") vegetation, which
produces fruits and wood, but also, and mainly, game and cattle.
Recent studies confirm that the critical factor, in that case, is
the use of fire in management of these areas. If fire is regularly

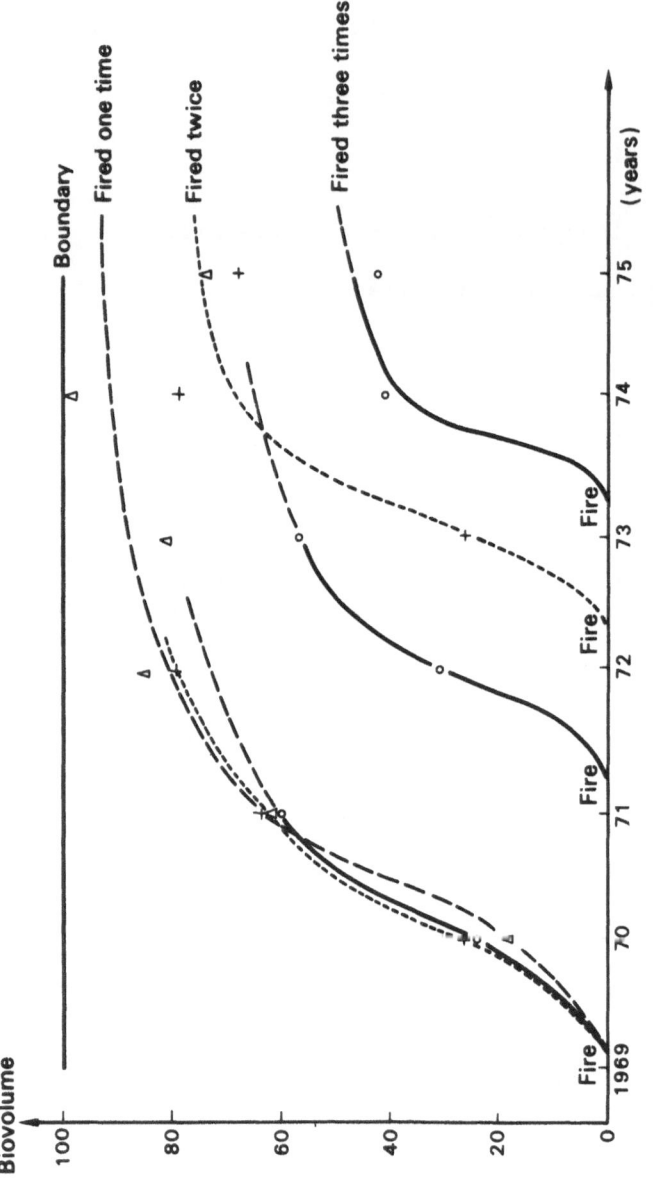

Fig. 1. Relative evolution of biovolume, in function of frequency of fire
        Ext. L. Trabaud (1973)

used at a period of about ten or fifteen years, three peculiar
phenomena occur:

(a) An "adaptated" flora develops, and vegetation gives a
     "regular" response.

(b) Erosion may modify the soil profile.

(c) Vertical distribution of biomass becomes peculiar.

The first two are classical, but also provide the simplest
way in order to define degradation, so some new information about
them will be presented.

CHANGES IN VEGETATION

The difference in flora and vegetation (L. Thurmann, 1849) is
particularly well illustrated by calciphile garrigue. L. Trabaud
(1973) shows that the flora remains constant in the regularly fired
areas. In one of the most typical cases, the Quercus coccifera
garrigue, Trabaud found that among the 83 species recorded in an
experimental area observed during four years with 6 types of burning,
28 are apparently indifferent to the action of fire, 41 have a
tendency to increase after fire, and only 14 have a tendency to
decrease after fire. In this case, it is not possible to demonstrate that fire diminishes floristic richness; on the contrary,
it furthers the specific diversity of the community, mainly because
it permits the arrival of annual (or biennal) species such as
Bromus madritensis, Convolvulus cantabricus, Euphorbia serrata,
Hippocrepis comosa, Lagoseris sancta, Scabiosa maritima, Silene
italica, Sonchus oleraceus and Sonchus tenerrimus. The species
which seem "indifferent" to fire are generally perennial plants
which have strong roots and which resprouts already some weeks
after the fire.

To obtain more precise figures, the presence of the species
has been observed by L. Trabaud along 10 m. transects, and the
results were:

Year	1969	1970	1971	1972	1973	1974
Number of species	12,5(fire)	11,0	10,0	11,7(fire)	10,7	13,0

If we consider now the biomass, we see a "regular" response
(Fig. 1, from L. Trabaud, in prep.) which is very close to a logistic one. Do we have to conclude, in this case, that the biocenosis is degraded? As long as the response of the system is
regular, it may be accepted that it works in good balance with its

environment, as a field of cereals does, showing the same type of
response (with an annual periodicity).

These periodical variations occur also in some quite "natural"
systems, like marshes or mountain grassland. We, men, know that
the periodical variations are caused, in the garrigue, by man;
but, for the plants, the cause of the variation has no importance
and the system, in itself, has no means of "knowing" the causes.
So, strictly speaking, when looking on twenty, or fifty, or some
hundred years, neither floristic richness nor overall functioning
provide criteria of degradation, and it is necessary to look back-
wards for millenia to find better criteria.

EROSION

Mid-european people are accustomed to think that their country was
covered by dense forests till the great clearings of the Middle
Ages. In the Mediterranean region it is different, because earliest
cultivation began eight thousand years ago: the "Cardial" people,
who knew cereals, were present near Marseille about 5,700 BC, at
the end of the Boreal period. So dense forest was probably not able
to cover all the country, and it is difficult to know what were
the equilibria between vegetation and soil, the depth of soil and
the type of humus.

We incline to think that soils were then more thick than now,
and that erosion has truncated them, but we have no decisive proof
of this. In fact, it is rather surprising to see that in these
garrigues erosion today is very feeble, and not directly induced
by fire. There is, on the surface of the soil, a "dallage" of
small stones which, when undisturbed, provides a very efficient
protection against rain. This "dallage" may be the result of
past erosion, which would have eliminated loam and clay from upper
soil horizons and left stones concentrated at the surface. One
could try to compute the thickness of soil which has been eroded,
but, unhappily, these soils have generally been ploughed and the
"model" of erosion, in this case, also gives a concentration of
stones at the surface of the soil.

Consideration of erosion thus does not help us to define the
nature of degradation in the Mediterranean region, so we need to
seek for another criterion.

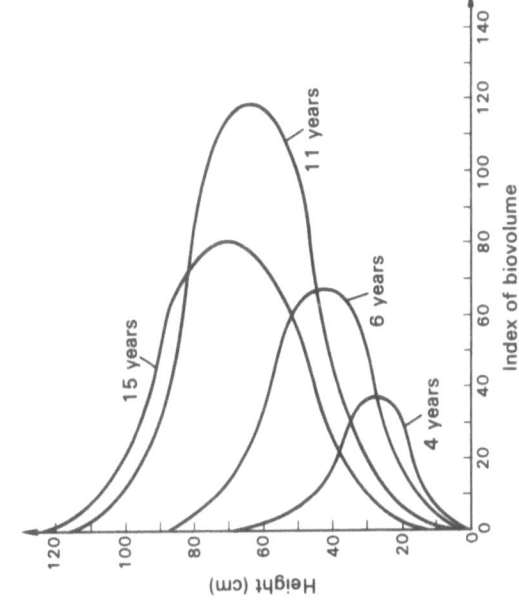

Fig. 3.   Evolution of the vertical profile of biovolume of *Cytisus*
purgans (L.) Benth.
Ext. of M. Debussche (in prep.)

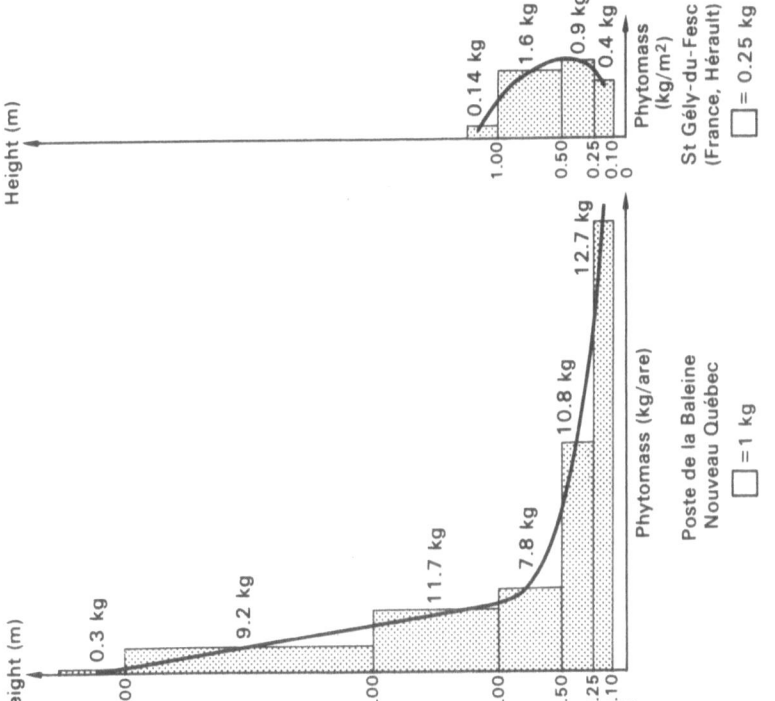

Fig. 2.   Vertical profiles of phytomass

VERTICAL DISTRIBUTION OF BIOMASS

The development process shown above (Fig. 1) seems to lead the
system towards a constant level with a rich flora and appearance
of a "climax" steady state. Everyone knows that it is not an eu-
climax, but we must try to characterise this type of para-equilibrium.

        The simplest way to find a solution may be to remember that the
main difference between para-climax and eu-climax lies in their
respective height above ground, and so, to look at the vertical
profile of biomass (Fig. 2). The comparison with a very old
"natural" forest, at the left of the same figure, leads to an
hypothesis: if the profile has the shape of the capital of a
column, the system is metastable; while the equilibrium between
the tendency of vegetation to grow towards the zenith and the
necessity for it to remain close to the soil (for water and nutrient
supply) would give a profile more like a pedestal (Godron, 1975).

        This hypothesis has recently been confirmed by the studies of
M. Debussche (in prep.; cf. Figs. 3 and 4), in the frequently burnt
communities of Aigoual (Cevennes, France) dominated by Cytisus
purgans. It may consequently be suggested that a general model
would be the succession of perhaps five or six "waves" of communities of increasing height; each of them seeming metastable when
its profile has the shape of a pedestal, being replaced by a
more developed one when new species become widespread in the canopy.
Such is the case of Pinus halepensis in some Quercus coccifera
garrigues, or of Quercus lanuginosa in some Quercus ilex old
coppices.

        In more arid parts of the Mediterranean region it is much more
easy to characterise degradation at this level of artificialisation
(C. Floret et al., 1973). In this part of the Mediterranean region,
aridity is the main character, but there is also a good relationship
between vegetation and milieu, this including grazing by sheep and
goats. Exclosures have often been made in order to see what happens
when the influence of cattle ceases. During the first years the
biomass increases but tends towards a maximum, and the centres of
the "bushes" of the main species often die.

        The type of equilibrium is then often an alternating one when
the phanerophytic "climax" species are absent, because they have
been used by the population for centuries. For example, Rhantherium
suaveolens is often in alternation with Aristida pungens, and the
vertical profile would become like a pedestal if Retama retam and,
subsequently, Acacia raddiana occurred, Long (pers. comm.). The
shape of the profile of vertical distribution of biomass could then
be a criterion of degradation of the phytocenose, within the third
step of artificialisation.

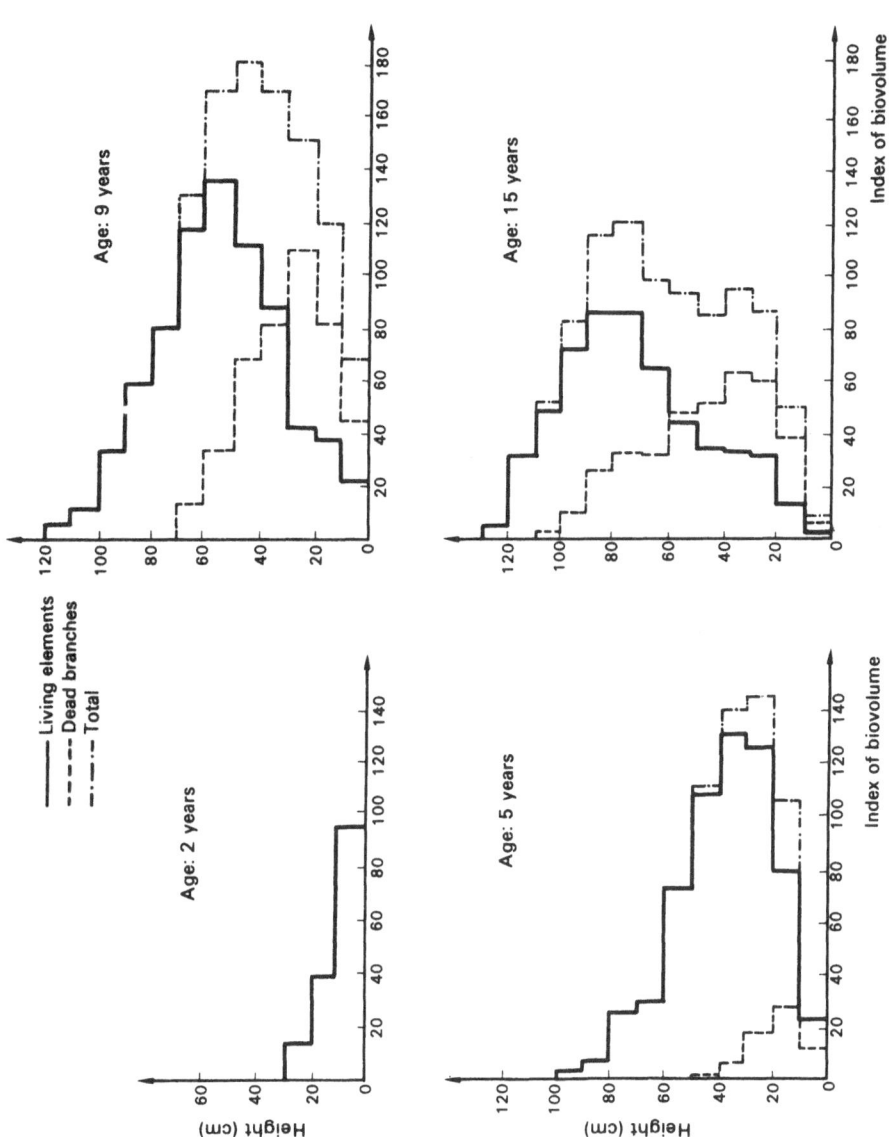

Fig. 4. Vertical profile of biovolume of <u>Cytisus purgans</u> (L.) Benth.
Ext. of M. Debussche (in prep.)

SUBSTITUTION OF SPECIES

The fourth step begins when man introduces chosen species, and
favours them in competition with naturally occurring species.
The diagnosis of degradation then becomes more classical, and the
criteria of floristical richness, concentration of the biomass
in the cultivated species, etc. become more useful; the problems
are then not much different in the Mediterranean region from those
in other regions, and they will not be discussed in this paper.

CONCLUSION

In the Mediterranean region, the geological history produced
extensive areas of hills and mountains (often calcareous) where
the long influence of man has led to a paraclimax, where erosion
has often been important, but is not now catastrophic if vegetation,
"adapted" to the disturbing action of man, is a "paraclimax".

    The problem which remains is to know why these biocenoses stay
in this paraclimax. A part of the answer is the rarity of phanero-
phytes able to colonise, and the hysteresis of their arrival (perhaps
linked to the shape of the vertical profile of biomass) but we must
also consider the peculiar characters of the Mediterranean climate.
This is rather well known (di Castri and Mooney, 1973), but it may
be useful to emphasise its unpredictability, for example by
approaches such as the pluviothermic expectation (Lepoutre, 1964;
Yi, 1974) which helps to indicate the probability of regeneration
of some important species.

    It seems, in conclusion, that the general model of Mediterr-
anean degradation is far from being built, but that we may, and
perhaps must, try to build it.

REFERENCES

Bolos O. de, 1954. Essai sur la distribution géographique des
    climax dans la Catalogne. Vegetatio, V-VI, 45-49.

Debussche M., (in prep). Éléments pour l'etude de la dynamique
    de la végétation. Exemples pris dans le massif de l'Aigoual.

di Castri F. and Mooney, H. A. 1973. Mediterranean type ecosystems.
    New York, Springer, 405p.

Dumont R. 1959. Agriculture comparee. Cours de l'Inst. Nat.
    Agron. (inedit).

Floret C., et al, 1973. Production, sensibilité et évolution de
    la végétation et du milieu en Tunisie presaharienne. Document
    C.N.R.S./C.E.P.E. N° 71, 45p.

Godron M. et al., 1968. Code pur le releve méthodique de la
    végétation et du milieu. Principes et transcription sur cartes
    perforees. Paris, C.N.R.S., 292p.

Godron M. 1975. Preservation, classification et evolution des
    phytocenoses et des milieux. Biologia contemp., 26. 51-59.

Lepoutre B., 1964. Premier essai de synthèse sur le mécanisme
    de régéneration du Cèdre dans le Moyen Atlas marocain. Annls.
    Rech. for. Maroc, 7, 55-163.

Thurmann J. 1849. Essai de phytostatique appliqué au Jura et aux
    contrées voisines. Paris, Bailliere. 2 vol.

Trabaud L. (in prep). Comparaison de l'effet des feux contrôles
    et des feux sauvages sur l'evolution quantitative globale
    des garrigues de Chene kermes (Quercus coccifera L.).

Trabaud L., 1973. Experimental study on the effects of prescribed
    burning on a Quercus coccifera L. Garrigue: early results.
    Proc. Ann. Tall Timb. Fire Ecol. Conference, 97-129.

Yi B. G. 1974. Essai d'application de l'espérance pluviothermique
    en vue de l'estension du Cèdre dans le Languedoc-Roussillon.
    D.E.A. Univ. Montpellier, 20 p.

DISCUSSION: PAPER 8

N. S. MARGARIS   Mediterranean areas are by definition areas
                        of environmental stress. They are dry, hot and
                        sunny. Species must have adapted to these
                        conditions and at least five strategies can be
                        recognised: the evergrees sclerophyll maquis
                        type, seasonal dimorphism, and the systems of
                        therophytes, geophytes and Crassulaceae. The
                        climatic conditions may make true forest systems
                        impossible in much of the region.

                        Professor Godron's paper emphasised human
                        influence in the Mediterranean basin - but
                        mediterranean type climates also occur in Chile,
                        Australia, South Africa and California where
                        this must have been much weaker. Would not
                        comparison with these places aid understanding?

M. GODRON   Taking the last point first, I did not
discuss these other areas, but they have been
studied, (see, for example, the book edited
by di Castri and Mooney, 1973). Fire has had
a great influence in some of them, like
California and perhaps South Africa, where the
Portuguese discoverers recorded seeing the
coastlands ablaze.

On the point about the possibility of
evergreen sclerophyll systems developing into
forest, I agree: the Mediterranean climax is
not always a forest (it is quite clear in
Spain and Morocco). If the area is too arid,
or too cold in winter, the climax will not be
forest and it is foolish to try to establish
one. But my paper is about areas where
garrigue is thought to be a degradation product:
our aim was to test this concept, and find
out whether it was now in a steady state.

G. LUCAS   Where forests are possible, are native
or exotic species best?

M. GODRON   Pinus halepensis was originally confined
to a small part of the Mediterranean, as P.
insignis was at Monterey. Castanea also had
a very local distribution. These species are
now present much more widely: should we call
them native or exotic? In my view these are
correct species to use: it may be less necessary to import remote exotics like eucalypts.

G. W. DIMBLEBY   Dr. Vita-Finzi's recent work on the fill
of Mediterranean valleys shows two main periods
of aggradation - the older fill of Paleolithic
age and the younger fill which is mainly of the
historical period. Can you explain the lack
of such evidence of accumulation associated
with the primary forest clearance, at which
time, as you showed, soil erosion was occurring?

M. GODRON.   It is possible that cultivation was not
very intense in the early Neolithic period,
perhaps because of demographic factors. Recent
paleo-ecological research suggests that human
impact was strong but patchy in early historic
times, leaving refuges for species in small
areas.

L. M. TALBOT   I agree with the stress on fire in the
                                Mediterranean regions of California. When it
                                is stopped, vegetation dominated by trees
                                ultimately develops. Are there any areas in
                                the Mediterranean basin which escaped early
                                cultivation? If so, are regeneration patterns
                                different there?

M. GODRON   In the areas I know, there is no place
                                where we may be sure that there has never been
                                cultivation. Simple manual systems, often on
                                a shifting basis, were probably used, and this
                                allowed crops to be grown even in very hilly
                                areas.

G. VAN DYNE   May we see the diagram relating biovolume
                                to successional stage again? How do you
                                characterise the distribution? Which parameters
                                were measured?

M. GODRON   We measured not only living biomass, dead
                                biomass and species composition but also the
                                proportion of biovolume attributable to each
                                species. Generally 20 per cent of the species
                                provided 80 per cent of the biovolume and 80
                                per cent of the species provided 20 per cent
                                of the biovolume. The variations around these
                                mean values seems to be significant for the
                                maturity of the phytocenose. The third part
                                of our approach, which I presented here in
                                very simplified form, is concerned with spatial
                                heterogeneity (which also changes with success-
                                ion). I would emphasise that the problem we
                                are concerned with is not whether Mediterranean
                                vegetation has been degraded, but what type of
                                degradation has occurred and what is really
                                degradation- this is what we are seeking to
                                model.

R. J. W. KEAY   Professor Godron showed height: phyto-
                                volume curves which are very like those of
                                production or regrowth in an even-aged forest.
                                These are evidence of recovery after degradation,
                                not degradation itself. Corsican maquis
                                invades degraded'land, regenerating freely,
                                and can cure environmental problems: it is
                                not evidence of degradation.

R. W. WEIN   I doubt if there is any universal definition of degradation: perhaps we should come back to such matters at the end of the conference?

9. THE DEGRADATION OF ECOSYSTEMS IN THE RHINE

W. J. Wolff

Nederlands Instituut voor Onderzoek de Zee

Texel, The Netherlands

THE RIVER

The river Rhine has a length of about 1,230 km and drains an area of about 220,000 km^2 (Fig. 1). In The Netherlands river water is sluiced from the river branches into the low-lying polder areas in order to water the growing crops in summer and to flush the canals which tend to become brackish due to seepage. Thus about 60% of The Netherlands is influenced by Rhine water.

At present about 85% of the Rhine water reaches the North Sea near Rotterdam, either by way of the Rotterdam Waterway or through the sluices in the Haringvliet. Most of the remaining 15% flows into the western Wadden Sea by way of the Lake IJssel.

Through the lowering of salinities the presence of Rhine water can be demonstrated in a large part of the North Sea coastal waters. A pesticide occurring exclusively in Rhine water has been shown to occur in mussels as far south as at least the island of Schouwen and to the north-east as far as Delfzijl on the Ems estuary (Koeman, 1971). Rhine silt has been shown to deposit in the Wadden Sea eastward as far as the Ems estuary (De Groot, 1973). A pesticide discharged into the Rhine estuary near Rotterdam could even be demonstrated in the eggs of Common terns breeding at the German island of Oldeoog North of the Weser estuary (Koeman, 1971). Finally the westernmost part of the Wadden Sea receives a relatively large amount of Rhine water (Zimmerman & Rommets, 1974).

The mean river discharge at Lobith, before the river divides into a number of distributaries, is about 2,250 m^3 sec^{-1}. Recorded

169

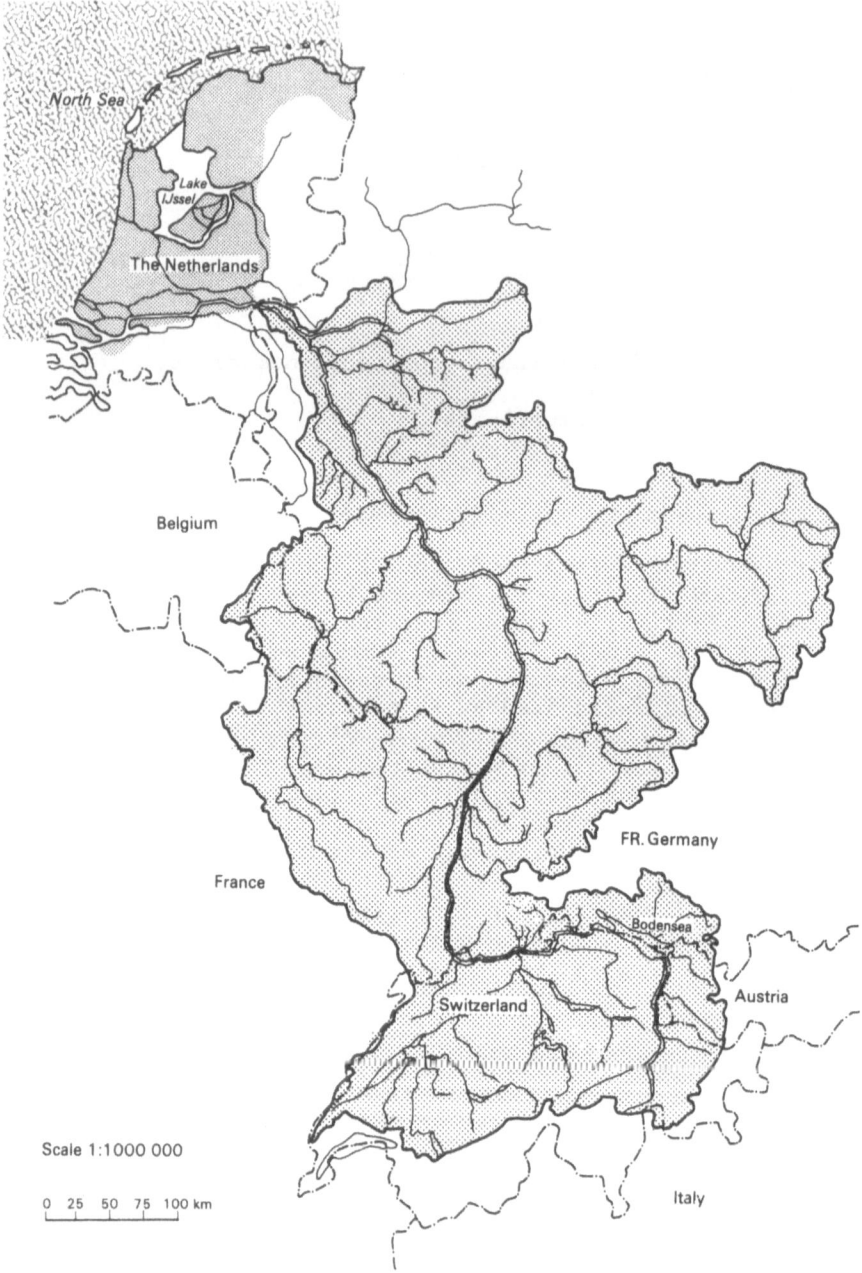

Fig. 1.   Catchment area of the Rhine and its tributaries and area
influenced markedly by Rhine water.

maximum and minimum values are 13,000 and 620 m^3 sec^{-1}, respectively.

MAN-INDUCED CHANGES OF THE ECOLOGICAL FACTORS IN THE RIVER ECOSYSTEM

Physical Changes

Especially during the last 150 years the shape of the river bed
has been altered greatly. In the interest of the extensive shipping
shallows have been dredged away and bends have been straightened.
Thus the length of the river between Basel and Mannheim has decreased
from 354 km to 273 km (Lauterborn, 1917). In the upper course of
the river as well as in a number of tributaries weirs have been
built. In the Netherlands all but one of the estuarine river mouths
have been dammed off or supplied with sluices.

Due to the discharge of heated cooling water the river presently
is about 1°C warmer than in the period 1910-1940 and a further increase is expected.

Chemical Changes

Within the catchment area of the Rhine over 50 million people live
in a heavily industrialised area causing the river to be the main
sewer of N. W. Europe. Table 1 summarises the major pollutants
and other dissolved compounds occurring at Lobith near the Dutch-
German border. Downstream of Lobith the pollution increases by
another 10-15%. A major part of this latter increase originates
from the industrial agglomeration of Rotterdam and, hence, does not
influence the river itself so much, but mainly the North Sea.

The BOD^2O_5-values (Biological Oxygen Demand) have increased
from about 4 $mg.1^{-1}$ in 1880 to nearly 9 $mg.1^{-1}$ in 1971-1975. Consequently the mean annual oxygen saturation percentages went down
from about 100% to about 49% in 1971-1975 with occasional minimum
values of 10-15%. In 1971 about 100 km of the river in Germany
became completely anoxic.

Among the 3,000 tons of "Other hydrocarbons" in Table 1
numerous toxic compounds occur, such as 7-10 tons of hexachlor-
benzene, 15-18 tons of benzene hexachloride, and 1 ton of dieldrin.
In 1969 a discharge of the pesticide endosulfan caused large-scale
fish mortality over a stretch of hundreds of kilometers. In
analyses of Rhine water many hydrocarbons of unknown structure have
been demonstrated. It is estimated that annually some 100-150 new
and unknown chemical compounds are discharged into the river.

Table 1. Pollution and chemical characteristics of the Rhine at Lobith near the Dutch-German border. The percentages indicate the share of human additions in the total discharges. The quantities are derived from the annual and quarterly reports of the Netherlands State Institute for the Purification of Sewage (RIZA) or from Ten Berge at al., 1973; all percentages are derived from calculations for the period 1970-72 by Ten Berge et al., 1973.

	DISCHARGE TOTAL (TONS YEAR^{-1})	HUMAN ADDITIONS (%)	CONTENTS TOTAL (PPM)	PERIOD	SOURCE OF DATA ON QUANTITIES
Cl	10.728.547	84	219	1971-75	RIZA
NO$_3$-N	167.141	80	2.9	1971-75	RIZA
NO$_2$-N	11.000	100	0.2	1970-72	Ten Berge et al.
NH$_4$-N	117.945	97	2.4	1971-75	RIZA
PO$_4$-P	19.048		0.34	1971-75	RIZA
total P	49.827	> 84	0.87	1971-75	RIZA
SO$_4$	4.572.720	47	86	1971-75	RIZA
HCO$_3$	7.000.000	17(?)	150	1970-72	Ten Berge et al.
Na	6.500.000	94	120	1970-72	Ten Berge et al.
K	490.000	35	8.5	1970-72	Ten Berge et al.
Ca	4.700.000	43	100	1970-72	Ten Berge et al.
Mg	470.000		10.5	1970-72	Ten Berge et al.
Fe	92.506	83	1.5	1973-75	RIZA
Pb	1.902	c. 85	34×10^{-3}	1973-75	RIZA
Cu	1.608	c. 75	32×10^{-3}	1971-75	RIZA
Zn	13.403	c. 85	247×10^{-3}	1971-75	RIZA
Cd	162		3.2×10^{-3}	1972-75	RIZA
Mn	10.512		165×10^{-3}	1973-75	RIZA
Be	38		0.7×10^{-3}	1973-75	RIZA
Cr	2.500	c. 80	54×10^{-3}	1970-72	Ten Berge et al.
Co	177	c. 70	3×10^{-3}	1973-75	RIZA
Hg	72	96	1.6×10^{-3}	1971-75	RIZA
As	312	c. 85	5.5×10^{-3}	1973-75	RIZA
Ni	788	c. 65	14×10^{-3}	1971-75	RIZA
oil	32.324	> 99	0.53	1972-75	RIZA
phenoles	2.200	100	48×10^{-3}	1970-72	Ten Berge et al.
detergents	18.291	100	0.32	1971-75	RIZA
other hydrocarbons	3.000	100	56×10^{-3}	1970-72	Ten Berge et al.

Biotic Changes

Some changes in the fauna of the Rhine due to direct human interference result from overfishing and the introduction of non-indigenous
species.

Overfishing, together with deterioration of the habitat, are
held to be responsible for the disappearance or very strong reduction of a number of commercially important fish species before the
Second World War (Verhey, 1961). This may apply to sturgeon
(Acipenser sturio), Allis shad (Alosa alosa) and Twaite shad (Alosa
fallax) (Fig. 2). At present practically no fisheries are left,
but the return of these species is prevented by the effects of
river regulation and pollution.

Purposeful introduction of species into the Rhine has been
rather an exception. The most important introduction has been that
of the Pike-perch (Stizostedium lucioperca) in Germany in 1885. The
species has spread over the whole river system since then.

The Chinese mitten crab (Eriocheir sinensis), the Zebra mussel
(Dreissene polymorpha), and other species have been introduced
accidentally. Due to the overriding effects of the changes of the
abiotic factors in the river ecosystem, the effects of these introduced species are currently very limited.

DEGRADATION OF THE RIVER ECOSYSTEM

Migratory Fishes

The Rhine system counted formerly at least eleven different species
of migratory fishes.

Sturgeon (Acipenser sturio), houting (Coregonus oxyrhynchus)
and Allis shad (Alosa alosa) disappeared before the Second World
War, probably due to overfishing and river regulation (Verhey, 1961).

Salmon (Salmo salar) and Twaite shad (Alosa fallax) also decreased considerably in that period (Fig. 2). From the 19th
century until 1942 hatchery-reared young salmons were introduced
in the river to compensate for the decrease. Chemical pollution,
dredging, the construction of weirs in the upper course of the
river, and possibly overfishing were factors causing virtually the
disappearance of salmon (Verhey, 1961). Twaite shad almost disappeared when the Haringvliet-Hollands Diep estuary was dammed off
in 1970 (Ir. B. Steinmetz, pers. comm.).

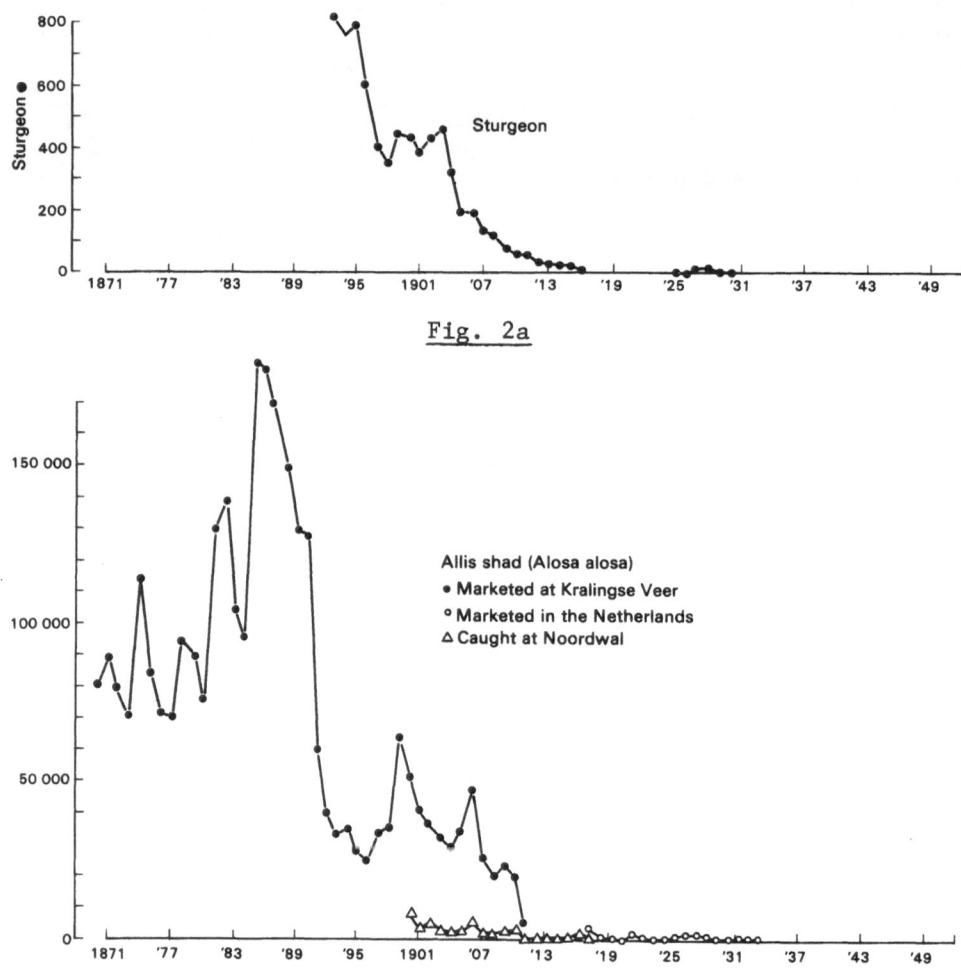

Fig. 2a

Fig. 2b

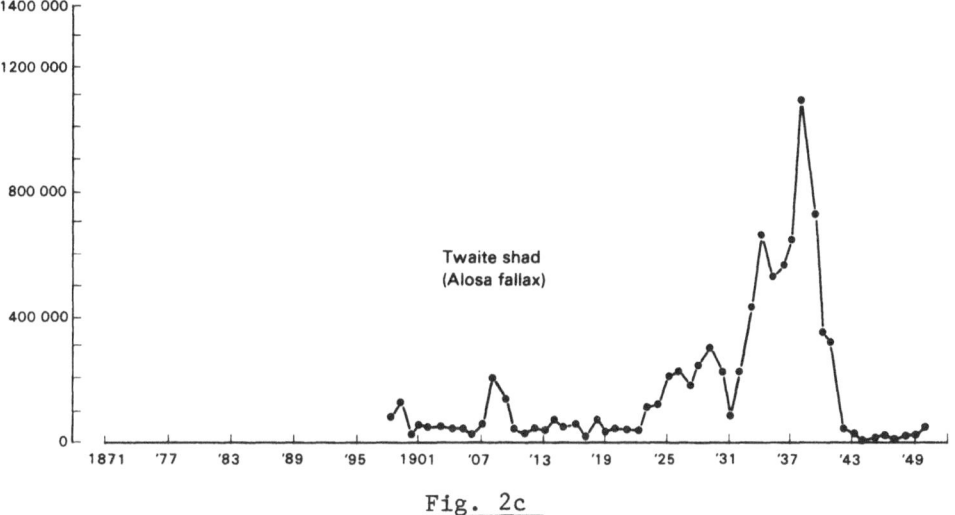

Fig. 2c

<u>Fig. 2.</u>   Catches of Sturgeon (a), Allis shad (b), and Twaite shad (c), made in the lower course of the Rhine in The Netherlands (after Verhey, 1961).

The Sea trout (Salmo trutta) occurred in low numbers at least until 1963 (Van Wijck, 1971). It probably has disappeared nearly completely.

Smelt (Osmerus eperlanus), lampern (Lampetra fluviatilis), Sea lamprey (Petromyzon marinus) and flounder (Pleuronectes flesus) were normal to abundant visitors of the lower course and the estuarine part of the river until recently (Vaas, 1968, Van Wijck, 1971). The closure of the Haringvliet estuary in 1970, however, caused a nearly complete disappearance of the first two species and a decline of the latter two (Steinmetz, 1974, 1975, pers. comm.). An abundant non-migratory smelt population still occurs in Lake IJssel, however.

Only the eel (Anguilla anguilla) is still a common species in the Rhine, whereas young flounders still occur in some years in the lower course (Ir. B. Steinmetz, pers. comm.).

Summarising it can be said that of the eleven migratory species once occurring in the Rhine, nine probably have disappeared or nearly so due to overfishing, river regulation and pollution.

Plankton

Peelen (1975) reviews a number of publications on the plankton of the Rhine. During the last sixty years the plankton composition did not change appreciably. Only some blue-green algae, e.g. Aphanizomenon flos-aquae and Microcystis aeruginosa, showed a considerable increase. Since the time of residence of the water between the Swiss Bodensee and the North Sea is only 14 days during average river discharge, Peelen, following Lauterborn (1916), Redeke (1948), and Leentvaar (1963), explains the lack of change by assuming that the plankton composition in the Rhine is determined mainly by the plankton in the large Swiss lakes and other stagnant waters in the catchment area of the river. This diverse input then will be flushed rapidly to the North Sea without the development of a characteristic river plankton.

Benthos

Knöpp (1957) compared the benthic fauna of the German stretch of the Rhine with the data of Lauterborn (1917, 1918) (Table II). The decrease of the total number of species is very marked, but the changes in the composition of the fauna are even more outstanding. The decrease of the share of clean water species is not less than 64-71%.

Table 2.   Composition of the benthic and littoral fauna
of the Rhine before 1918 (after Lauterborn,
1916-1918) and in 1955 (after Knöpp, 1957)

	Upper Rhine 1918	Upper Rhine 1955	Middle Rhine 1918	Middle Rhine 1955	Lower Rhine 1918*	Lower Rhine 1955
Spongillidae	5	2	1	2	1	2
Vermes	7	10	5	11	2	15
Crustacea	3	3	3	4	1	2
Plecoptera	12	0	12	0	1	0
Ephemeroptera	9	1	3	1	5	0
Neuroptera	0	0	0	0	1	0
Odonata	1	2	0	0	0	0
Rhynchota	1	0	1	0	1	0
Diptera	5	3	3	2	1	3
Trichoptera	9	3	11	1	5	2
Coleoptera	2	1	2	0	0	0
Bryozoa	3	1	2	1	1	3
Mollusca	20	14	18	12	17	8
TOTAL	77	40	61	34	36	35

* According to Lauterborn (1916-1918) the Lower Rhine has
been investigated insufficiently.

Peeters & Wolff (1973) investigated the soft bottom fauna of the Waal, the major Rhine branch in The Netherlands. In 76 0.1 m^2 Van-Veen grab samples they found in extremely low densities the molluscs Bithynia tentaculata, Limnaea cf. peregra, Sphaerium corneum, and Pisidium casertanum, the isopod Asellus aquaticus, and unidentified leeches and chironomid larvae. Maximum densities were 1.0/m^2 for Sphaerium corneum, 0.8/m^2 for chironomid larvae and 0.3/m^2 for Bithynia tentaculata. Only unidentified tubificids (Oligochaeta) reached densities up to several thousands per m^2. Compared to these figures may be the densities in the relatively clean Meuse, where 6.9 specimens per m^2 of Bithynia tentaculata and nearly 1,000 chironomid larvae per m^2 were found. The situation in the Rhine between Strasbourg and Koblenz appeared to be very similar to that in the Dutch Rhine (Arbeitsgemeinschaft Umwelt Mainz, 1972).

A number of invertebrate species has disappeared altogether from large parts of the Rhine. This concerns the large freshwater mussels Anodonta anatina, Unio tumidus, U. pictorum, U. crassus and Pseudanodonta complanata (Wolff, 1968), the fingernail clams Sphaerium rivicola, Pisidium amnicum, P. moitessierianum, and P. supinum (Wolff, 1970; Kuiper & Wolff, 1970), the isopod Proasellus meridianus (Wolff, 1973), and nearly all Plecoptera, Ephemeroptera, and Trichoptera (Knöpp, 1957). Many other species probably went unnoticed.

The Nekton

Van Wijck (1971) and Peeters and Wolff (1973) together record 20 species of fishes and one hybrid from the Waal, the main branch of the Rhine in The Netherlands (Table III). From their papers it becomes evident that characteristic river species still recorded by Redeke (1948), such as barbel (Barbus barbus), the cyprinid Chondrostoma nasus, chub (Leuciscus cephalus), bullhead (Cottus gobio), and Stone roach (Neomacheilus barbatulus) have disappeared or become extremely rare. Dace (Leuciscus leuciscus) and gudgeon (Gobio gobio) are the last representatives of this group. The remaining species are ubiquitous freshwater fishes. Koeman (1971) found that roach living in Rhine water showed much higher PCB-contents than roach living in other Dutch fresh waters.

Also the Rhine fisheries declined dramatically. A century ago thousands of people took part in the Dutch Rhine fisheries. In 1947 120 fishing boats were still active on the Waal branch, in 1961 16 and currently only 1 or 2 (Leentvaar, 1963; Peeters, pers. comm.).

Table 3. Occurrence of fish species in the Waal according to Van
Wijck (1971) and Peeters & Wolff (1973). The species
for which a percentage has been calculated, occurred in
the sample recorded in the heading; crosses denote
species observed at other occasions.

	Van Wijck (1971) n = 1449 (1970)	Peeters & Wolff(1973) n = 2203 (1971)
Roach (Rutilus rutilus)	57.7%	84.8%
White bream (Blicca bjoerkna)	13.6%	1.6%
Bream (Abramis brama)	11.7%	8.1%
Bleak (Alburnus alburnus)	6.2%	0.1%
Eel (Anguilla anguilla)	4.5%	0.0%
Dace (Leuciscus leuciscus)	4.2%	0.0%
Gudgeon (Gobio gobio)	1.8%	4.3%
Perch (Perca fluviatilis)	0.2%	0.0%
Pike (Esox lucius)	0.1%	-
Lamprey (Petromyzon marinus)	+	-
Lampern (Lampetra fluviatilis)	+	-
Salmon (Salmo salar)	+	-
Carp (Cyprinus carpio)	+	-
Bitterling (Rhodeus amarus)	-	-
Orfe (Leuciscus idus)	+	-
Rudd (Scardinius erytrophthalmus)	+	-
Bream x Roach (Abramidopsis leuckarti)	-	0.2%
Three-spined stickleback (Gasterosteus gasterosteus)	-	0.1%
Pike-perch (Stizostedium lucioperca)	+	-
Pope (Acerina cernua)	+	0.5%
Flounder (Pleuronectes platessa)	-	0.0%

Aquatic Phanerogams

Aquatic phanerogams formerly common (Lauterborn, 1918; Redeke, 1948), such as pondweeds (Potamogeton pectinatus and P. perfoliatus), Flowering rush (Butomus umbellatus) and arrowhead (Sagittaria sagittifolia), have disappeared almost completely from the shallow parts of the lower course river.

Aquatic Birds

The populations of many bird species living along the Rhine have changed considerably in numbers. However, in most cases it is difficult to demonstrate that relationship with the changes in the river exists. Only the complete disappearance of breeding colonies of cormorants (Phalacrocorax carbo) along the branches of the Rhine in The Netherlands may probably be related to the deterioration of the river.

The River Ecosystem

The present foodwebs in the Rhine appear to be simple. The phyto-plankton is probably fed on by the rather scarce zooplankton, although the latter category also may take the abundant detritus. However, neither phytoplankton nor detritus are apparently exploited to any extent by a filter-feeding benthic fauna, notwithstanding the fact that transport of detritus across a cross-section of the river is in the order of magnitude of 50 g. dry weight m^{-1}. sec^{-1}.

The only benthic animals of any quantitative importance are the deposit-feeding Oligochaeta (Tubifex tubifex and other species). Dependent on the amount of deposition of silt and detritus (up to 1 m. $year^{-1}$), their numbers can amount to many thousands per m^2. This source of food most probably is being exploited by several fish species, for instance roach (Van Wijck, 1971), and diving ducks such as pochard (Aythya ferina) and Tufted duck (Aythya fuligula). Only in this way can the abundance of roach and pochard (40,000 along a few kilometers of the river) be explained.

All other steps in the foodchains are nowadays absent or very unimportant. Thus arises a picture of a very simply structured ecosystem in which only the foodchain sewage – detritus – oligochaetes – cyprinid fishes or diving ducks is of any importance.

DEGRADATION OF TERRESTRIAL ECOSYSTEMS IN THE FLOOD PLAINS

The flood plains are influenced by the Rhine during the annual inundations in winter and early spring. During these inundations silt contaminated with toxic compounds settles. This induces manganese deficiency in different arable crops through high levels of zinc. High arsenic contents of the soils in these flood plains render potatoes undesirable for human consumption (De Groot, 1973). High copper values of the grass in the flood plain meadows have been demonstrated to influence the health of grazing sheep. Remarkably enough the high mercury contents of the grass do not seem to influence grazing animals, probably because this metal is not taken in as methylmercury (Heidinga et al, 1971; Ten Berge et al, 1973).

Data on the influence of the polluted river water on the terrestrial wild flora and fauna are scarce. A marked increase of nitrophile and even of halophile plant species along the river has been demonstrated.

DEGRADATION OF FRESHWATER LAKES

Introduction

In the Netherlands a number of man-made lakes are fed with Rhine water, viz. Lake IJssel (since 1932), Lake Brielle (since 1950) and Lake Haringvliet (since 1970). All of them increasingly show a negative influence on the Rhine water.

Eutrophication

All lakes show signs of increasing eutrophication. In Lake IJssel and Lake Brielle blooms of blue-green algae are a regularly occurring phenomenon during the last ten years. Also in Lake Haringvliet blue-greens begin to show up. However, it has been shown that in Lake Brielle a local sewage outfall also contributed to the eutrophication.

In Lake IJssel suspended sediment and algal blooms probably have led to disappearance of submerged phanerogams. This in its turn caused a decline of many diving ducks and swans feeding on the submerged vegetation. Among these is the rare Bewick's swan, Some years ago the larger part of this bird's world population wintered on this lake.

Toxic Compounds

When the river water enters the lakes the riverborne silt tends
to settle. Sedimentation rates up to 1 m per year have been
observed in the entrance of Lake Haringvliet (Nieuwe Merwede).
Since many toxic compounds tend to adsorb on the silt particles,
the concentration of e.g. heavy metals in the bottom sediments can
be very high.

Possibly the very sparse occurrence of benthic animals in Lake
Haringvliet five years after its origin (Wolff, unpublished data)
can be related to this fact.

Also bird and fish species living in these lakes have been
shown to be contaminated. Cormorants fishing in Lake IJssel have
been shown to die because of excess contamination with polychlorin-
ated biphenyls. Eel (Anguilla anguilla) from Lake IJssel had an
average total mercury content of 0.49 ppm, which is significantly
higher than 0.14 ppm in eel from other inland waters in the Netherlands. Also Pike-perch (Stizostedium lucioperca) and perch (Perca
fluviatilis) from Lake IJssel showed high mercury levels (Greve &
Wit, 1971).

INFLUENCES ON MARINE AND ESTUARINE ECOSYSTEMS

Eutrophication

Van Bennekom et al. (1975) demonstrated a noticeable influence of
the pollution of the Rhine on the Southern Bight of the North Sea,
resulting in rapidly increasing eutrophication. For instance, in
the period 1970-1975 the concentrations of dissolved phosphate in
the coastal waters of the Netherlands have nearly doubled (Tijssen
& Van Bennekom, 1976). Since phytoplankton production in the Dutch
coastal waters tends to be light-limited, the effects probably have
been comparatively small along the coast, but further offshore
phytoplankton production might have increased considerably (Postma,
1973; Gieskes & Kraay, 1975).

This is probably one of the sources of the increase of organic
matter imported into the Wadden Sea (De Jonge & Postma, 1974).
Postma (1954) estimated the annual import of organic matter from
the North Sea towards the Wadden Sea at 80 g particulate organic
carbon m^{-2} $year^{-1}$, but De Jonge & Postma (1974) arrived at an
estimated import three times larger. Hence, this import has trebled
during the past 20 years. Postma (1973) indicates the possibility
that this process may lead to a disturbance of the oxygen balance
in estuarine areas through the mineralisation of too large amounts

of particulate organic matter. Tijssen & Van Bennekom (1976)
found evidence for this in the western Wadden Sea.

Toxic Compounds

Koeman (1971) showed how the discharge of waste-water from a pesticide factory on the Rhine estuary caused the mass mortality of birds
in the western Wadden Sea. This occurred through northward transport
of river water with the North Sea residual current and through
amplification in the food-chains leading to these birds. Most
probably this pollution caused the numbers of Sandwich tern (Sterna
sandviciensis) in the Netherlands to decline from about 40,000
breeding pairs in 1954 to 650 breeding pairs in 1965. Since the
pesticide plant has been closed, the numbers of terns are increasing
again. Mortality also was observed in spoonbill (Platalea leucorodia)
eiderduck (Somateria mollissima), Herring gull (Larus argentatus),
Common tern (S. macrura), and Great crested grebe (Podiceps
cristatus).

Also PCB's discharged by the Rhine have been demonstrated to
be responsible for bird mortality. Cormorants (Phalacrocorax carbo)
found dead, very probably dead from the effects of high levels of
PCB's (Anonymous, 1975).

Porpoises (Phocaena phocaena) have virtually disappeared from
Dutch coastal waters, and Harbour seals (Phoca vitulina) are decreasing at an alarming rate (Fig. 3). Although high concentrations
of mercury have been found in liver and brain tissue of these species,
it is possible that the toxic action of methylmercury is counteracted by accumulation of selenium (Koeman et al., 1975). High
levels of PCB's also have been reported in these animals, but an
explanation of their decrease is still a matter of conjecture
(Anonymous, 1975).

Koeman et al. (1971) concluded that many organisms in the Dutch
coastal waters are contaminated with mercury, the larger part of
which probably originates from the Rhine. This effect even extends
to the English Wash where wintering waders show elevated concentrations of mercury probably taken up during their stay in autumn
in the Wadden Sea (Parslow, 1973).

CONTROL OF THE POLLUTION OF THE RHINE

The effects of the pollution of the Rhine are relatively well known.
Since 1885 attempts to restore the quality of the river have been
made. In this respect the following points may be mentioned.

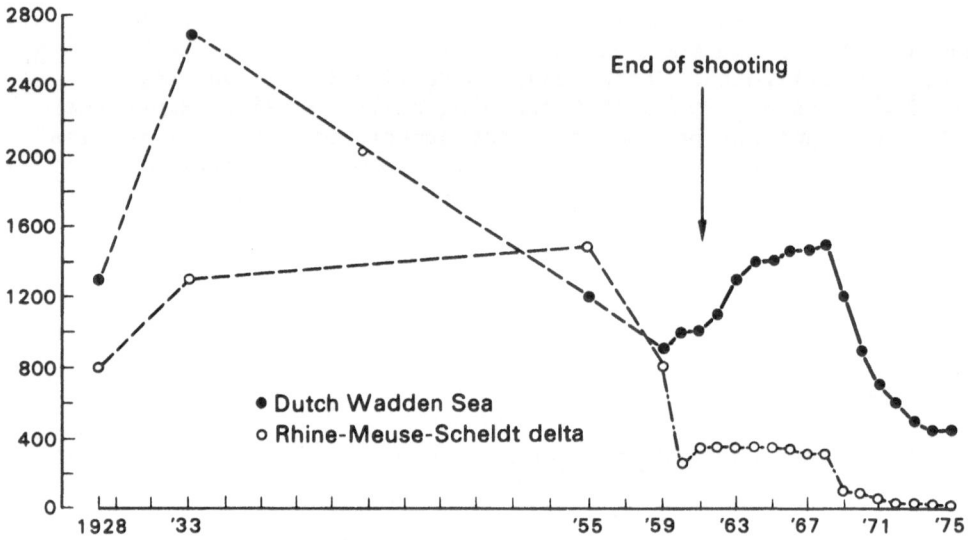

<u>Fig. 3.</u>   Decrease of Harbour Seals in The Netherlands (after Van
          Haaften, 1974).

There exists an International Commission for the Protection of
the Rhine against Pollution. In this Commission Switzerland, France,
Germany, Luxembourg and the Netherlands take part. The Commission
monitors the quality of the river water and advises the five governments on steps to be taken. It has no executive powers however.

The Council of Europe is at present preparing a European
Convention for the Protection of International Watercourses against
Pollution. For this convention a Committee of Technical Experts is
assembling criteria for the quality of the Rhine water in order to
protect the wild flora and fauna and the capacity for self-purifica-
tion of the river.

Roughly 50% of the organic sewage discharged into the river
is being purified, either mechanically or biologically. This percentage has increased during the last few years as a result of the
enforcement of special water protection laws in some of the countries
involved. This is probably the cause of the recent decrease of the
ammonia concentration in the river. Also the concentration of
phenols has decreased since about 1971.

Detergents causing excessive quantities of foam have been
banned completely from the river in 1964-1967. The levels of
mercury and beryllium in the river have dropped considerably since
it became known that these elements could have disastrous effects
on human health.

In Germany a fleet of small tankers has been created to receive
the oil-polluted bilge water of the many barges on the river. A
system of land stations has been set up in the Netherlands to serve
the same purpose.

The Governments of the countries in the watershed of the Rhine
regularly meet to discuss measures to be taken to decrease pollution,
as do various international organisations. Agreement has been
reached on a number of specific measures, e.g. control of thermal
pollution and restriction of chemical pollution and discharge of
sodium-chloride.

Despite these promising developments it cannot yet be concluded
that the condition of the Rhine is improving. Many pollutants maintain their high levels and others even increase.

ACKNOWLEDGEMENTS

Valuable comments on the first draft of this paper were received
from Messrs. A. J. van Bennekom, Dr. D. Eisma, J. L. Koolen, Dr.
D. J. Kuenen, P. Leentvaar, Dr. A. Lelek, J. C. H. Peeters, Dr. H.
Postma, P. Reijnders, J. Rooth, B. Steinmetz, Dr. C. W. Stortenbeker,
and Dr. J. J. Zijlstra. Mr. T. A. W. van Rossum helped with the
translation.

REFERENCES

Anonymous, 1975. Ecological aspects of water pollution in specific
    geographical areas: study of sublethal effects on marine
    organisms in the Firth of Clyde, the Oslo Fjord and the Wadden
    Sea. Report on a Working Group, Wageningen, 2-4 Dec. 1974,
    World Health Org., 38p.

Arbeitsgemeinschaft Umwelt Mainz, 1972. Bestandsrückgang der
    Schneckenfauna des Rheins zwischen Strassburg und Koblenz.
    Natur Mus., Frankf., 102, 197-206.

Bennekom, A. J. van, Gieskes, W. W. C. & Tijssen, S. B. 1975. Eu-
    trophication of Dutch coastal waters. Proc. R. Soc. B, 189,
    359-374.

Berge, W. ten, et al. 1973. Rijn Nota. Veren. Milieudefensie,
    Amsterdam, 116p.

Gieskes, W. W. C. & Kraay, G. W. 1975. The phytoplankton spring
    bloom in Dutch coastal waters of the North Sea. Neth. J. Sea
    Res., 9, 166-196.

Greve, P. A. & Wit, S. L. 1971. (Totaal) kwikgehalte van zoetwater-
    en zeevis. TNO-nieuws, 26, 395-399.

Croot, A. J. de, 1973. Occurrence and behaviour of heavy metals
    in river deltas, with special reference to the Rhine and Ems
    rivers. In: North Sea Science, edited by E. D. Goldberg, 308-
    325. (Mass.), MIT.

Haaften, J. L. van, 1974. Zeehonden langs de Nederlandse kust.
    Wet. Meded. K. ned. natuurh. Veren., 101, 36p.

Heidinga, M. C., et al. 1971. Onderzoek naar de accumulatie van
    kwik in de uiterwaarden van de Rijn. TNO-nieuws, 26, 382-384.

Jonge, V. N. de, & Postma, H. 1974. Phosphorus compounds in the
    Dutch Wadden Sea. Neth. J. Sea Res., 8, 139-153.

Knöpp, H. 1957. Die heutige biologische Gliederung des Rheinstroms zwischen Basel und Emmerich. Dr. gewässerk. Mitt., 1, (3), 65-63.

Koeman, J. H. 1971. Het voorkomen en de toxicologische betekenis van enkele chloorkoolwaterstoffen aan de Nederlandse kust in de periode van 1965 tot 1970 (In Dutch with English summary). Thesis, Utrecht, 136 pp.

Koeman, J. H., et al. 1971. Kwik in het Nederlandse kustmilieu. TNO-nieuws, 26, 402-409.

Koeman, J. H., v.d. Ven, W. S. M., de Goeij, J. J. M., Tjice, P. S. & van Haaften, J. L. 1975. Mercury and selenium in marine mammals and birds. In: The Science of the total environment. 279-287. Amsterdam: Elsevier.

Kuiper, J. G. J. & Wolff, W. J. 1970. The Mollusca of the estuarine region of the rivers Rhine, Meuse, and Scheldt in relation to the hydrography of the area. III The genus Pisidium. Basteria, 34, 1-40.

Lauterborn, R. 1916, 1917, 1918. Die geographische und biologische Gliederung des Rheinstromes I, II, III. Sber. heidelb. Akad. Wiss. B., Jahrgang 1916; 6 Abh, 1-16, 1917; 5 Abh, 1-70, 1918.

Leentvaar, P. 1963. Stiefvader Rign, kunstmoeder Maas. Natura, Amst., 60, 50-56.

Parslow, J. L. F., 1973. Mercury in waders from the Wash. Environ. Pollut., 5, 295-304.

Peelen, T. 1975. Changes in the composition of the plankton of the rivers Rhine and Meuse in the Netherlands during the last fifty-five years. Verh. int. Verein. theor. angew. Limnol., 19, 1997-2009.

Peeters, J. C. H. & Wolff, W. J. 1973. Macrobenthos and fishes of the rivers Meuse and Rhine, the Netherlands. Hydrobio. Bull. Amst., 7, 121-126.

Postma, H. 1954. Hydrography of the Dutch Wadden Sea. Archs. neerl. Zool., 10, 405-511.

Postma, H. 1973. Transport and budget of organic matter in the North Sea. In: North Sea Science, edited by E. D. Goldberg, 326-334, Cambridge, (Mass.): MIT.

Redeke, H. C. 1948. Hydrobiologie van Nederland. De zoete wateren.
    Amsterdam: De Boer, 580p.

Steinmetz, B. 1974. Orienterend onderzoek naar de visstand van
    het Haringvliet en het Hollands Diep. Visserij, 27, 113-128.

Steinmetz, B. 1975. Resultaten van het visserijkundig onderzoek
    in het Haringvliet en het Hollands Diep in 1974 en 1975.
    Visserij, 28, 474-486.

Tijssen, S. B. & Bennekom A. J. van, 1976. Lage zuurstofgehaltes
    in het water op het Balgzand. H_2O, 9, (2), 28-31

Vaas, K. F. 1968. De visfauna van het estuariumgebied van Rign
    en Maas. Biol. Jaarb. Dodonaea, 36, 115-128.

Verhey, C. J. 1961. De vissen en de visvangst. In: De biesbosch,
    land van het levende water, edited by C. J. Verhey, 139-164.
    Zutphen, Thieme.

Wijck, C. J. A. van, 1971. Onderzoek naar de visfauna in de
    omgeving van Nijmegen. Report Zool. Lab. Afd. Dieroecologie.
    Kath. Univ. Mijmegen, 40p.

Wolff, W. J. 1968. The Mollusca of the estuarine region of the
    rivers Rhine, Meuse and Scheldt in relation to the hydrography
    of the area. I. The Unionidae. Basteria, 32, 13-47.

Wolff, W. J. 1970. The Mollusca of the estuarine region of the
    rivers Rhine, Meuse and Scheldt in relation to the hydrography
    of the area. IV. The genus Sphaerium. Basteria 34, 75-90.

Wolff, W. J. 1973. The distribution of Asellus aquaticus (L.)
    and Proasellus meridianus (Rae.) in the southwestern part of
    the Netherlands. Hydrobiologia, 42, 381-392.

Zimmerman, J. T. F. & Rommets, J. W. 1974. Natural fluorescence
    as a tracer in the Dutch Wadden Sea and the adjacent North
    Sea. Neth. J. Sea Res., 8, 117-125.

DISCUSSION: PAPER 9

J. CHRISTIE   In the Great Lakes there has been a
                        sequence of changes parallel to those in the
                        Rhine. Up to 1920 overfishing caused a prog-
                        ressive loss of fish. Thereafter pollution
                        has had an increasing impact: most recently
                        we have been concerned with pesticide levels

in birds and fish. Fisheries, however, persist
but take species smaller in size and lower in
the trophic series, although maintaining a
comparable volume to that caught previously.

There is an added stress to those of overfishing and pollution, namely the invasion of
exotic fish species, like smelt and white
perch.   Such new species may increase explosi-
vely in response to removal of predators.
Control measures are now being applied to
pollution - it remains to be seen how the
system will respond.

L. M. TALBOT   I was struck by the diagram showing
apparently cyclic changes in fishery catch:
was this due to changes in river conditions?

W. J. WOLFF   I don't know. I have never seen the fish
that used to be caught in the Rhine!

N. POLUNIN   Can we hear more about the celebrated fish
kill in the Rhine six years ago, and the process
of recovery from it?

W. J. WOLFF   Six years ago a pesticide, found (or
known?) to be endosulphan, was accidentally
discharged into the Rhine in Germany. Many
fish were killed - and indeed it was this
that led to the detection of the pollution.
Stocks of these fish were depleted but they
have now recovered.

E. M. NICHOLSON   What is the effect of the closure of the
delta likely to be? What regime will exist
behind the barriers?

W. J. WOLFF   The effect will be complicated. In the
past there were two main estuaries. One was
the shipping channel for the port of Rotterdam,
the other was once a very large estuary now
closed by a row of sluices impounding a big
freshwater lake (which became fresh very
quickly because of the high river flow). The
environment in this lake is poor, and it has
been colonised by small numbers of a few
species and is now exhibiting eutrophication,
with blooms of blue-green algae. Sedimentation is heavy and over the years many metres

of polluted sediment are likely to accumulate.
I cannot predict just how the biocenosis will
develop, but suggest it will be poor.

N.K. JACOBSEN   How far are recent trends in marine life
detectable in the Frisian Wadden Sea?

W. J. WOLFF   The influence of the Rhine is clear in
the Dutch Wadden Sea and in the German area
residues probably brought down the Rhine are
present in the eggs of seabirds breeding in
the Weser estuary, but no effect is detectable farther north.

H. KÖPP   The seal population decline is an international problem and there are other factors
than just pollution from the Rhine. Bans on
hunting and shooting did not coincide in
Germany, Denmark and the Netherlands. Now
there are such bans the threat now is of
disturbance by tourists throughout the year.

W. J. WOLFF   The data, however, do show a major decline
in the Dutch Wadden Sea since 1968 and it is
hard to relate this to hunting or tourism.

G. VAN DYNE   How far can the variation in pollutant
loading over time be ascribed to changes in
rainfall, and hence to dilution effects?

W. J. WOLFF   Variation in river flow, and in sediment
load, which can take up pollutants by ad-
sorption, certainly occurs: levels of
industrial discharge also vary.

J. N. R. JEFFERS   It is striking that hydrologists have
precise models of river flow, yet biologists
are reluctant to use these. Are there such
models for the Rhine? Many fluctuations could
be explained by physical components in the
models. Also, many fluctuations will have a
multivariate cause - graphs, isolating single
factors and projecting relationships on a
single plane may distort our appreciation of
the true relationships. Much more intensive
analysis is clearly needed where so many data
exist and the situations are potentially so
complex.

B. ULRICH

In a terrestrial ecosystem the soil is not only a source of nutrients for plants but a place where decomposition and recycling goes on. In the last century we decided not to use the soil to decompose and recycle human body wastes: instead we put this waste into water and air, and worse, we mixed domestic sewage and industrial effluents at our treatment works. Now we have watercourses polluted with both organic wastes and heavy metals. We should be developing new technology that will again make use of the soil for waste disposal and breakdown rather than water or air.

C. GELIN

Surely one essential measure is to treat sewage properly before it enters the river. Will the Rhine Commission stimulate action on this and similar matters in time?

W. J. WOLFF

The schedules agreed by the Commission are reasonable enough, and there is plenty of information and ideas - but intergovernmental agreements take time to become effective in practice.

H. KÖPP

This whole story illustrates in a classic fashion the way international efforts can work. Progress has been very slow - only in 1976 have some practical agreements been reached, after 10 years of debate. The Council of Europe Convention has also been discussed for over a decade. Meantime, damage has increased. Yet the Rhine is far better known than most rivers. The delay has been caused because of the very different economic interests in Switzerland, Germany, France and the Netherlands. There has been fear of upsetting economic competition and it is this, not a lack of data or of the technical means for control, that has held things up.

10. THE ROLE OF FIRE IN THE DEGRADATION OF ECOSYSTEMS

Ross W Wein

Department of Biology, University of New Brunswick

Fredericton, New Brunswick, Canada, E3B 5A3

INTRODUCTION

Man has long been in conflict with wildfire yet as Sauer (1975) argues, early man retained fire and utilised it to dominate his environment. Lightning and volcanic activity are older causes of fire over much of the world and one cannot deny their importance (Komarek, 1973) but man in his somewhat more systematic approach has changed the frequency of fire. Although until recently man has considered fire to be a very destructive agent, fire is now considered such a positive force by some researchers that its negative effects may be ignored.

The objective of the present paper is to examine the fire literature to draw attention to cases where fire has lead to ecosystem degradation or where the potential for long-term degradation exists.

It is important to establish my definition of degradation for this paper. Degradation occurs when the ecosystem components do not recover to the expected degree during the normal fire frequency period. Obviously, I do not equate species change with degradation. For this paper I try to avoid identifying fire as bad or good as have many other authors. To establish any level of agreement of degradation requires identical criteria of degradation among those discussing a test case. This rarely occurs and the result is more confusion than clarification.

HISTORIC AND PRESENT ATTITUDE TOWARD FIRE

When European man colonised eastern North America, fire was used to
remove vegetation so that agriculture could be developed, but fire
was also a fearful power. The White pine of New Brunswick were cut
to provide ship masts for the British Royal Navy and later, lumber
of other tree species supported a thriving export trade and ship-
building. In 1825 the dry forest refuse, plus possibly insect-
killed trees, combined with a year of abnormally low rainfall and
numerous escaped clearing fires, led to a fire that burned between
15,500 and 20,700 km^2 across New Brunswick and into Maine. One
hundred and sixty lives were lost in what was to be called the
Great Miramichi Fire (Ganong 1906). This was the largest fire
recorded in North America (Holbrook, 1943) but there is some evidence
that even larger fires have occurred (Ganong, 1902).

Spectacular fires plus the continued prevalence and threat of
fire as the remainder of North America was being settled, shaped
the belief that fire must be controlled. Fires were fought, and
as equipment and fire fighting techniques improved, fires were more
effectively controlled whether they were in grassland, commercial
forests, parks or wilderness areas. It is obvious that fires
threatening human life, dwellings and equipment, and causing forest,
wildlife and recreational losses must be controlled but should fires
be eliminated to protect ecosystems which are to be maintained in
a "natural state"?

In the late 1960s there were only a few researchers calling
for the restoration of fire to these areas but recently there has
been a dramatic increase in the study of fire in the ecosystem. The
increased number of symposia (Tall Timbers Research Station 1962
to 1974; Slaughter et al. 1971; USDA Forest Service 1972; Heinselman and Wright 1973) and books (Kozlowski and Ahlgren 1974)
attest to this interest. Today many ecologists view fire as a
necessary factor for the maintenance of many conifer forest ecosystems (e.g. Heinselman, 1973; Wright, 1974 and many others).

THE PRESENT STATE OF FIRE FREQUENCY KNOWLEDGE

It is important that prehistoric fire frequencies be duplicated
in ecosystems, such as wilderness areas, where natural ecological
controls are to prevail because ecosystems have evolved under this
fire regime. In other ecosystems where man is utilising resources
and wishes to use fire as a management tool, prehistoric fire
frequencies could serve as a standard with which to compare present
frequencies. If differences in frequencies exist, long-term ecological problems may develop.

Unfortunately historic and prehistoric fire frequencies are known for only a few geographic areas. These frequencies have been developed primarily from fire scars on trees and from paleoecological techniques which provide a longer period of record. Paleoecological researchers have long recorded charcoal in lake sediments and a few researchers have linked the charcoal in sediment varves to years of large fires (e.g. Davis, 1967). Swain (1973) reported on a comprehensive record of fire frequencies in lake sediments of the Boundary Waters Canoe Area of northern Minnesota. This study will serve as a standard and has already stimulated a number of similar studies in other vegetation types.

In New Brunswick, there is increasing pressure to prescribe fire on clearcut forest land to facilitate tree planting. These prescribed burns will generally be less intense than natural fires because they will be conducted in the spring or autumn when the fire hazard is not as great. The prescribed burn frequency will be on a 40 to 60 year harvest rotation basis. At present we do not know the prehistoric fire frequency under which the forest evolved but we do know that the 40-60 year fire rotation far exceeds the rate burned since 1930. Wein and Moore (1977) have calculated recent fire rotations ranging from about 230 years to over 1,000 years for the different vegetation types in New Brunswick.

Table 1 presents some idea of the range of fire frequencies found in North American vegetation types and the information on which the frequencies are based.

The reasons for the variability are many but basically they include different ignition frequencies and different fuel characteristics of the biomass. The main point is that the frequency of fire has shaped the evolution of the vegetation and it would seem logical that information should be examined critically before managment plans established in one area are implemented in another area with a very different natural fire frequency. Thus, even though there are many easily demonstrated effects of fire in areas with high fire frequency (Wright and Heinselman, 1973), fire may not have similar effects in vegetation types with low fire frequencies.

FIRE AND THE PHYSICAL COMPONENT OF THE ECOSYSTEM

Erosional cycle and land stability: It must be remembered that fire will not cause erosion, it only functions as a "triggering mechanism". Accelerated erosion will only occur on areas that are already geologically unstable and the amount of erosion is dependant on erodibility of the soil, steepness of slope, rainfall characteristics

Table 1.   Fire frequencies from selected vegetation types in North America.

Vegetation type and location	Record type and period	Fire frequency (years)	Reference
Prairie, Missouri	–	1	Kucera and Ehrenreich (1962)
Long-leaf pine and blue stem ranges, Southeast U.S.	–	3	Duvall and Whitaker (1964)
Mixed conifers, California.	Fire scars on trees (1581 – present)	7 – 9	Wagener (1961)
Ponderosa pine, Arizona	Fire scars on trees (1700 – present)	5 – 12	Weaver (1951)
Mixed forest, Minnesota	Lane sediments (1,000 years)	60 – 70	Swain (1973)
Boreal conifers, Northwest Territories	Fire report records (1966–1972)	110	Johnson and Rowe (1975)
Mixed forest, New Brunswick	Fire report records (1920 – 1975)	230 – 1,000	Wein and Moore (1977)

and the amount of protecting vegetation and soil surface organic
matter left after the fire.

Much of the available literature on soil erosion in forested
land has focused on road building and logging because these two
activities frequently initiate much more soil movement than fire
(Swanston, 1971). It should also be noted that the building of
fire guards can cause much more erosion than the fire alone
(Bolstad, 1971; Biswell, 1974). The present author would suggest
that land managers would do well to establish the frequency of
natural fires and the erosion rate following fires for a given
area and then use these data as an index to permissible erosion
rates for road building, logging and other activities of ecological
concern.

A spectacular example of erosion following fire has been
studied by Anderson et al. (1959) and Krammes (1965) in the
chaparral-dominated San Gabriel Mountains, near Los Angeles,
California. On steep slopes of this highly erodible watershed
an annual average of 8,000 kg/ha of debris was lost downslope.
After a fire swept through the watershed, debris movement
accelerated by a factor of ten.

Another factor that contributes to soil movement on slopes
in chaparral vegetation is the development of water-repellent soils
resulting from compounds volatilised or leached from resinous
vegetation (DeBano, 1969). Fire releases compounds which diffuse
downward and condense on the cooler soil particles. During rains
the soil above the water-repellent soil layer becomes saturated,
the soil mass overcomes the force of gravity and mudflows result.

Mudflows and sheet erosion are also important in fine-textured,
high ice-content soil of arctic and sub-arctic regions. In this
case, vegetation and organic matter provide thermal insulation
which protects the ice-rich soil. After fire more energy penetrates
the soil and the surface thawed layer (active layer) thickens
(Mackay, 1970; Heginbottom, 1971; Viereck, 1973; Wein and Bliss,
1973). With increasing ice content there is an increasing amount
of subsidence and silt flows can result on slopes (Heginbottom, 1971).

As a result of erosion, the physical condition of streams
change. High runoff peaks may scour the channel and even low
erosion rates of fine-textured soils may fill the small openings
between rocks or completely cover gravel beds. In essence, a stable
stream ecosystem may be so changed by erosion that aquatic organisms
must recolonise the substrate. To the author's knowledge few data
exist on the rates of this recolonisation but erosion and sedimentation probably are more important biologically in clear water rather
than where water already has a high sediment load.

Nutrient cycles: It has been well documented that nutrient pools
and nutrient fluxes among pools change dramatically during and
after fire. Volatilisation of nitrogen usually is the greatest
loss from the system but there is considerable disagreement as to
the quantity of losses by particulate matter to the atmosphere and
by leaching of the non-volatile nutrients (Grier, 1975; Knight,
1966; Smith, 1970; Evans and Allen, 1971; Lewis, 1974).

On the soils that are coarse-textured with low clay content,
and therefore low cation exchange capacity, a relatively large
proportion of the nutrient pool can be held in the vegetation biomass and in the soil organic layer. Following fire, released
nutrients may not be held by the soil and subsequent revegetation
is slow (Viro, 1974). Fire-barrens can result. Fire-barrens are
not uncommon in Fenno-Scandi (Viro, 1974) and Canada (Damman, 1971;
Strang, 1972) and some foresters believe that these areas cannot be
reclaimed economically for forest production (Strang, 1972).

Gimingham, (1972 p. 204) raises another potential nutrient
regime problem. He hypothesises that Calluna monocultures maintained by fire in northern Britain may be contributing to an increased
rate of podzolisation. The long-term effects of this could be the
development of iron pans in the B horizon, impeded drainage and
development toward acid bog conditions.

Hydrologic cycle: Water is the transport medium for nutrients and
sediment so the erosion, nutrient and hydrologic cycles cannot be
readily separated. Water movement through the terrestrial ecosystem
becomes a major physical component of the freshwater ecosystem and
the quality of runoff water affects the functioning of the aquatic
ecosystem.

Stream flow increases following removal of deep-rooted, woody
vegetation from such vegetation types as chaparral, oak savannah
and pinyon-juniper woodland (Brown, 1965; Pase and Ingebo, 1965;
Lewis, 1968). Runoff immediately after storms can be very high
due to greater overland flow on burned areas (Ursic, 1969).

Burning or cutting vegetation that shades streams may also
lead to an increase in water temperature above the lethal temperature for trout and salmon. Mean monthly maximum increased as much
as 8°C (14°F) in summer for a watershed in Oregon (Brown and Krygier,
1967). Stream water temperature can also be raised by sedimentation
because pools become shallow and suspended sediments absorb more
radiation per unit volume of water (Cordone and Kelley, 1961).

Many researchers have found that nutrient and organic matter movement into aquatic ecosystems increases following fire (Brown et al. 1973; Snyder et al. 1975; McColl and Grigal, 1975). One would assume that this would increase the primary production or increase the detritus feeders but to the author's knowledge this has not been examined in any detail.

FIRE AND THE BIOLOGICAL COMPONENT OF THE ECOSYSTEM

<u>Fire, the individual and the population:</u> Individual plants or animals survive fire by avoidance through movement away from the fire or by being passively located in, or actively selecting, a habitat that is sufficiently insulated from the fire. No living tissue can survive normal fire temperatures without some degree of insulation. In other words, fire exerts selection pressure at the level of the individual by determining which die or survive. A second aspect of individual species success following fire is that of invasion when lethal temperatures no longer exist on the burned area. A great amount of literature is available on plant and animal adaptations related to fire that will not be reviewed here. The reader is referred to reviews such as Daubenmire,(1974); Gill, (1975) and Bendell, (1974).

Man is more often concerned with populations lost by fire and with populations that result from surviving or invading species. Fire does not burn the vegetation of complete landscapes and small segments of populations usually survive in unburned refugia, such as along streams, on islands or in isolated valleys. To my knowledge, fire in itself has not caused the extinction of a species although populations can be readily depleted. A great volume of literature gives comparisons of populations before and after fire and there are some longer-term studies which determine the length of time required for populations to return to prefire levels. Only a few studies of populations will serve as examples.

Although my topic is the degradation aspect of fire I wish to briefly mention a few positive effects on populations. There are many examples of the reduction of plant parasites such as Dwarf mistletoe on Black spruce (Irving and French, 1971) and plant diseases such as Brown needle spot (<u>Septoria alpicola</u>) on Long-leaf pine (<u>Pinus palustris</u>) (Siggers, 1934), Blueberry red leaf (<u>Exobasidium</u> sp.), and leafspot of blueberry (<u>Septoria</u> sp.) (Markim, 1943). Komarek (1970) reviews the role of fire in reducing insets that damage food and fibre crops. Burning crop residues is standard practice to control alfalfa weevil, stem borers of rice and many other insects. The positive relation between fire and moose, Black-tailed deer, Bobwhite quail, Scottish red grouse and Blue grouse

are well known and have been reviewed by Bendell (1974). Finally
it is well known that fires have been instrumental in the establishment of spruce and pine populations that are extremely valuable
to the forest industry.

Negative effects of fire on populations occur when desirable
populations are lost or unwanted populations expand. Fire-
damaged aspen and pine are susceptible to infection by Fomes sp.
(Schmitz and Jackson, 1927; Balsham, 1957), grasshoppers and leaf-
hoppers increase in burned grasslands (Rice, 1932; Hurst, 1970),
animals and birds that appear to be dependent on late stages of
Boreal Forest succession such as caribou, marten, Red squirrel,
Grizzly bear, wolverine, fisher and Spruce grouse may suffer
population declines or may be displaced following fire (Grange,
1948; Edwards, 1954; Cringan, 1958; Hayes, 1970; Scotter, 1971).
More recent work on this last example has shown that these organisms
are not entirely restricted to later stages of vegetation and may
indeed be better adapted to a mosaic of differing age vegetation
types.

Changes in population size often result from changes due to
fire in the physical and biological habitat. Similar changes are
found in aquatic systems but are not as well documented. The
effects of increased stream sediment loads on young salmon survival
and populations changes reviewed by Phillips (1971) illustrate
detrimental effects that could result from burning.

Another aspect of population-fire interactions is concerned
with plant monocultures that commonly develop after fire and that
are inherently susceptible to outbreaks of a species that utilises
the monoculture. A good example to illustrate this point is the
periodic population increase of Heather-beetle (Lochmaea suturalis)
on Calluna heathland (Gimingham, 1972, p.204).

Fire and the community: All fires change the general composition
of the biotic community but in fire-dominated ecosystems such as
grasslands and shrublands, the shift in composition is quite rever-
sible and often rapid.

There is evidence of irreversible shifts in community composition following fire for areas where fire frequency is naturally
low. Fire is frequently confounded with other activities of man
and examples are not easy to identify. In the wettest parts of
western Tasmania where fires do not occur naturally the forest is
dominated by trees of Nothofagus and Atherosperma. With increased
fire frequency to once in 250 years three species of Eucalyptus
invade after fire and overtop the two tree species already mentioned.
A fire frequency of once or twice per century causes lower-statured

trees such as <u>Pomaderris, Olearia</u> and <u>Acacia</u> to replace <u>Nothofagus</u>. If fires occur every 10-20 years monocultures of low-statured <u>Eucalyptus</u> dominate. Areas with more frequent fires develop a tree-less moor condition (Jackson, 1968; Walter, 1973, p. 142-143).

Fires have also changed tree-dominated communities to tundra communities near treelines in alpine and arctic areas (Clark, 1940; Ritchie, 1962, p. 25; Billings, 1969; Douglas and Ballard, 1971) but it is not known if these changes are irreversible. Paleoeco-logical evidence given by Nichols (1976) suggest that treelines shift over thousands of years because climatic conditions change. Reproduction may no longer be possible and loss of the relic trees is inevitable. Fire only hastens their demise.

SUMMARY

Fire has been an evolutionary force in many ecosystems and it has been recognised only recently that some ecosystems depend on fire for long-term stability. Fire is generally considered to be a positive force but it must be noted that most fire research has been conducted in areas with high fire frequency. The long-term effects of increased fire frequency for areas with naturally low fire frequency are not as well documented.

ACKNOWLEDGEMENTS

The author wishes to acknowledge the assistance of Ms J M Moore in the preparation of this paper. Fire research in the Acadian Forest of New Brunswick has been supported by National Research Council of Canada Operating Grant A-6878.

REFERENCES

Anderson, H. W., Coleman, C. B. and Zinke, P. J., 1959. Summer slides and winter scours - dry-wet erosion in southern California mountains. <u>Tech. Pap. Pac. N.W. for. & Range Exp. Sta.</u>, PSW 36.

Basham, J. T. 1957. The deterioration by fungi of jack, red, and white pine killed by fire in Ontario. <u>Can. J. Bot.</u>, <u>35</u>, 155-172.

Bendell, J. F., 1974. Effects of fire on birds and mammals. 73-138. In: <u>Fire and ecosystems</u>, edited by T. T. Kozlowski and C. E. Ahlgren, New York: Academic Press.

Billings, W. D. 1969. Vegetational pattern near alpine timberline as affected by fire-showdrift interactions. <u>Vegetatio</u>, <u>19</u>, 192-207.

Biswell, H.H, 1974. Effects of fire on chaparral. In: Fire and
    ecosystems, edited by T. T. Kozlowski and E. C. Ahlgren, 321-
    364. New York: Academic Press.

Bendell, J.F, 1974. Effects of fire on birds and mammals. 73-138.
    In Fire and ecosystems, edited by T.T. Kozlowski and C.E. Ahlgren,
    New York: Academic Press.

Bolstad, R. 1971. Catline rehabilitation and restoration. p. 107-
    116. In Fire in the northern environment, edited by C.W.
    Slaughter, R.J. Barney and G.M. Hansen. Pac. N.W. For. Range
    Exp. Sta. 107-116.

Brown, H.E. 1965. Preliminary results of cabling Utah juniper
    Beaver Creek Watershed project. Proc. Arizona Watershed
    Symposium, 16-21.

Brown, G.W., Gahler, A.R. and Marston, R.B. 1973. Nutrient losses
    after clear-cut logging and slash burning in the Oregon coast
    range. Water Resour. Res., 9, 1450-1453.

Brown, G.W. and Krygier, J.T. 1967. Changing water temperatures in
    small mountain streams. J.Soil Wat. Conserv., 22, 242-244.

Clark, C.H.D. 1940. A biological investigation of the Thelon Game
    Sanctuary. Bull. natn. Mus. Can., No 96.

Cordone, A.J. and Kelly, D.W. 1961. The influence of inorganic
    sediment on the aquatic life of streams. Calif. Fish Game,
    47 189-228.

Cringan, A.T. 1958. Influence of forest fires and fire protection
    on wildlife. For. Chron. 34, 25-30.

Damman, A.W.H. 1971. Effect of vegetation changes on the fertility
    of a Newfoundland forest site. Ecol. Monogr.,41, 253-270.

Daubenmire, R.F. 1974. The fire factor. In: Plants and environment -
    a textbook of plant autecology. (3rd edn.), 320-337. New York:
    John Wiley and Sons.

Davis, R.B. 1967. Pollen studies of near-surface sediments in Maine
    lakes. In: Quaternary paleoecology, edited by E.J. Cushing and
    H.E. Wright, Jr. 143-173. New Haven, Conn: Yale Univ. Press.

DeBano, L F. 1969. Water repellent soils; a worldwide concern in
    management of soil and vegetation. Agric. Sci. Rev., 7, 11-18.

Douglas, G.W and Ballard T.M. 1971. Effects of fire on alpine
    plant communities in the North Cascades, Washington. Ecology
    52, 1058-1064.

Duvall, V.L.and Whitaker, L.B.1964. Rotation burning: a forage
    management system for longleaf pine-bluestem ranges. J.
    Range Mgmt., 17, 322-326.

Edwards, R.Y. 1954. Fire and the decline of a mountain caribou
    herd. J. Wildl. Mgmt., 18, 521-526.

Evans, C.C.and Allen, S.E.1971. Nutrient losses in smoke produced
    during heather burning. Oikos 22. 149-154.

Ganong, W.F. 1902. Great fires in New Brunswick. Bull. nat. Hist.
    Soc. New Brunswick 20, 434-435.

Ganong, W.F. 1906. On the limits of the Great Fire of Miramichi of
    1825. Bull. nat. Hist. Soc. New Brunswick,5, 410-418.

Gill, A.M. 1975. Fire and the Australian flora: A review. Aust.
    For.,38, 4-25.

Gimingham, C.H. 1972. Ecology of heathlands. London: Chapman and
    Hall.

Grange, W.B. 1948. Wisconsin grouse problems. Wisc. Conserv.
    Dept., Madison, Wisc. (See Bendell 1974).

Grier, C.C 1975. Wildfire effects on nutrient distribution and
    leaching in the coniferous ecosystem. Can. J. For. Res.,5,
    599-607.

Hayes, G.L. 1970. Impacts of fire use on forest ecosystems. In:
    The role of fire in the Intermountain West, 99-118. Missoula,
    Montana: Intermountain Fire Res. Council.

Heginbottom, J.A. 1971. Some effects of a forest fire on the perma-
    frost active layer at Inuvik, N W T. In: Proc. Permafrost
    Active Layer Conf. 1971, edited by R.J.E. Brown,31-36, (C.N.R.C.
    Technical Memorandum 103).

Heinselman, M.L. 1973. Fire in the virgin forests of the Boundary
    Waters Canoe Area, Minnesota. Quaternary Res. 3, 329-382.

Heinselman, M.L.and Wright, H.E.Jr. (Eds.). 1973. The ecological
    role of fire in natural conifer forests of western and northern
    North America. Quaternary Res. 3, 317-513.

Holbrook, S.1943. Burning an empire. New York: Macmillan.

Houston, D.B. 1976. Wildfires in northern Yellowstone National
     Park. Ecology,54. 1111-1117.

Hurst, G.A. 1970. The effects of controlled burning on arthropod
     density and biomass in relation to bobwhite quail brood
     habitat on a right-of-way. In: Proc. Tall Timbers Conf. Ecol.
     Animal Contr. Habitat Manage. No. 2. 173-183. Tallahassee,
     Florida: Tall Timbers Res. Sta.

Irving, F.D.and French, D.W. 1971. Control by fire of dwarf
     mistletoe in black spruce. J. For.,69, 28-30.

Jackson, W.D. 1968. Fire, air, water and earth - an elemental
     ecology of Tasmania. Proc. Ecol. Soc. Aust.,3, 9-16.

Johnson, E.A.and Rowe,J.S. 1975. Fire in the sub-arctic wintering
     ground of the Beverly Caribou Herd. Am. Midl. Nat., 94,
     1-14.

Knight, H. 1966. Loss of nitrogen from the forest floor by
     burning. For. Chron.,42, 149-152.

Komarek, E.V.Sr. 1970. Insect control - Fire for habitat manage-
     ment. In: Proc. Tall Timbers Conf. Ecol. Animal Contr. Habitat
     Manage. No. 2, 157-171. Tallahasse, Florida: Tall Timbers
     Research Sta.

Komarek, E.V, SR. 1973. Ancient fires. In: Proc.
     Tall Timbers Fire Ecol. Conf. No. 12. 219-241. Tallhasse,
     Florida: Tall Timbers Res. Sta.

Kozlowski, T.T and Ahlgren C.E.(Eds.). 1974. Fire and ecosystems.
     New York: Academic Press.

Krammes, J.S. 1965. Seasonal debris movement from deep mountain-
     side slopes in southern California. Misc. Publs. U.S. Dep.
     Agric., No. 970, 85-89.

Kucera, C.L and Ehrenreich,J.H. 1962. Some effects of annual
     burning on central Missouri prairie. Ecology,43, 334-336.

Lewis, D.C. 1968. Annual hydrologic response to watershed conversion
     from oak woodland to annual grassland. Wat. Resour. Res.,
     4, 59-72.

Lewis, W.M.Jr. 1974. Effects of fire on nutrient movement in a
     South Carolina pine forest. Ecology,55, 1120-1127.

Mackay, J.R. 1970. Disturbances to the tundra and forest tundra environment of the western Arctic. Can. geotech. J., 7, 420-432.

Markim, F.L. 1943. Blueberry diseases in Maine. Bull. Me. agric. Exp. Stn. no. 419, 395-417.

McColl, J.G and Grigal, D.F. 1975. Forest fire: Effects on phosphorus movements to lakes. Science, N.Y., 188, 1109-1111.

Nichols H. 1976. Historical aspects of the northern Canadian treeline. Arctic,29, 38-47.

Pase, C.P and Ingebo, P.A. 1965. Burned chaparral to grass: early effects on water and sediment yields from two granitic soil watersheds in Arizona. Proc. Arizona Watershed Symposium, 8-11.

Phillips, R.W. 1971. Effects of sediment on the gravel environment and fish production. In: Forest land uses and stream environment, edited by J. T. Krygier and J. H. Hall, 64-74. Corvallis: Oregon State Univ.

Rice, L.A. 1932. The effect of fire on the prairie animal communities. Ecology,13, 392-401.

Ritchie, J.C. 1962. A geobotanical survey of northern Manitoba. Arctic Instit. N. Amer. Tech. Paper No. 9. 47p.

Sauer, C.O. 1975. Man's dominance by use of fire. In: Grasslands ecology - A symposium. Geoscience and man. Vol. 10, edited by R. H. Kesel, 1-13. School of Geoscience, Louisiana State Univ., Baton Rouge.

Schmitz, H. and Jackson, L.W.R. 1927. Heart rot of aspen with special reference to forest management in Minnesota. Tech. Bull. Univ. Minn. agric. Exp. Stn. no. 50.

Scotter G W. 1971. Fire, vegetation, soil and barren ground caribou relations in northern Canada. In: Fire in the norther environment, edited by G. W. Slaughter, R. J. Barney and G. M. Hansen, 209-230. Portland, Oregon. Pacific Northwest Forest & Range Exp. Stn.

Siggers, P.V. 1934. Observations on the influence of fire on the brown-spot needle blight of long-leaf pine seedlings. J. For., 32, 556-562.

Slaughter, C. W., Barneym R.H. and Hansen, G. M. (Eds.). 1971. Fire in the northern environment. Portland, Oregon: Pacific N.W. For. & Range Exp. Stn.

Smith, D. W. 1970. Concentrations of soil nutrients before and after fire. Can. J. Soil Sci., 50, 17-29.

Snyder, G. G., Haupt, H. F. and Belt, G. H., Jr. 1975. Clearcutting and burning slash alter quality of stream water in northern Idaho. U.S.D.A. For. Serv. Res. Paper, INT-168.

Strang, R. M. 1972. Ecology and land use of the barrens of western Nova Scotia. Can. J. For. Res., 2, 276-290.

Swain, A.M. 1973. A history of fire and vegetation in northeastern Minnesota as recorded in lake sediments. Quaternary Res. 3, 383-396.

Swanston, D. N. 1971. Principal mass movement processes influenced by logging, road building, and fire. In: Forest land uses and stream environment edited by J. T. Krygier and J. D. Hall, 29-39. Corvallis: Oregon State Univ., 252p.

Tall Timbers Research Station. 1962-1974. Proc. Tall Timbers Fire Ecology Conferences Nos. 1-13. Tallahassee, Florida.

U.S.D.A. Forest Service, 1972. Fire in the environment - Symposium Proceedings. U.S.D.A. For. Serv., FS-276.

Ursic S. J. 1969. Hydrologic effects of prescribed burning on abandoned fields in Northern Mississippi. Res. Pap. U.S.D.A. For. Serv., S.O.-46.

Viereck, L. A. 1973. Wildfire in the taiga of Alaska. Quaternary Res., 3, 465-495.

Viro, P. J. 1974. Effects of forest fire on soil. In: Fire and ecosystems, edited by T. T. Kozlowski and C. E. Ahlgren, 7-45. New York: Academic Press.

Vogl, R. J. 1970. Fire and the northern Wisconsin Pine Barrens. Proc. Tall Timbers Fire Ecol. Conf., 10th, 175-209.

Wagener, W. W. 1961. Past fire incidence in Sierra Nevada forest. J. For., 59, 739-748.

Walter, H. 1973. Vegetation of the earth. New York: Springer.

Weaver, H. 1951. Fire as an ecological factor in the Southwestern
    ponderosa pine forests. J. For.,49, 93-98.

Wein, R.W. and Bliss, L.C. 1973. Changes in Arctic Eriophorum tussock
    communities following fire. Ecology,54, 845-852.

Wein, R.W. and Moore, J.M. 1977. Fire history and rotations in the
    New Brunswick Acadian Forest. (Manuscript form.).

Wright, H.E, Jr. 1974. Landscape development, forest fires, and
    wilderness management. Science, N.Y., 186, 487-495.

Wright, H.E Jr. and Heinselman, M.L.1973. Introduction. Quaternary
    Res.,3, 319-328.

DISCUSSION: PAPER 10

C. GELIN   Have you experience with the effects of
                                fire on the inflow of nutrients to adjacent
                                aquatic systems?

R. W. WEIN   Not personally. There have been many
                                studies of nutrient inflow into aquatic
                                systems. What has not been determined is the
                                importance of these extra nutrients flushed
                                into streams and lakes, to the long term
                                development of the lake systems. For how long
                                will the effects of fire be felt? The altering
                                of river systems, leading to deposition of
                                sediments and even stones that affect spawning
                                beds are much more spectacular, and thus more
                                studied, effects of fire.

V. GEIST   I have studied the effects of fire on the
                                fauna in Canada. The reproductive biology of
                                caribou and moose cannot be understood without
                                taking fire into account. Caribou can in fact
                                benefit from, or be harmed by, fire. Faunas
                                in wet valleys, deltas, and on avalanch slopes
                                move into burned areas after fire. The fauna
                                are clearly adapted to the periodic burning
                                of the system: some National Parks may lose
                                fauna because of fire control.

R. W. WEIN   Data from Western Canada and the United
                                States show this. There are also other
                                influences between fire and fauna. We are told

that one-half of the Balsam fir forest of New
Brunswick has been killed by Spruce budworm
forming a powder keg with potential to cause
an enormous fire, perhaps as big as that of
1825 which may have similarly been affected
by an epidemic. It would seem obvious from
an evolutionary point of view that with higher
fire frequency there should be a greater degree
of adaptation to fire.

L. M. TALBOT   This balance of "degradation" and "recovery" is basic to the conference. We are
concerned with the regeneration of ecosystems
for human benefit, but what are we trying to
re-create? Early man's influence has been so
pervasive and persistent we cannot return to
the original ecosystem.

In Yosemite, the Indians burned land to
help them hunt. The practice was stopped by
the Parks authority. The vegetation then grew
to the point where views were obscured and
there was extreme danger of catastrophic fire.
Now trees are being cut, and there is even some
controlled burning! We have to decide the
kind of ecosystem we want, and this determines
whether fire is a help or a danger.

R. W. WEIN   I would agree.

B. ULRICH   How far is fire needed as an influence on
the regeneration of forest? Each fire causes
nitrogen loss, and nitrogen is the nutrient
that controls productivity: fire can therefore reduce primary productivity.

R. W. WEIN   If the primary concern is to produce trees,
which fire destroys, stability of primary productivity decreases. If man applies sufficient
energies to establish trees, to prevent disease,
etc., fire may not be needed and should be
avoided to prevent fibre and nitrogen loss.
From the ecosystem standpoint, or where multiple
land-use is a concern, fire effects may be less
negative. Fire creates conditions which permit
species to invade or increase. This may create
stability in the long term, even if fires
cause short term instability.

L. M. TALBOT   There are two kinds of fire. The first
                          is periodic and changes the system by consuming
                          the standing crop of primary producers. The
                          second occurs every 4 years or so in the Sierra
                          Nevada, operating to consume standing dead
                          material that would otherwise build up to a
                          catastrophic fire. These frequent fires do
                          not harm the live standing crop: they thus
                          sustain stability. Another stabilising effect
                          applies when, as in redwoods, fire is needed
                          to ensure regeneration.

R. W. WEIN   I agree there are two basic types of fire –
                          the catastrophic fire which has a low fire
                          frequency but destroys almost all biomass and
                          what I would call the short term maintenance
                          fire which occurs frequently but with less
                          destruction of the biomass. Both types should
                          be considered to have a form of stability.

Part III: The Restoration of Degraded Ecosystems

Introduction

M. W. Holdgate

The programme of the Conference provided for a third session, consisting of case studies of the successful restoration of degraded ecosystems and a fourth and final session devoted to the discussion of general guidelines and methods. In practice, there was a substantial overlap between these sessions. Inevitably, the contributors of case studies drew more general conclusions from their work, while those seeking to define guidelines referred to specific examples. The papers have, therefore, been re-grouped for publication.

It became evident from many contributions to the conference that the restoration of soil stability and fertility was the key to the rehabilitation of many terrestrial ecosystems. The first two papers in this section of the volume deal with the general question of restoring soil fertility and with specific achievements in Iceland. Soil management is also stressed in several of the four following papers concerned with vegetation: especially paper 14 which describes the restoration of plant cover on British sites made derelict by mineral exploitation and paper 13 reporting the achievements of reafforestation in Iceland. General questions of forest management in the Federal Republic of Germany (15) and of the conservation of plant genetic resources lead on to the difficult (and at times emotive) issue of the role of predators in ecosystem management (paper 17). Successive groups of papers deal with practical experience and general principles of restoration and maintenance of freshwater systems and with broader questions of land use planning.

OPENING REMARKS BY SESSION CHAIRMAN

G. W. Dimbleby

Institute of Archaeology
University of London

These papers are inevitably concerned with many habitats and
organisms. But there are common threads. One general conclusion
is that to restore ecosystems we need to understand the steps in
the process by which they become degraded. Many of these are
mentioned in Part II as well as Part III; they include soil
nutrient depletion, the destruction of vegetation by fire and the
pollution of aquatic systems. It is not, however, a clear-cut
picture. Atmospheric pollution, for example, may supply nitrogen
or sulphur as well as cause damage. The interactions are vitally
important.

What is meant by "successful restoration?" To what are we
restoring the system? To something like its previous configuration
or to something new? It is probably impossible to recover the
preceding condition, even if we know what it was. But if we are
to evaluate our success, we need a baseline against which to assess
the degree of restoration. The final papers in this part of the
volume (22,23 and 24) emphasise that the objectives of ecosystem
management will be based upon a combination of ecological practica-
bility and social value judgement: the same goals will not be
appropriate everywhere, even for the identical system. The papers
indicate some of the options open to us: the things that can be
done, and some of the factors involved in making a choice - but
what is actually done in a new situation must depend on judgement
and resources at the time.

A. The Restoration of the Soil

11. THE RESTORATION OF SOIL PRODUCTIVITY

D. Parkinson

Department of Biology

University of Calgary, Calgary, Alberta, Canada, T2N 1N4

INTRODUCTION

The general subject of restoring soil productivity following devastation is vast, encompassing a wide variety of interacting pedological phenomena (biological, chemical and physical). This contribution concentrates on the restoration of the productivity of soil organisms for the following reasons:

    (i)   soil organisms are known to play vital roles in the maintenance of soil fertility through their activities in organic matter decomposition and nutrient cycling;

    (ii)   soil organisms (specifically the micro-organisms) exhibit great metabolic diversity (from chemolithotrophy to obligate parasitism), many species have great adaptive potential, and some species can survive and even exhibit metabolic activity in uncongenial environments;

    (iii)   the specific interests of the author.

Over recent years much data have been obtained on the productivity of undisturbed terrestrial ecosystems. This has been largely a result of projects set up under the aegis of the International Biological Programme, and has led to a better understanding of the interactions between the different components of ecosystems.

One important effect of integrated ecosystem studies has been that opportunities were provided for detailed studies on the interacting roles of microflora and fauna in decomposition processes

213

and nutrient cycling in soil. Whilst such studies have been frequently hampered by methodological problems (summarised in Phillipson, 1971 and Parkinson, Gray and Williams, 1971), they have clearly shown soil biological activity to be of greater complexity and a more important 'driving force' in ecosystems than hitherto appeared to have been realised by many ecologists.

The complexity of the interactions between different components of the soil biota has been represented diagrammatically by various authors (e.g. Mitchell, 1974). Over recent years there has been considerable speculation on the relative roles and contributions of the soil microflora and fauna in organic matter decomposition. From these has emerged the concept of the components of the micro-flora (bacteria and fungi) being the major agents of decomposition, responsible for 80-90% of the energy flow in litter-soil systems. However, such a statement should be treated with caution, for the role of soil fauna is of at least indirect importance - an importance far greater than the bare energy flow figures suggest.

Over a number of years data have been obtained on the effects on soil populations and processes of eliminating various groups of organisms from the soil system. Partial sterilisation of soil using heat (steam) or volatile chemicals (e.g. chloropicrin, methyl bromide etc.) has been used as a means of control of pathogens of greenhouse or field crops of high commercial value. One of the dangers of such methods is that not only are pathogens killed but so also is a large proportion of the saprophytic microflora and fauna ; thus any pathogen recolonising the treated soil faces considerable decreased microbial competition. Garrett (1970) summarised the effects of partial sterilisation of soil by both methods as: (1) consistent increases in crop growth; (2) accumulation of ammonia in the soil; (3) with the killing and subsequent autolysis of microbial and animal cells there is a quick release of nitrogen, phosphate and other nutrients; (4) inoculum potential of a range of organisms (pathogenic, weakly pathogenic and saprophytic) is lowered.

Exclusion experiments (Edwards & Heath, 1963) and chemical killing treatments (Witkamp & Crossley, 1966) have been used to examine the effects of eliminating soil fauna. Considerable decreases in rates of organic matter decomposition followed fauna elimination.

Selective inhibitors have been used (Anderson & Domsch, 1973; 1975) in an attempt to assess the relative contributions of bacteria and fungi to organic matter decomposition in forest soils. Such studies have indicated the fungi to be the dominant decomposers. With respect to nutrient cycling phenomena, the numerous transformations involved in the nitrogen cycle have been thoroughly investigated and therefore prediction of the effects of changes in environmental factors (e. g. available organic matter, pH, aeration, trace elements etc.) on such phenomena as N_2 fixation, ammonification, nitrification and denitrification can be predicted.

Much information has also been obtained about the effects of pesticides on soil organisms and this has been reviewed by Edwards (1973).

Thus it has been demonstrated in a number of ways that relatively small perturbations of the below ground ecosystem may have considerable effects on abundance, species diversity and activities of soil organisms. However, when one turns to the effects on soil organisms of soil devastation (as recognised in this symposium), it can only be said that the information published in the open literature is at best fragmentary. Such information in no way approaches the quantity or quality of data available for the effects of devastation on higher plants or animals.

Various degrees of soil devastation result from a variety of causes both natural and man-made. With these come a variety of effects on the chemical and physical nature of the soil environment, which in turn affect the nature and activities of the soil biota, and must determine the methods used for amelioration of the devastated condition. This present contribution does not attempt any complete survey of the possibilities but aims, through choice of a small number of examples, to expose the need for much more comprehensive studies.

BRIEF COMMENT ON METHODOLOGY

In any study of soil devastation one of the first steps must be to assess the effects of that devastation not only on soil chemical and physical factors but also on the soil organisms. It would appear that, until recently, an "out of sight, out of mind" attitude has existed.

Much has been written on the methods available for studying soil organisms in general, specific taxa, or specific physiological groups (e. g. Phillipson, 1971; Parkinson et al., 1971). Therefore the only general (platitudinous) comment which must be made is that methods must be chosen which are appropriate to attempt to answer the considered (not haphazard) questions being asked e.g. if one were attempting to isolate fungi present in the soil as hyphae one would not use the soil dilution plate technique (which effectively isolates fungi present as spores).

If a study of soil organisms is to be made, it should be
properly based statistically and the data obtained subjected to
appropriate analysis. This entails replicate sampling, preferably
at more than one point in time. Again, this comment may appear
so obvious as to be unworthy of reiteration, but one frequently
gains an impression that groups interested in environmental impact
assessments have the erroneous belief that significant soil
biological data can be obtained in a short time using small numbers
of soil samples.

The methods which are available for assessing the nature and
amounts of soil micro-organisms allow assessments of standing crop
but do not allow critical study on the rates of cell production in
organic matter or soil (Gray & Williams, 1971). Attempts have
been made to achieve rough estimates of both fungal and bacterial
productivity by regular replicate biomass estimates over varying
time periods (e.g. Parinkina, 1974).

ROOT-MICROBE ASSOCIATIONS

Root-microbe interactions range from specific symbioses (mycorrhizal,
root nodules) and obligate parasitism to the much less specific
rhizosphere effects. The soil-root interface is the real action
zone with respect to nutrient and water uptake, and therefore it
is here that micro-organisms can have their most potent effects
(beneficial or harmful) on higher plant development.

The mycorrhizal condition is widespread in the plant kingdom,
in fact non-mycorrhizal plants (e.g. aquatic plants, sedges,
crucifers) are exceptional. Mycorrhizal associations have been shown
to be important in reforestation of barren areas such as avalanche
slopes (Bjorkman, 1970), and their presence on plants colonising
devastated soils allows significantly better growth and survival
than if mycorrhizae are absent. Schramm (1966) in an extensive
account of plant colonisation of anthracite wastes in Pennsylvania
concluded that early ectomycorrhizal development was a pre-requisite
for the establishment of seedlings of a variety of tree species
and for the continued colonisation of devastated areas (application
of nitrogen fertilisers allowed only temporary plant growth). His
work also indicated that, because of the adverse conditions existing
on mine wastes (e.g. high temperatures, xeric conditions, adverse
pH), many potentially mycorrhizal fungi were incapable of active
growth. However Pisolithus tinctorius was shown to be one mycorrhizal fungus capable of active development under such rigorous
conditions. Marx (1975) reported results of his work on the significance of ectomycorrhizal formation on tree seedling growth on
a range of strip-mine waste areas in the eastern U.S.A. He demonstrated the widespread and predominant occurrence of P. tinctorius

on roots of several Pinus and Betula species growing on coal and
kaolin wastes, some of which had pH regimes as low as 2.9.

A good deal more work is required on the general problem of
the role of ectomycorrhizal fungi (some of which are host specific)
in the rigorous environments seen in devastated soils. More information is required, for example, on the rate of mycorrhizal formation
on seedling trees, and the ways of ensuring inoculum of the appropriate mycorrhizal fungus in the root region of the tree seedling.

With respect to this last point (mentioned above), Marx and
Barnett (1974) demonstrated improvement of tree seedling growth
in containers when the appropriate ectomycorrhizal fungus is
incorporated into the containers. Thus, since container grown tree
seedlings will probably be used for re-vegetation programmes on
devastated land, inoculation of containers with the appropriate
mycorrhizal fungus (using vegetative hyphae or spores as inoculum)
could solve the problem of soil inoculation.

Endomycorrhizal associations with plants growing on devastated
areas have received limited study. Schramm (1966) in his detailed
study observed that endomycorrhizal native plants did not colonise
coal wastes. However, other workers (Daft & Nicholson, 1974; Daft,
Hacskaylo & Nicholson, 1975) have found abundant endomycorrhizae
on grasses and other herbaceous species growing on coal waste in
Scotland and Pennsylvania. Much more research is required on
methods of inoculating devastated soil with endomycorrhizal fungi,
their survival and growth onto seedling roots, and on the effects
of endomycorrhizal fungi on the subsequent growth and survival of
higher plants.

The role of symbiotic nitrogen fixation by both legumes and
nodulated non-legumes in ameliorating rigorous soil conditions
in early stages of succession is well-known. Therefore the potential
values of such plants in re-vegetation of devastated areas has
received attention. Schramm (1966) showed that plants which were
hosts to endophytic nitrogen fixing micro-organisms were successful
colonisers of mine waste areas. Such nitrogen fixing species as
alder and acacia have been used for erosion control, and it has
been shown (e.g. Hashimoto et al., 1973) that growth of alder can
bring about amelioration of surface soil conditions (increasing
carbon and nitrogen contents, cation exchange capacity, and moisture
holding capacity). Such ameliorated soil conditions allow the
subsequent development of a more balanced saprophytic soil microflora.

As with mycorrhiza, much work is required on the problem of
achieving efficient nodulation of appropriate plants growing on
devastated soils. A great deal is known on the ecology and host

specificity of Rhizobium species, Rhizobium strain efficiency, and
the general biology of the Rhizobium – legume symbiosis (Quispel,
1974) and similar data are accumulating for the non-legume nitrogen
fixing systems. These data have been obtained in laboratory, greenhouse and non-devastated field experiments but they have provided
a body of data which must be used in examining the special problems
of barren soils. One of these problems is the means of host inoculation with the specific, efficient micro-organism, and Rothwell
(1973) provided data on the use of plastic planting bullets for
seedling growth and examination for nodulation following inoculation.
Once again there is need for detailed data on the ecology of the
nitrogen fixing microbial symbionts in devastated soils.

With successful plant growth there is the development of a
diverse and active saprophytic microflora in the general root
region (rhizosphere and root surface). This root region microflora may have considerable effects on root development, nutrient
uptake and defence against root pathogens. On root death, rhizosphere micro-organisms are capable of decomposing the added organic
matter. Thus roots growing in devastated soil can represent the
developing centres of microbial activity essential for organic
matter decomposition and nutrient cycling.

NATURAL SOIL DEVASTATION

Flood and fire have been chosen as two examples of natural factors
of soil devastation which have received study from a soil biological
viewpoint.

Flooding

Flooding has been used in a controlled way in agricultural practice
and for the control of soil-borne pathogens (Brown, 1933; Stoner
& Moore, 1953; Stover, 1954). However severe flooding with fresh
or sea water brings about alterations in soil moisture, aeration
and nutrient status, and these lead to qualitative and quantitative
changes in the soil microflora.

An example of a quantitative study of microbiological effects
of fresh water flooding is that of Mitchell and Alexander (1962).
Working with several Central American soils they showed an immediate
drop in soil fungal populations following flooding (as indicated
by soil dilution plate counts). These fungal propagule counts remained low as long as there was free water above the soil, and re-

covery of the fungi following drainage was slow. Data obtained
for bacteria and actinomycetes did not show such pronounced effects
of flooding. It was suggested that lowering of oxidation-reduction
potential, accumulation of sulphides and manganous or ferrous ions,
CO_2 concentrations, accumulation of toxic substances, digestion of
fungal spores and hyphae by bacteria, and competition for nutrients
were the factors affecting fungal populations during flooding.

Detailed studies on polder reclamation in the Netherlands
give a classic example of detailed microbiological studies accompanying other ecological and engineering investigations. Much of
this, primarily bacteriological work, has been summarised by van
Schreven and Harmsen (1968).

Damming the Zuider Zee (1932) produced the IJsselmeer, which
in time became a freshwater lake. However in the construction
of the Wieringermeer polder (early 1930's) salt water was a factor
of importance in land reclamation. Immediately following drainage,
bacterial numbers were very low (about 1/1,000th of those in 'old'
soil). Within two years total bacterial numbers had increased
particularly where cultivation had begun. Increased bacterial
activity was correlated with decreases in moisture and salt contents.
With respect to bacteria concerned in nitrogen transformations, it
was found that nitrifying bacteria only increased in numbers after
good drainage was established, Azotobacter populations remained low
(but could be increased by green manuring), whilst Rhizobium spp.
showed a complex pattern of appearance (for Pisum and Vicia spp.,
inoculation of soil with appropriate Rhizobium strains was unnecessary; for Trifolium spp. inoculation, whilst not necessary,
did produce better results; for Phaseolus, Medicago, Lupinus and
Serradella there was failure without inoculation).

By the time of construction of the North-east polder (early
1940's) the IJsselmeer was effectively a freshwater lake. After
pumping out most of the water, drainage and cultivation, total
bacterial numbers increased significantly (an increase which was
paralleled by increasing in amylolytic, cellulolytic and proteo-
lytic forms). Again numbers of nitrifying organisms rose rapidly
following drainage. Azotobacter populations showed a slow rate
of development depending on soil type and cultivation, and Rhizobium
spp. showed the same pattern of occurrence as for the Wieringermeer.

Studies of unaerated, sub-aqueous soils from IJsselmeer showed
the presence of substances toxic to bacteria. These substances
could impede the development of populations of aerobic bacteria
during the early stages of polder occlamation.

Fire

Fire exists both as a man-induced factor for the maintenance of
vegetation (e.g. grassland, forest, heath) and as a destructive
natural environmental factor in which the degree of soil devastation depends on the intensity of the fire and a number of edaphic
factors. Detailed reviews on the effects of fire on soil chemistry
and biology can be found in Kozlowski and Ahlgren (1974).

During fire the temperatures generated in the upper layer of
soil vary greatly but most are sufficient to kill at least a
proportion of the micro-organisms. Apart from the rapid oxidation
of organic material producing ash and charcoal (which is resistant
to decomposition), fire effects physical and chemical changes in
the soil (e.g. reduction in pore size, aeration and water holding
capacity, increase in pH etc.). Accompanying these changes is a
reduction in biological activity. The subsequent re-development
of soil biological activity appears to be much affected by moisture
conditions - rainfall following fire frequently causes an increase
in soil microbial populations probably as a result of the leaching
into soil of minerals from the surface ash.

The majority of the work on bacterial response to fire has
been quantitative. However, work with nitrogen transforming
bacteria has indicated that, in coniferous forest soils, free-
living nitrogen fixing bacteria (Azotobacter and Clostridium)
increase following burning (Lunt, 1951), however reduction in the
rate of nitrogen fixation and nitrification have been reported
from dry Kenyan soils (Meiklejohn, 1955). It has also been shown
that actinomycetes are less affected by heat and drought than are
bacteria.

With respect to fungi, considerable qualitative data are
available. The development of characteristic fungi on burned ground
is well-known (e.g. Wicklow, 1975): many of these are either dis-
comycetes or pyrenomycetes whose ascospores appear to require heat
or chemical activation before germination. With respect to the
general soil fungal flora several detailed studies have been made.
Jorgensen and Hodges (1970) indicated that the soil mycoflora remained unchanged after fire, but most studies have shown a decrease
in fungal propagules. Widden and Parkinson (1975), in a comparison
of fungi in burned and unburned areas of a Pinus contorta forest,
found reduction species of Trichoderma and Penicillium in soil
following forest burning, whilst Cylindrocarpon destructans appeared
unaffected by fire. They suggested that in the absence of Trichoderma and Penicillium it was possible that C. destructans could
achieve a high inoculum and thus could perhaps result in high
seedling disease incidence.

It would appear that recolonisation of surface soil layers affected by burning occurs by means of wind-blown microbial propagules, wind-blown organic debris and growth from the unaffected subsurface layers. Meiklejohn (1955) reported that an Aspergillus species replaced a Penicillium species, and Jalauddin (1969) reported that Trichoderma and Penicillium spp. were the first to recolonise burned ground. Little appears to be known on the reduction and/or change in the populations of mycorrhizal fungi as a result of fire. Presumably the litter inhabiting ectomycorrhizal basidiomycetes are decimated by fire and populations are subsequently re-developed from wind-transported basidiospores.

There are very few assessments of the amounts of fungal hyphae in burned soils. Widden and Parkinson (1975) showed that two weeks after a forest fire there was more mycelium in the H layer of a burned forest plot than in the unburned control whereas in the A horizon no significant differences in mycelial lengths were recorded.

With respect to the soil fauna, again data are limited and generalisations difficult. Initial heat followed by post-fire environmental changes appears to have greater effects in forest than grassland.

MAN-MADE DEVASTATION

In contemplating the environmental implications of man's industrial activities, a number of spectacular examples of land devastation spring to mind. However, it should be remembered that the overgrazing and over-cropping of land and indiscriminate land clearance techniques have slowly, but effectively, devastated vast areas of land (through loss of organic matter, nutrients, and soil structural aggregations, and the consequent massive erosion). In relation to the problem of improving soil structure and stability it has been shown that the micro-scale of soil crumb formation micro-organisms (particularly filamentous fungi and capsule-forming bacteria etc.) can be effective, but at the macro-scale of soil stabilisation large input of organic matter is required for amelioration to be achieved.

The expanding demands for energy and mineral resources have been responsible for the creation of areas of soil devastation. The two chosen here for cursory survey are crude oil spillage and coal mining.

Oil Spillage

Extensive studies have been carried out on the effects of crude
oil spillage on the biological, chemical and physical properties
of soils. As a result of these, and allied, studies it has been
shown that hydrocarbon-utilising organisms (a wide range of bacterial and fungal genera) are present in a wide range of environments.

It appears that when petroleum hydrocarbons are added to
soil at least a part of the microbial population is stimulated.
E. g. Plice (1948), showed an approximately 10-fold increase in
bacterial numbers in a 3 month period following application of
oil at a level of 1% by weight.

Gossen (1973), in a detailed study of the effects of crude
oil spillage on microbial biomass in several northern Canadian
soils (Mackenzie Delta, N.W.T.), demonstrated increased bacterial
numbers following oil spillage, whilst fungi showed an initially
decreased hyphal biomass followed by a rapid re-development of
fungal biomass. Measurements of soil respiration also indicated
greater activity in oil-treated as compared with untreated soils.
However, GLC analyses showed a slow rate of oil degradation in these
soils. This last observation pointed out the dangers of inferring
rates of oil degradation using solely microbiological data. Another
significant observation made in this study was the effect of micro-
topography on rates of oil degradation. The study site was characterised by marked, small (< 1 m diam.) hummock formation, and
marked temperature differences were recorded in the upper soil
layers of the north and south-facing hummock slopes, whilst oil
degradation proceeded significantly faster in the soil of the
south-facing hummock slopes.

The chemical composition of crude oil varies considerably,
and this composition has significant effects on degradability and
on the general soil biota. Crude oils with high contents of n-
alkanes and low molecular weight aromatics are more easily attacked
by micro-organisms.

Schwendinger (1968) summarised the techniques for reclaiming
terrestrial oil spills as:

1.   Direct removal of oil

   a)   ignition of oil residues, followed by intense discing
        or ploughing;

   b)   rapid, on-site, mixing of oil-contaminated soil with
        uncontaminated soil.

2.   Indirect removal of oil

    a)   deep ploughing (to achieve deep soil aeration without
         too much surface disturbance);

    b)   microbial 'seeding';

    c)   fertiliser treatment.

The methods which are feasible for use depend on the geographic location of the study area, i.e. in tundra regions only microbial seeding and fertiliser treatment appear practical for use (because any disturbance of the surface vegetation mat causes uncontrollable damage to the tundra). In many geographic zones biological or chemical treatments appear (to the author) preferable to the more drastic physical-mechanical means of reclaiming soil.

'Seeding' of oil spills with micro-organisms capable of active oil degradation is an immediately attractive method for speeding soil reclamation. However it has been shown (Atlas & Bartha, 1973; Jobson et al., 1974) that such 'seeding' was not effective in accelerating crude oil degradation in various environments. In this respect, it has been pointed out (Gossen, 1973) that even if an hydrocarbon-degrading bacterial species were found effective in such 'seeding' studies it would be necessary to have data on survival and future consequences (i.e. effects on future develop-ment of balanced soil biological activity) of such an organism in soil following exhaustion of the oil residues.

When oil is added to soil (temperate and tropical) there is an input of much carbon and the supplies of nitrogen and phosphorus are rapidly depleted as a result of the stimulated microbial activity.

Schwendinger (1968) demonstrated that fertiliser treatment (nitrogen and phosphorus) increased the rate of oil degradation, and that application of fertiliser to soil prior to oil-spillage allowed microbial activity to achieve higher levels than after post-spill application. Gossen (1973), working with tundra soils in both laboratory and field studies, confirmed that nitrogen and phosphorus treatments were necessary for accelerating oil degradation in soil. Similar data have been obtained from a range of Canadian soils (Jobson et al., 1974; Cook and Westlake unpublished data).

Coal Mining

Strip-mining for coal has economic attractions but considerable environmental hazards. The vast physical disruption of terrain, piling of 'tailings' and inefficient replacement of overburden create

problems which have not been systematically tackled from the viewpoint of re-developing soil biological activity. Apart from desul-
tory studies on bacterial numbers and soil respiration, little
general work has been done. The scene is lightened by the excellent
work which has been done on the importance of plant-microbe symbioses
in re-vegetation of mine waste areas (see previously), and on
specific problems related to the activity of individual physiological groups of soil micro-organisms.

An example of this latter type of study was given by Duggan
(1975), who studied the activity of acidophilic bacteria (e.g.
Thiobacillus spp.) in sulphuric acid production from iron pyrite
exposed during strip mining. He estimated that, as a result of
this phenomenon, over 1 million lbs of acid are discharged daily
into the streams in Ohio - a discharge which must have drastic
environmental consequences. Therefore an investigation was carried
out on ways of inhibiting the activity of Thiobacilli in spoil
heaps. It was found that anionic detergents and low molecular
weight organic acids inhibited acidophilic Thiobacilli, and it
was noted that organic acids were present in sewage sludge in
significant amounts i.e. sewage sludge additions to spoil heaps
could have the two-fold benefit of increasing the organic matter
content and inhibiting the iron-oxidising bacteria. However, the
addition of organic matter could have the deleterious effect of
producing the environmental conditions of high organic matter
plus high sulphate which could result in anaerobiosis and allow
development of sulphate-reducing bacteria (e.g. Desulphovibrio)
with the consequent evolution of H_2S (toxic and objectionable).

This example indicates the necessity for detailed studies of
individual physiological groups of micro-orgainisms associated
with reclamation processes. Much more work is required on the
relative values of different organic matter applications in helping
to restore an active general soil microflora and to supplement
the restricted centres of activity around mycorrhizal or nodulated
plants colonising mine waste areas without creating environmental
conditions facilitating the development of micro-organisms whose
activities will have deleterious effects. In mine waste reclamation, at least in some areas, it would seem essential to
escape the simplistic view that seed plus inorganic fertiliser
is the formula for reclamation.

The foregoing account is, by necessity, at best fragmentary,
and it could be said that due account has not been taken of much
relevant published work. Certainly, much valuable work has been
accomplished on specific physiological groups of soil micro-
organism, the role of micro-organisms in rock weathering, lichen
symbioses etc. However, it must be clear that it is strongly
felt that much more statistically critical general soil biological

work is required in any study of the effects of amelioration
processes in devastated land reclamation. Through such detailed
studies will come the greater possibility for making predictions
on the possibilities for maintaining the essential nutrient cycling
processes in the below-ground ecosystem.

REFERENCES

Anderson, J. P. E. and Domsch, K. H. 1973. Quantification of
    bacterial and fungal contributions to soil respiration. Arch.
    Mikrobiol., 93, 113-127.

Anderson, J. P. E. and Domsch, K. H. 1975. Measurement of
    bacterial and fungal contributions to respiration of selected
    agricultural and forest soils. Can. J. Microbiol., 21, 314-322.

Atlas, R. M. and Bartha, R. 1973. Effects of some commercial
    oil herders, dispersants and bacterial inoculation of oil in
    sea water. In: The microbial degradation of oil pollutants,
    edited by D. G. Ahearn and S. O. Meyers, 283-389. Baton Rouge:
    Louisiana State Univ. (LSU-SG-73-01).

Bjorkman, E. 1970. Forest tree mycorrhiza – the conditions for
    its formation and the significance for tree growth and
    afforestation. Pl. Soil, 32, 589-610.

Brown, L. N. 1933. Flooding to control root-knot nematodes.
    J. agric. Res., 47, 883-888.

Daft, M. J. and Nicolson, T. H. 1974. Arbuscular mycorrhizas in
    plants colonising coal wastes in Scotland. New Phytol., 73,
    1129-1138.

Daft, M. J., Hacskaylo, E. and Nicolson, T. H. 1975. Arbuscular
    Mycorrhizas in plants colonising coal spoils in Scotland and
    Pennsylvania. In: Endomycorrhizas: proceedings of the
    symposium held at the University of Leeds, July, 1974, edited
    by F. E. Sanders, B. Mosse and P. B. Tinker, 561-580. London:
    Academic Press.

Duggan, P. R. 1975. Bacterial ecology of strip mine areas and
    its relationship to the production of acidic mine drainage.
    Ohio J. Sci., 75, 266-279.

Edwards, C. A. and Heath, G. W. 1963. The role of soil animals
    in breakdown of leaf material. In: Soil organisms, edited
    by J. Doeksen and J. van der Drift, 76-84. Amsterdam:
    N. Holland Pub.

Edwards, C. S. 1973. Persistent pesticides in the environment. 2nd Edition. Cleveland, Ohio: CRC Press.

Garrett, S. D. 1970. Pathogenic root-infecting fungi. London: Cambridge Univ. Press.

Gossen, R. G. 1973. Microbial degradation of crude oil. Ph.D. thesis, Univ. of Calgary.

Gray, T. R. G. and Williams, S. T. 1971. Microbial productivity in soil. In: Microbes and biological productivity, edited by D. E. Hughes and A. H. Rose, 255-286. London: Cambridge Univ. Press.

Hashimoto, N., Kojima, T., Ogawa, M. and Suzuki, T. 1973. Effects of alder and acacia on devastated land. In: Ecology and reclamation of devastated land, Vol. 1. edited by R. J. Hutnik and Grant Davis, 357-366. London: Gordon and Breach.

Jobson, A., McLaughlin, M., Cook, F. D. and Westlake, D. W. S. 1974. Effect of ammendments on microbial utilisation of oil applied to soil. Appl. Microbial. 27, 166-171.

Jorgensen, J. R. and Hodges, C. S. 1970. Microbial characteristics of a forest soil after 20 years of prescribed burning. Mycologia, 62, 721-726.

Kozlowski, T. T. and Ahlgren, C. E. (eds.) 1974. Fire and ecosystems. New York: Academic Press.

Lunt, H. A. 1951. Liming and twenty years of litter raking and burning under red and white pine. Proc. Soil Sci. Soc. Am., 15, 381-390.

Marx, D. H. 1975. Mycorrhizae and establishment of trees on strip-mined land. Ohio J. Sci. 75, 288-297.

Marx, D. H. and Barnett, J. P. 1974. Mycorrhizae and containerised forest tree seedlings. In: Proc. N. American Containerised Forest Tree Seedling Symposium, Denver, Colorado, August, 1974; edited by R. W. Tinus, W. I. Stein and W. E. Balmer, 85-92. U.S.D.A. Forest Service. (The Great Plains Agricultural Council Pub. no. 68).

Meiklejohn, J. 1955. The effect of bush-burning on the microflora of a Kenya upland soil. J. Soil Sci., 6, 111-118.

Mitchell, M. J. 1974. Ecology of soil oribatid mites. Ph.D. thesis, Univ. of Calgary.

Mitchell, R. and Alexander, M. 1962. Microbiological changes
    in flooded soils. Soil Sci., 93, 413–419.

Parinkina, O. M. 1974. Bacterial production in tundra. In: Soil
    organisms and decomposition in tundra, edited by A. J. Holding
    et al., 65–77. Stockholm.

Parkinson, D., Gray, T. R. G. and Williams, S. T. 1971. Methods
    for studying the ecology of soil micro-organisms. I.B.P.
    Handbook No. 19. Oxford: Blackwell.

Phillipson, J. (ed) 1971. Methods of study in quantitative soil
    ecology. I.B.P. Handbook No 18. Oxford: Blackwell.

Plice, M. J. 1948. Some effects of crude petroleum on soil
    fertility. Proc. Soil Sci. Am., 13, 413–416.

Quispel, A. (ed) 1974. The biology of nitrogen fixation.
    Amsterdam: N. Holland Pub.

Rothwell, F. M. 1973. Nodulation of various strains of Rhizobium
    with Robinia pseudoacacia seedlings planted in strip-mined
    soil. In: Ecology and reclamation of devastated land, Vol. 1.
    edited by R. J. Hutnik and Grant Davis, 349–356. London:
    Gordon and Breach.

Schramm, J. 1966. Plant colonisation on black wastes from
    anthracite mining in Pennsylvania. Trans. Amer. phil. Soc.,
    56, 1–94.

Schwendinger, R. 1968. Reclamation of soil contaminated with oil.
    J. Inst. Petrol., 54, 182–197.

Stoner, W. N. and Moore, W. D. 1953. Lowland rice farming, a
    possible cultural control for Sclerotinia sclerotiorum in
    the Everglades. Pl. Dis. Reptr., 37, 181–186.

Stover, R. H. 1954. Flood-fallowing for eradication of Fusarium
    oxysporum f. cubense. Soil Sci., 80, 397–412.

van Schreven D. A. and Harmsen, G. W. 1968. Soil bacteria in
    relation to the development of polders in the region of the
    former Zuider Zee. In: The ecology of soil bacteria, edited
    by T. R. G. Gray and D. Parkinson, 474–499. Liverpool:
    Liverpool Univ. Press.

Wicklow, D. T. 1975. Fire as an environmental cue initiating
    ascomycete development in a tall-grass prairie. Mycologia.
    67, 852–862.

Widden, P. and Parkinson, D. 1975. The effects of a forest fire
    on soil microfungi. Soil Biol. & Biochem., 7, 125-138.

DISCUSSION: PAPER 11

G. VAN DYNE   Professor Bradshaw shows that the best
                      results in reclaiming a limestone quarry were
                      given by sewage sludge treatment which would
                      contribute both nitrogen and phosphorous.
                      But could micro-organisms in fact have been
                      introduced at the same time and account for
                      his results?

D. PARKINSON   I don't know. Most of the people who
                      have examined sewage sludge for micro-organisms
                      have looked for enteric pathogens rather than
                      forms likely to be of ecological value.

A. D. BRADSHAW   The sewage sludge we used was incorporated
                      in the surface layer and had a textural effect
                      on the soil as well as contributing nutrients.

B. ULRICH   We are examining the chemistry of phosphate
                      in soils. Solubility appears to be a critical
                      factor - the phosphate ion has to be taken in
                      by the soil in a form that is available to
                      plants. Various phosphate compounds need to
                      be modified to a more soluble form if they
                      are to be useful, and microbial enzymes and
                      plant roots are important in these chemical
                      modifications within the soil.

D. PARKINSON   In my paper I mentioned that Thiobacillus
                      spp. could produce acid in the presence of
                      sulphur; this ability has been used to bring
                      about dissolving of rock phospate in agricul-
                      tural soils (i.e. after top dressing with
                      sulphur).

J. N. R. JEFFERS   Professor Parkinson's paper has emphasised
                      that the soil must not be omitted when ecolog-
                      ical processes are defined. In the design of
                      I.B.P. projects there was a great emphasis on
                      large herbivores, but one of the more startling
                      findings of that programme was that micro-
                      organisms were much more important! They had
                      not been recognised when the research projects

were designed as such major components in the
system and considerable adjustments were needed
as projects proceeded.

A. D. BRADSHAW   What is the relationship between nitrogen
fixation and the rhizosphere system? Could
it be more important than we have thought?

D. PARKINSON   There are free living nitrogen fixers in
the rhizosphere soil of various crop plants,
and it is likely that in this region (where
simple carbon compounds are available for such
bacteria), N_2 fixation could be very important.

12. SOIL CONSERVATION IN ICELAND

S. Runolfsson

Landgraedsla Rikisins

Gunnarsholt, Rang, Iceland

THE ESTABLISHMENT OF THE ICELANDIC
SOIL CONSERVATION SERVICE

At the close of the cold period from 1860 to 1890 a good deal was written and a lively discussion maintained on the problem of wind erosion in Iceland. The great damage it had caused, particularly in the previous decades, seemed bound to continue on an ever escalating scale if protective measures were not undertaken. The drift-sand was closing in on most farms in the middle of the South and a great many had already been evacuated.

About the beginning of the 20th century the Agricultural Society of Iceland showed a growing concern for the problem of erosion in general and the drift-sand in particular. On its initiative a number of specialists, including Danes, were sent to undertake a field study of the problem. The Icelandic Althing granted a small sum of money for fighting the sand-drift. Some farmers took some protective measures, which consisted mainly of erecting walls of stones or timber right across the direction of the erosion at the junction of the eroded areas and the vegetated land. The results of these individual and public attempts to stop the erosion were however negligible.

The State Soil Conservation Service was previously named the Icelandic Sand Reclamation and the oldest legislation on sand re-clamation dates back to an 1895 "Act for Resolution on Sand Erosion and Reclamation". This contained authority for District Commissions, but the Act proved a dead letter with no implementation so it made no mark in its field.

231

In 1907 an Act on "Forestry and Prevention of the Erosion of Land" was enacted and a special Sand Reclamation representative was engaged. He was Gunnlaugur Kristmundsson of Hafnarfjord and for 40 years he fought a serious shortage of funds and a disbelief in the importance of his function.

The link between sand reclamation and forestry, however, remained very weak and was abolished by a new "Act on Sand Reclamation" in 1914 whereby the Governor was charged with the administrative supervision of sand reclamation with the Agricultural Society of Iceland which was to look after sand reclamation affairs.

Since then there has at all times been some legal connection between the Agricultural Society of Iceland and Sand Reclamation. The act on Sand Reclamation was thereupon augmented and amended in 1923 and again in 1941. Then again 1965 marks the end of an era with the passing of the Act on Land Reclamation. The name of the Institute was then amended to the State Soil Conservation Service and the operation and task of this Institute was vastly extended and is now as follows:

1) To stop and prevent destruction of vegetated areas and soil and further the reclamation of eroded areas;

2) Protection of growth, which is achieved by obstructing the excessive utilisation of plant growth anywhere in the country;

3) Supervision of all grazing areas in Iceland;

The organisation began its activities by importing barbed wire for fencing in areas which were being eroded and protecting them completely from grazing. Sowing of sand lymegrass or "Melgras", in Icelandic, (Elymus arenarius) was undertaken on a very small scale, mainly on a trial and error type basis and it was even tried to plough down the seed.

The farmers living in the neighbourhood of the eroding areas did not have much faith in the experiments and considered their usefulness of a very limited value. Most of the farmers accepted erosion as a fact of life and would not have anything to do with interfering with God's will. According to the Act of 1975 the Service can take any land and protect it as long as is needed in order to reclaim it and make it fit for controlled grazing (under the supervision of the S.C.S.). However, rather than enforcing the previously mentioned law legal contracts have been made between the landowners and the Service.

It soon became evident that some regrowth took place within

the protected areas, even if the plants grew slowly to begin with.
Elymus arenarius took root first, then came creeping red fescue,
Agrostis spp. and Poa. Systematic gathering of Elymus seed was
undertaken fairly early on, for the spring sowing. Simultaneously
with the sowing, erection of windbreak fences was undertaken. Such
were the main activities of the organisation until 1920. A considerable number of small desert reclamation fences had been erected
in many places in the country. People were beginning to believe
that wind erosion and drift-sand could be stopped by protecting
the land from grazing, by windbreaks and by sowing of Elymus in
the worst drifts.

After 1920 the organisation began to expand, more money was
allocated in the budget for this purpose, and a number of larger
soil conservation fences were erected. The greatest effort was
made in Gunnarsholt which is in the middle of the volcanic area
of the South.

In 1975 over 190,000 hectares (nearly 2% of the total area of
Iceland) were enclosed by reclamation fences, but a number of older
areas had been given back to the farmers for controlled grazing.
The activities of the organisation have been confined almost
entirely to the volcanic areas in Iceland, which indicates that the
basic reason for the incredible erosion which has taken place in
the past is the volcanic origin of the soil (discussed later).
The main activities of the S.C.S. will be discussed independently
below.

SAND DUNE STABILISATION

The total area of the so called inland dunes has been estimated to
have covered in the past up to 2,000km^2. They were only to be
found in the volcanic areas. The texture of the volcanic loess on
the palagonite formation consists mainly of silt and fine sand plus
some considerable percentage of coarser sand, but very little clay
material. When the vegetative cover on these areas was weakened,
e.g. by overgrazing so that the roots no longer managed to keep
the topsoil in place, the finer soil fractions (silt, fine sand
and organic matter) were carried away by wind, leaving the coarser
fraction behind. This sorting action removed the most important
material from the standpoint of plant productivity and water
retention.

Eventually a soil condition was created in which plant growth
was minimised and in extreme situations the sand began to drift and
form unstable dunes. These encroached on better surrounding lands
and in the last two centuries seriously threatened and even caused
the evacuation of many farming areas.

The only way this problem could be tackled at the time was
to fence them off and protect them completely from grazing. Extensive sowings of the Icelandic Elymus arenarius had to be done as
soon as possible. This plant grew wild in many places and its seeds
were formerly used to some extent as a human food. In severe cases
the establishment of the melgras had to be aided by the erection
of windbreaks. These were either made of lumber or stone walls and
seedlings of Elymus managed to survive on the leeward side of the
walls. Elymus does not thrive well except in drift sand and then
only when there is a fresh annual addition of loose sand. In
general Elymus does not form a continuous vegetative cover, however,
but collects the sand around its roots in small heaps which gradually
grow into hills or stable dunes. Hardy grass species like red
fescues (Festuca rubra) and Agrostis spp. were able to start
growing in the sand once it had stopped drifting. Nowadays the
areas in between the dunes are seeded down with red fescue and the
whole lot top dressed with fertilizer by aircraft and the areas
can be exposed to strictly controlled grazing. One area in particular which was reclaimed in this manner 20 years ago and has been
completely protected from grazing ever since, shows a very good
example of plant succession. The area which was tackled first
is now covered by low growing birches and willows, showing the
tendency towards a climax vegetation.

Today most of these moving inland dunes are now being brought
under control by the Icelandic S.C.S.

Coastal sand dunes are very common in Iceland – the total area
of them being approximately 1,500 km^2. The origin of these dunes
is quite different from those discussed previously and they are
also found on the coastal basalt formations in addition to the
palagonite formations.

The sand which is washed ashore by wave action is blown inland
on drying and deposited in mounds or small dunes. The initial
product of this process is the formation of a foredune just beyond
the high-tide mark. Halophytic (those plants which are able to
withstand periodically high concentration of salt in the sand)
grass species, mainly Elymus, gain foothold in these but the dunes
are by no means stable especially if not protected from grazing,
since halophytic plants are usually succulent and sought after by
livestock, and particularly sheep. As the size of the foredunes
increases, the sand at the crest is blown further inland where it
accumulates into another dune etc.

The way in which the reclamation of the coastal sand dunes has
been tackled by the Icelandic S.C.S. is first of all by fencing the
areas off, then sowing of Elymus in long continuous strips just
beyond the scouring action of the tides. Gradual build-up of dunes

amongst the shoots of Elymus then takes place, the upward part of
the plant being renewed repeatedly after burial. The whole ridge
is bound together by the buried stems of the Elymus, but at first
the surface sand is loose, giving the name "Mobile Dunes". Later
however colonisation of the surface takes place, often initially
by mosses, later on by red creeping fescue and this plant community
is later invaded by Agrostis and Poa spp.

Because of the continuous supply of sand from the sea and in
some cases sediment from river banks, these areas are extremely
vulnerable to grazing, and those within a S.C.S. fence are in most
cases not grazed at all.

These areas are a considerable threat to the farming communities
situated inland from the sand dunes, and quite often large amounts
of sand are deposited on grass fields and blown into farm buildings
etc. As is often the case with such sand dune areas, a dune slack
or very wet area with a constantly very high watertable forms behind
the dunes, being fed by rain and drainage from inland. Farmers
have drained this land by open ditches and got good grazing pastures
out of it; however there have been cases where these ditches have
been filled up by the sand blown from the sand dunes.

Once the sand dunes have been reclaimed or stabilised it is
usually necessary to protect them from grazing for a long time, and
probably grazing should never be allowed there. Experience has
shown these areas to be extremely vulnerable and probably not even
fit for recreation purposes.

The reclamation work on the dunes is often hampered by the
partial flooding of the sand, especially in the winter, with subsequent killing of the seedlings.

RECLAMATION OF ERODED AREAS

The disastrous erosion which has taken place in Iceland left a
great many scars in the volcanic areas where all the vegetation
and most of the topsoil have been completely stripped off, leaving
in many instances level fine gravel or fluvioglacial sands. Old
lavas were often uncovered in a similar manner, but usually quite
a lot of aeolian sand is moving within the lava. Loss of vegetated
areas in the rangelands did not result in a lowering of the total
number of sheep in this century, so this in fact meant increased
grazing pressure on the remaining rangelands.

The staff of the Sand Reclamation Service realised this, and
in order to relieve the grazing pressure on the most vulnerable
grazing areas the Service started reclaiming eroded areas, even

though they were past the stage in the process of erosion where
they did harm to farmland. This work was begun in Gunnarsholt in
1946 with various grass seeds imported mostly from the U.S. and
Canada. Later on varieties from Norway, Denmark and Alaska were
also tried. Over 50 different grass species and a great many
varieties have been tried, both for reclaiming eroded land and also
for stabilising moving sand and fixing the fine volcanic loess.

Of the grasses tested Festuca rubra, Poa macranthe, Poa
pratensis, Bromus inermis (varieties from Canada, Minnesota and
Alaska), Phleum pratensis and Agrostis tenuis, have given best
results. Other varieties have often managed to survive for one or
two years and then died or were overtaken by low yielding Icelandic
creeping red fescue. The grasses mentioned above usually last for
5 or 6 years and then the native creeping red fescue dominates in
most cases.

Some legumes were tried without any promising results although
lupins will survive and grow extremely well where there is no sand-
drift and wild white clover grows in lawns and sheltered areas.

In the year 1948 the cultivation of grasses for agricultural
use was started on the sands at Gunnarsholt. This cultivation has
been kept on and expanded on a growing scale and today there are
over 1,200 hectares of fields utilised for hay and silage on this
farm where before had been practically bare sands and fine gravel.
To begin with these fields required more fertiliser in order to
yield the same as other meadows, but the incidence of winter kill
has been far less than on other soils in Iceland. The difference
may lie in the higher organic matter of many Icelandic soils,
(average 10%) and the fact that these "sands" drain more easily.

The results of this reclamation of the "sands" at Gunnarsholt
were so promising that some farmers recognised this and in 1954
the S.C.S. assisted the farmers from a district in a volcanic area
of the South to reclaim 300 hectares of eroded land. This is all
in one field and the farmers manage the fertilisation, haymaking
etc. co-operatively. Since then the Service has assisted farmers
in other parts of the country to reclaim for hay making purposes
over 3,000 hectares of level eroded areas. These grass fields
are grazed in the spring, then cut for hay in the summer and grazed
again in the autumn, thus reducing greatly the grazing pressure
on the rangelands.

The reclamation of the older lavas, once covered by vegetation,
did not become a reality on a big scale until the Service bought
its first fertiliser aeroplane in 1957. Usually there is quite a
lot of volcanic loess and sand on top of these lavas but very little
vegetation. By sowing Danish creeping red fescue from the air and

applying fertilisers for the first three years these lavas could
be turned into valuable grazing lands. However, this land had to
be fenced off to control the grazing pressure. Hopefully this
land will be an important factor in reducing the grazing pressure
on the commons in the volcanic areas in the future.

In the past few years vast areas on the boundary between common
grazing areas and homeland have been reclaimed. The farmers take
active part in this work and once the vegetation has established
itself these areas are used for grazing in the autumn. This re-
lieves the grazing pressures on the commons.

IMPROVEMENTS OF RANGELANDS

Range research has revealed that most of the commons in the volcanic
areas are being overgrazed, with consequent changes in botanical
composition to unpalatable and lower yielding species. Many of
these rangelands are already badly damaged by wind erosion. Prolon-
ged overgrazing has left the soil impoverished, but fertilisation
of the rangelands results in rapid, extensive changes in the
vegetation. Grasses become dominant and plant density increases.
Generally an application of 70 kg nitrogen and 70 kg phosphate per
hectare for two successive years has proved satisfactory. As yet
it is somewhat uncertain how stable these changes are, and whether
it is possible to maintain, without further fertilisation, the new
composition and increased herbage yield under proper grazing.

In the Soil Conservation Act of 1965 provisions were made for
enforcing the farmers to reduce the number of grazing livestock
according to the grazing capacity of the commons (a difficult thing
to accomplish in reality) and to assist the farmers to improve their
rangelands by aerial topdressing of fertilisers. This method of
applying fertilisers is especially applicable for the hummocky
surface in Iceland and the relative inaccessibility of Icelandic
Highlands.

The use of aircraft in soil conservation started here in 1957
and has continued on an ever increasing scale. The organisation
now owns two fertiliser aeroplanes. A Piper Pawnee with carrying
capacity of 500 kg of fertiliser and grass seeds and a DC-3 which
carries 4,000 kg. The DC-3 was given to the S.C.S. by the Iceland
Air and the distributing mechanism was imported from New Zealand.
The pilots from Flugleidir h/f fly this aeroplane completely free
of charge which indicates the tremendous interest and willingness
amongst Icelanders to help the reclamation work.

OTHER SOIL CONSERVATION SERVICE OPERATIONS

In addition to the main activities of the S.C.S. already described,
some of its other tasks should be mentioned. The service runs the
Gunnarsholt farm, which is the Service's headquarters and main
experimental farm. The livestock there has provided the Service
with valuable information on the grazing capacity of reclaimed
areas.

The S.C.S. always has some fertiliser observations and seeding
trials going, although the Agricultural Research Institute at
Keldnaholt does the required research work.

Other aspects of the S.C.S. include erection of dams in order
to get some sort of a surface irrigation. This is particularly
important on the very porous pumice soils, where the establishment
of the first vegetation is especially difficult but once some
vegetated cover is achieved it helps retain the moisture and the
area can eventually be reclaimed for grazing purposes.

FUTURE OUTLOOK

Today one half of the forage consumed by large herbivorous animals
comes from the rangelands. Unfortunately range research in Iceland
has revealed the fact that about one half of the rangelands are
overgrazed, and what is even more serious is that most of this
land is on the volcanic areas. Furthermore a quarter of the rangelands are perhaps properly utilised and one fourth underused.
Increasing population places increasing demands upon the vegetated
areas for food production. Therefore, there is an urgent need for
land reclamation and range improvement by aerial topdressing of
fertilisers. Proper allocation of grazing livestock according to
grazing capacity of the land is of particular importance.

On the 1,100th anniversary of Icelandic settlement, the Icelandic parliament provided 1,000 million Icelandic kronur for reclamation purposes i.e. soil conservation, forestry and research.
This was to be spent according to a five years programme and has
vastly increased the activities of the Soil Conservation Service
and the aim is to stop all accelerated erosion in the inhabited
areas of Iceland before 1979.

Even if we are gaining slightly in our fight against soil
erosion, it is a meagre fact in the face of the blatant truth
that Icelanders still owe their country over 20,000 km^2 of
vegetated land.

DISCUSSION: PAPER 12

H. P. BLUME   It is clear that there has been severe erosion in Iceland. Why has not this occurred in other countries that are also over-grazed? Could it be because there is very little clay in Icelandic soils?

On our excursion yesterday we saw meadows surrounded by drainage ditches. How is this likely to modify the erosion of the soil by water?

S. RUNOLFSSON   As Dr. Fridriksson's paper states, the digging of such channels leads to the drying of the soils and changes the wetland flora towards grasslands which provide better grazing. It could also lead to some water erosion but generally such soils are stable. As to your first point, there is no doubt that erosion is very much influenced by inherent soil characteristics like the proportion of clay.

V. GEIST   I have not seen this kind of erosion in Arctic Canada. But erosion does occur over white volcanic ash in sheep grazed areas of the St. Elias range. I would also comment that the wild lupin that is native in Canada does not appear to be grazed there and moose in Canada rarely graze birch, in contrast to the sheep in Iceland which appear to feed on it freely. Indeed, I have heard it said that Icelandic sheep graze everything except the rocks! Is it generally true that domestic livestock do more damage than wild species?

S. RUNOLFSSON   Many lupin species are poisonous, although some are taken by sheep in Iceland. The sheep in this country only graze birch in early spring and again in autumn when other forage is not available. The real damage in the past occurred because of the Icelander's habit of grazing the sheep in the winter in shrublands when they eat the bark from the birches thus killing the trees.

G. VAN DYNE   I have two questions. First, what perc-
                      entage of the land surface in Iceland is suf-
                      fering from erosion? Second, are the re-
                      clamations today exceeding the losses?

S. RUNOLFSSON   There is some debate over the extent of
                      erosion. Mr. Bjarnason estimates in his
                      paper that we had lost about half of the orig-
                      inal vegetated areas at a rate which would
                      average 2,000 hectares per year since settle-
                      ment. At the present time we are treating
                      about 6,000 to 7,000 hectares each year. In
                      addition some enclosed land recovers without
                      treatment. We appear to be gaining although
                      there is still a great deal of erosion still
                      to cure.

M. W. HOLDGATE   There are areas in the northern Pennines
                      of England which show erosion patterns very
                      comparable with those we have seen in Iceland,
                      although the soil is a deep peat rather than
                      a volcanic loess. These are also areas subject
                      to heavy sheep grazing. I believe there is
                      a relationship between heavy sheep grazing
                      and erosion because in another region of
                      deep blanket peat which I have seen in southern
                      Chile, the only areas of open gully erosion
                      were also heavily sheep grazed. What controls
                      are you able to impose on sheep grazing in
                      Iceland, especially in the period immediately
                      following restoration?

    S. RUNOLFSSON   There is strict control of grazing in
                      areas where the top soil is being restored.
                      This is less essential on stony eroded areas
                      but here top dressing with fertilisers is
                      applied at a rate which is related to the
                      grazing level. Where the soil is actively
                      eroding we either prevent grazing or reduce
                      it to minimal levels. There is also a pro-
                      hibition of grazing on most of the sand dunes
                      and we may need to prohibit grazing of the
                      coastal zone for a very long time.

B. The Restoration of Vegetation and the Conservation of Plant Diversity

13. EROSION, TREE GROWTH AND LAND REGENERATION IN ICELAND

H. Bjarnason

Skograekt Rikisins

Reykjavik, Iceland

INTRODUCTION

Iceland is, without doubt, the most eroded land in Europe, if not in the world. For this reason it is very appropriate that this conference should be held here.

More than half of the vegetation cover, which occurred here at the time of settlement some 1,100 years ago, has been destroyed. Furthermore the fertility of the remaining soil is but a fraction of the original.

Birch forest and shrubland, which is the natural climax vegetation, covered then about 40,000 km^2 but has been reduced to about 1,000 km^2. Half of this area of remnant birch is still deteriorating in quality and in imminent danger of destruction. Other vegetation types, apart from wetlands, are also rapidly deteriorating under heavy grazing and from erosion. Such is the position here in the final decades of the enlightened 20th century.

This desperate situation must be remedied so that the land can be used in the most efficient way in the future. But for any improvement to be possible, we must solve the following problems:

1. What has caused the tremendous devastation of soil and vegetation?

2. What is the true biological potential of the country, and what type of vegetative cover will give us the most productive use for each unit of land?

241

THE CAUSES OF DEVASTATION

With regard to the first question it is obvious that man's settlement of the country is the prime cause of the devastation. The
destructive effects of man's intrusion have been magnified by
the harsh climate and are, not least, due to the properties of the
soil itself, which is loess of basaltic origin. This type of soil
has little resistance to the eroding effects of wind and water
after the protecting cover of vegetation, in this case birch woodland, is removed.

Once erosion has begun wind and water are far more destructive
than the innocent looking sheep and work a thousand times faster.
However, had the vegetation cover not been totally removed, then
neither wind nor water unaided would have been able to devastate
the vegetation.

It has been claimed by some people that volcanic eruptions
have played a major role regarding soil erosion, whilst others have
blamed the various periods of climatic deterioration. Sigurdur
Thorarinsson has shown that volcanic acitvity has only played a
negligible part, and periods of climatic deterioration, which may
have occurred two or three times after the settlement, would have
had very little effect if the devastation had not already begun.

One point regarding land devastation and regeneration, which
has emerged during recent years, must be mentioned. Many enclosures of varying areas have been established in most parts of
the country and protected from grazing, some of them 50 to 70
years ago. When we compare these protected areas with the grazing
land outside we see how the soil erosion comes to a standstill
without any further action taken. The totally eroded sites inside
the fences are first colonised by Festuca vivipara which gradually
spreads over the bare ground. If there is woodland nearby, birch
will seed itself when the ground is partially vegetated. This
process usually takes several decades and the speed of regeneration
depends very much on the annual rainfall.

Besides this, remarkable changes occur in the composition of the
vegetation. Heath turns gradually into grassland and grassland
into a herb community, which shows that the soil recovers its
fertility within a few decades. If birch has survived hidden in
the sward, or woodland is nearby, then birch regeneration is soon
noticeable.

These examples show better than anything else that not only
is the grazing the prime cause of erosion but it is the main reason

why the vegetation cannot recover and therefore a major factor
for continuing erosion.

It is fortunate that neither volcanic activity nor climatic
deterioration is the main cause of devastated land in Iceland:
instead it is grazing, the only factor over which we could have
complete control. If all grazing outside cultivated areas came to
a stop, Iceland would become revegetated within a century or two.

THE BIOLOGICAL POTENTIAL OF ICELAND

Let us now consider the second question. The biological potential
of the country is not so obvious or as easy to determine as the
causes of the erosion.

The Icelandic flora is poor in species, with only some 440
vascular plants. This is due to the isolation of the country
after the last glacial period. In Northern Scandinavia, where
the climate is similar to here, there are over 700 species, and in
the coastal area of South-Alaska there are about 1,200.

It is now generally accepted that most of the Icelandic flora
is derived from those species which survived the last glaciation
Steindorsson, (1963). Therefore it is not surprising that the
flora has definitive Arctic and Subarctic affinities.

The biological potential of an area is usually evaluated
from the existing plant species and vegetation types growing there.
This method was practised in Iceland during the early part of this
century, and people were very certain about what species could
grow and thrive. Among other things they were sure that conifers
would never be able to grow here. But the only way to find the
real biological potential of the country, and to evaluate it
correctly, is to introduce both trees, herbs and grasses from
those areas of the world where there are similar climatic conditions to what is here and follow up their growth in various regions
in the country. Trees are especially good indicators as it is
possible to measure both annual growth rings and height from year
to year.

The first forest plantings in Iceland were made 77 years ago
and several tree species were imported. This work has continued
without a break except an interlude from 1913 to 1933. The
beginnings were small and planting progressed at a slow rate until
the end of last war. After that we were able to obtain a considerable amount of tree seed of many species from different parts of

the world, and in addition a large variety of other plants, such
as shrubs, herbs and grasses, some of which have given valuable
information.   Care has been taken to avoid mixing the different
provenances.

Altogether about 60 tree species have been introduced from
well over 400 different areas, mainly from Alaska and North Europe,
but also from many mountain regions in America, Europe and Asia.
Of the 60 species tried, 27 can be safely cultivated in the
country, several are still at the trial stage and some have failed.
Conifers make up 15 of the "safe" list and are used mainly for
forestry purposes.   The 12 hardwood species are mostly used for
amenity purposes or in shelterbelts.

The different provenances of the species have been planted
in sufficiently large stands to enable measurement plots to be
established.   Many of the provenances have also been planted at
several different enclosures around the country in order to compare
the growth conditions within regions.

The climate varies often considerably between different
regions in Iceland.   Thus the South and West regions have a high
rainfall while it is low in the North and North east Districts, and
in the South the growing season is markedly longer and warmer
than in the North.   Therefore the various species thrive differently in the different regions, and it must be added that there are
vast areas within the country where neither forests nor trees
can grow.

A country-wide survey was made of stands established before
1960 in 1974 and 1975, during which about 1,000 plots were laid
out.   The object was to assess the growth and suitability of the
various species and provenances.   Unfortunately the results of the
survey have not been completed yet so I can only give a few examples
on the performance of the introduced species.

The earliest planting with a species of known provenance is
at Hallormsstadur in North east Iceland with Engelmann blue spruce
(Picea engelmanni).   It is raised from seed collected at about
3,000 above sea level in Colorado and planted in 1905.   There are
only 5 trees left in this stand and they are between 14 and 15 m.
in height with a breast diameter of 35 - 45 cm.   Since 1947 they
have born fertile seed at regular intervals and several thousand
progeny have been raised and planted.   It is difficult to make
accurate assessments of mean annual increment from so few trees,
but a conservative estimate gives a mean annual growth of $5 - 6$ m^3
per hectare.   No climatic damage has been observed on these trees,
but their progeny have proved to be unsuited to the unstable winter
climate in South Iceland.

A stand of Siberian larch of an Archangelsk provenance was planted in 1938. At the age of 36 years the mean height was 11.6 m. and mean diameter at breast height 18.3 cm. The mean increment was 6.8 m^3 annually per hectare. Such a yield is very profitable in a wood-starved country like Iceland and this stand has given a high rate of return on the investment. These larches have produced fertile seed and a fair amount of self sown seedlings can be found near the stand. It is interesting to note that the same provenance (or at least one very similar) has been planted at two localities in North Sweden and that its annual growth is similar to the Icelandic stand.

So far we only have a single stand of Sitka spruce for comparison and it is a very young one, only 20 years. However, by judging from the growth of older individuals one can expect the Sitka spruce to be at least as productive as the larch on suitable sites. The same can be said for Lodgepole pine (Pinus contorta), which is an even more recent introduction. Both species have yielded fertile seed and some natural regeneration of the spruce has been observed. When a tree species produces fertile seed at regular intervals, not to say annually as Pinus aristata does here, it would generally be considered to have gained a secure footing in its new environment.

From this we can deduce that the factors determining growth are not unlike those at the original site. And the outcome of this comparison is that in the South and West regions the climatic conditions are akin to the coastal area of Alaska from Cordova in the east to Homer in the west.

It was pointed out above that, the North and North east regions have a different type of climate. Formerly it was classified as a continental type but now we are of the opinion that it is more comparable to a subalpine climate as is found at high elevations in the mountain ranges of the Northern hemisphere where the growing season is short and there are big fluctuations in the diurnal temperature range. This is shown by the successful growth of Engelmann spruce, Bristlecone pine and Alpine fir as well as Siberian larch from high elevation. Furthermore, it even applies if the seed comes from latitudes as far south as 40° - 50°N, providing it is collected from the subalpine zone.

As said before we have also collected seed from various herbaceous plant species and brought them to Iceland. Two species have awakened special interest and provoked some speculation. One is the Alaska lupin, Lupinus nootkatensis, the other a lyme grass, Elymus mossis. Both were collected at Point Pakenham in the Prince William Sound in Alaska.

Although natural regeneration sets in when eroded areas are fenced off, this regeneration is often very slow. If the lupin is sown or planted, even at wide spacings, in such eroded land, it will completely colonise such areas within ten years. The lupin has a massive root system with nitrogen-fixing bacteria, which have been found to penetrate down a half a metre or more within three years after sowing. It flowers early in the summer and disperses seed from early August to October. Furthermore, it is hardy at 400 m. elevation, where it both flowers and sets fertile seed annually. Vegetative growth is luxuriant and the plant reaches a height of one metre after flowering. Both leaves and stem rot quickly after falling, and a fertile soil is rapidly formed, as is easy to see by rapid invasion of earthworms. In addition a rich insect fauna is associated with the lupin. After a few years a wide variety of plants begin to appear under the lupin canopy, and gradually it gives way to many of them. All these factors leave us with no doubt that this lupin must have a great future in reclaiming devastated land.

Elymus mollis is not indigenous to Iceland, but we have Elymus arenarius throughout the country. Both species flower and disperse their seed at the same time, and their growth period seems to be equally long. Yet E. mollis yields at least 30% more grass under the same growth conditions. May be this is due to a difference of the species, but it also gives one much food for speculation.

For various reasons, both these species and several others which have been brought in from climatically similar areas, are far more productive than analogous native species. The same applies for trees although no comparable species are native to Iceland. On many exposed sites it is obvious that Sitka spruce is much hardier than the native birch.

It is possible that the Icelandic flora, most of which is derived from species surviving the glaciation, is unable to utilise the improved climate which has followed the glacial period as well as plants imported from localities that now have a climate similar to that of Iceland.

If that is the case then we must introduce more "economic" plant species from similar areas where we have sought tree species, in the belief that their growth and productivity will be more than the native plants.

But in this case we should remember the advice Francis Bacon gave us four centuries ago "Nature can not be commanded except by being obeyed".

REFERENCE

Steindorsson, S. (1963). On the age and immigration of the
    Icelandic flora. Visindafjelag Islandinca (Societas Scientarium
    Islandica).

DISCUSSION: PAPER 13

M. W. HOLDGATE	May I thank Dr. Bjarnason for a very clear paper. It seemed to me to establish a situation not unfamiliar from other areas, namely one in which we have the scientific knowledge to rehabilitate damaged environmental systems, but are limited by the need to control the human component of the system. In Iceland there is obviously a conflict between sheep grazing and soil and vegetation stability: how far can grazing be controlled?
H. BJARNASON	It is almost impossible to control grazing under the present circumstances. Farmers are difficult people to persuade and there are policy problems.
B. ULRICH	Does solifluxion contribute to soil break-up and instability in Iceland? Especially around the ice-caps?
H. BJARNASON	There is some permafrost above 600 metres around the ice-caps. Stone stripes, polygons and small terraces are present on many hill slopes, where soil is certainly moved by solifluxion. Wind erosion and deposition, and also water transport, are more important.
G. VAN DYNE	Grazing is a very important aspect of Iceland's economy. I have travelled widely in this country in the last ten years and agree there is much severe erosion with a metre or more of lost soil, but equally many more areas are well managed. As I understand it, many highland areas were not forested originally. It may not be right to try and restore forest there: such places may be best maintained under a grazing economy. I think we should guard against an emotional anti-grazing feeling though evidently we need to manage it.

H. BJARNASON   But we must not overgraze. Today Iceland
is overstocked with sheep and ponies. When
the island was settled 50,000 km^2 was vegetated.
There are about 17,000 km^2 under 200 m. and a
very large area under 400 m. Shrubs range to
300 m. in the south, 450 m. at Gullfoss, and
as high as 540 m. in the north. Birch was
the climax vegetation almost everywhere
except in the wetlands at the time of colonisation. Deforestation took place not only
to create grazing but to meet human needs
for firewood and charcoal (used in iron
smelting).

H. KOPP   What is the land ownership pattern?
Clearly this is important to land management.

H. BJARNASON   About 70 - 75% of land is privately
owned. The state is the largest single land
owner, managing some land on behalf of the
Church. Some land is leased to private farmers,
but these leases confer almost as much freedom
as ownership.

D. PARKINSON   The ameliorating power of the Alaska
lupin highlights the great importance of
symbiotic associates of plants. Have
mycorrhiza been looked for in trees? It is
possible tree production could be enhanced by
inoculation with the right fungi.

H. BJARNASON   Not enough is known about this. Siberian
larch is always followed by specific associated
fungi. Concerning the lupin: last year Sitka
spruce planted in open soil, soil manured
with horse manure and soil with lupins was
analysed. The leaves in the first site had
1.2% nitrogen against 1.4% in the second and
1.7% in the third, demonstrating the effects.

G. W. DIMBLEBY   Was your assessment of the original
wooded state of Iceland based on documentary
or other types of evidence?

H. BJARNASON   Both kinds are available - I have used
documentary sources, but pollen analysis
has provided information although this needs
to be collated to provide an overall picture.

14. THE RESTORATION OF VEGETATION ON DERELICT LAND PRODUCED BY

INDUSTRIAL ACTIVITY

A.D. Bradshaw, R.N. Humphries, M.S. Johnson & R.D. Roberts

University of Liverpool

England

INTRODUCTION: DERELICT LAND

The theme of this conference is the rehabilitation of severely damaged ecosystems. "Severely damaged" implies that the damage is considerable. In this case the question we must ask is "can it be repaired?".

If the answer is that the damage is irreparable then there are likely to be serious problems. Destruction of the eocsystems of the world, because of mining, quarrying, and other forms of industrial activity, is proceeding at a considerable rate. In Britain it is estimated that 4,800 ha of derelict land are being produced every year and in the USA 61,000 ha are being produced from strip mining alone. To these must be added in Britain the 58,000 ha, and in the USA the 1.6 m ha of land already derelict from past industrial activity (Goodman and Bray, 1965). If all this must remain derelict then we should soon see the permanent destruction of much of our environment.

If, however, the damage can be repaired then we must ask:

i) how much will the repairs cost?

ii) will the ecosystem be as good as new afterwards?

It is possible that the repairs could be too expensive for either industry or society to bear, and if the repaired ecosystem is not as good as new after repair, we may not wish to allow the damage to take place at all. But there is the possibility that

the repaired ecosystem, although it may not be as good as new
(or in other words restored to its original state), could become
something attractive or valuable for a new use.

The derelict and degraded land produced by industrial activity
is characterised by its lack of vegetation and productive ecosystems.
Either recolonisation takes place slowly, or hardly at all, and
whatever happens the ecosystem that develops in no way resembles
the ecosystem that was present before disturbance. It is obvious
that the major determining factor lies with the soil, and it is
this that we must consider.

PREVENTION

Prevention is better than cure. This is very obvious, and yet the
quantity of new derelict land produced every year suggests that
prevention is not occurring. Industrial activity must continue,
but steps must be taken to ensure an immediate and complete restoration so that an accumulation of derelict land is prevented.

This can only take place if (i) the original soil is removed
and replaced in its original form carefully and rapidly, and (ii)
the original ecosystem is re-established carefully and quickly.
The outstanding example of this is the restoration of the coastal
heathlands mined by the mineral sand industry in Australia. Here
the soil is removed and replaced with great care in a matter of
weeks, and the natural vegetation re-established by vegetative
fragments, seed and planted seedlings (Coaldrake, 1973).

In the temperate regions the restoration carried out in
Britain of open cast coal workings by the National Coal Board is
equally outstanding. By careful storage of top soil and subsoil
and its subsequent replacement, highly productive farmland is
reinstated immediately after mining is complete (National Coal
Board, 1974). So effective is the restoration that five years
later it is not possible to make out where the mining took place
except by the absence of mature trees and hedges.

CURE

However until we, as a society, require full and immediate restoration there will always be a large amount of derelict land requiring rehabilitation using means that do not involve the replacement
of top soil, since this has been lost. There are three steps in
the process:

i) <u>The restoration of soil fertility and structure</u>. The factors to be overcome can be divided into three: (a) physical, (b)
nutritional, (c) toxicity. For each there are now a wide range
of solutions which are not necessarily elaborate or expensive.
They are listed in Table 1. Some of the solutions are obvious,
such as the use of fertilisers, but the use of waste products
such as sewage, sludge or even other derelict land soils is
perhaps rather novel.

ii) <u>The establishment of species</u>. There are a number of ways
in which species can be introduced into a site; ranging from a
passive acceptance of natural colonisation to a careful controlled
planting programme. They are listed in Table 2. Some of the
methods are novel but are extremely promising and could be used
on a wide scale. Of particular interest is the fact that we can
now establish not only agricultural species but also wild species,
and therefore can set about recreating natural as well as artificial
ecosystems.

iii) <u>The subsequent management for ecosystem development</u>. Once
the soil has been rehabilitated and plants have been established,
the ecosystem could be left to develop on its own. But the starting
point for this development is very low, whatever component of the
environment is considered, such as soil nitrogen and other nutrients
or plant cover and species diversity. It is possible to let natural
processes of development occur, but inevitably in this case the
build up of a complex mature ecosystem is likely to take so long
a time as to be unacceptable. So some degree of management or
aftercare will be necessary, which must be related to the desired
land use (Table 3). Often the management may be directed towards
maintaining a desired subclimax such as grassland: in some reclaimed areas this may not be possible because of the nature of
the substrate.

In any one situation the particular problems and requirements
will be different and the steps in the process of restoration may
be in conflict. But in the end they will have to be built up into
a simple integrated programme involving as few operations as
possible in order to be economic. To see what is involved we can
examine three different areas which have very different end points.

SAND WASTE FROM KAOLIN MINING IN CORNWALL

Large amounts of coarse quartz sand waste are produced as a by-
product of kaolin extraction in Cornwall. The present rate of
production of waste is about 15 m tonnes/yr^{-1}. This cannot be
put back in the pits from which it was excavated because the deposits
are bottomless and extraction is taken progressively deeper with
time.

Table 1.   The problems of soil fertility and structure of derelict
land preventing the establishment of a vegetation cover,
and the ways in which they can be overcome.

PHYSICAL

Erosion a) because too steep   reduce gradients and establish
vegetation cover.

b) because too fine   establish vegetation cover, if
necessary using a cover crop,
stabiliser or mulch.

Texture a) too hard   rip by tining, and cultivate.

b) too soft   establish vegetation to
stabilise.

c) too coarse   apply fine material or organic
waste.

Water   a) too little   sow drought tolerant species.

b) too much   arrange for better drainage.

NUTRITIONAL

Deficiency of Nitrogen   apply nitrogenous fertiliser
or organic manure and sow
legume or other N fixing
species.

Deficiency of other Macro-   apply appropriate fertiliser.
nutrients

Deficiency of Micronutrients   apply appropriate micronutrients or complete mixture.

TOXICITY

Extreme a) too low   apply lime.
pH
b) too high   apply sulphate, pyritic coal
shale or organic matter.

Excess heavy Metals   apply organic matter or
phosphate or sow metal-
tolerant varieties.

Excess Salinity   leach or sow salt-tolerant
varieties.

Table 2. The problems of establishment of species on derelict
land, and the ways they can be overcome.

PHYSICAL

lack of microsites for
germination

cultivate surface or spread
coarse material or fibrous
mulch.

lack of water retaining medium

sow in wet season or apply
organic mulch.

excessive exposure

sow nurse crop of annual
species or cover with chopped
straw (or equivalent) or
spread branches.

BIOLOGICAL (LACK OF SUITABLE
COLONISING SPECIES)

need for species of any sort

introduce appropriate
agricultural species by seed.

need for wild species

introduce appropriate wild
species
    by hand collected seed
    or by hand planted
        vegetative material
    or by top soil containing
        vegetative material and
        seed
    or by seed-bearing branches
    or by flail harvested seed-
        bearing material.

MECHANICAL (METHODS OF DISTRIBUTION)

topography has normal gradients

use standard agricultural
machinery.

gradients too steep

use hydraulic seeding
technique
    or sow by hand (with
        riddle)
    or sow by air

Table 3.   The problems of management and aftercare of rehabilitated
           derelict land and the ways they can be overcome.

Agricultural end point

maintenance of soil fertility	monitor soil nutrients and pH regularly and add extra amounts of fertiliser and lime required over and above normal practice.
	maintain legume by grazing management and fertiliser treatment unless nitrogen can be added continuously.
maintenance of soil structure	maintain a vigorous growth of vegetation.
general management	apply rigorous agricultural standards.

Forestry end point

maintenance of soil fertility and structure	little need be done after initial fertilizing and soil treatment if this is properly carried out but some N + P fertilizing may be required.
general management	apply rigorous forestry standards.

Nature Conservation end point

if restoring original ecosystem	ensure nutrient levels are being restored to original level, by fertilising and growth of legumes.
	ensure re-establishment of all previous species (see table 2).
if initiating ecological succession	apply no treatments except minimum to ensure site stability etc.
	ensure species which could colonise have opportunity to do so, if necessary by purposeful introduction.

The surrounding countryside is dominated by moorland or small
dairy farms so that most of the area is heavily grazed permanent
grassland. There are small woods. The waste heaps are either
conical or now more usually long, flat and about 30m high. Although
they are new additions to the landscape and therefore there is no
possibility of restoring an original ecosystem, the present need
is to blend these heaps into the surrounding landscape and to reduce
dust blow, by the establishment of a grass cover.

The waste is coarse, and very deficient in all plant nutrients.
There is no toxicity. The essential characteristics and problems
to be overcome are given in Table 4. There is some natural colonisation giving a depauperate vegetation cover mainly of heath species
tolerating poor acid conditions. However, if acid tolerant legumes
such as gorse (Ulex europaeus) and the introduced Tree lupin (Lupinus
arboreus) invade, there is a build up of a much more vigorous vegetation confirming the importance of nitrogen to ecosystem development on the material (Dancer, Handley and Bradshaw 1976a).

A large number of experiments have been carried out which show
that major plant nutrients, particularly nitrogen are critical
(Bradshaw, Dander, Handley and Sheldon, 1975) (Figure 1). The basis
therefore for successful establishment of vegetation must be an
adequate supply of nitrogen which is most effectively provided by
the nitrogen fixation properties of White clover (Trifolium repens).
This can contribute 100 kg N/ha^{-1}yr^{-1} to the developing ecosystem
(Dancer, Handley and Bradshaw, 1976b). Nitrogenous fertilisers,
although important in the early stages before nitrogen fixation
occurs are not very effective because they are heavily leached in
the coarse well drained material (Dancer, 1975).

Since the objective is a grass sward in keeping with the
surrounding agricultural land the seed mixture used contains a
wide range of agriculturally available species including species
with rapid establishment and growth such as ryegrass (Lolium
perenne) but also species more tolerant of low fertility situations
such as Bent grass (Agrostis tenuis). Normal methods of establishment are not possible because of the steepness and instability of
the heaps. Hydraulic seeding is therefore used, although normal
agricultural methods are very effective and economical on flat
areas. Hydraulic seeding is however a technique which although
widely used has not been effectively researched. Considerable
failures were discovered in the establishment of clover, which
were subsequently found to be well known elsewhere but not explained. Investigations have shown that the failure is because
of interaction between clover and fertiliser at germination, clover
being particularly sensitive. If fertiliser levels, particularly
nitrogen, are kept low in the hydraulic seeding mixture but
applied later, clover establishment is excellent.

Table 4.   Characteristics of sand waste from kaolin extraction.

PHYSICAL	
<u>texture</u>	coarse sand with little finer material.
<u>stability</u>	unstable as tipped, liable to wind and water erosion.
<u>moisture</u>	can be extremely dry on surface but always moisture below.
NUTRITIONAL	
<u>nitrogen</u>	very low indeed and leaches readily.
<u>other macronutrients</u>	Ca P K Mg very low.
<u>micronutrients</u>	satisfactory.
TOXICITY	none if limed.

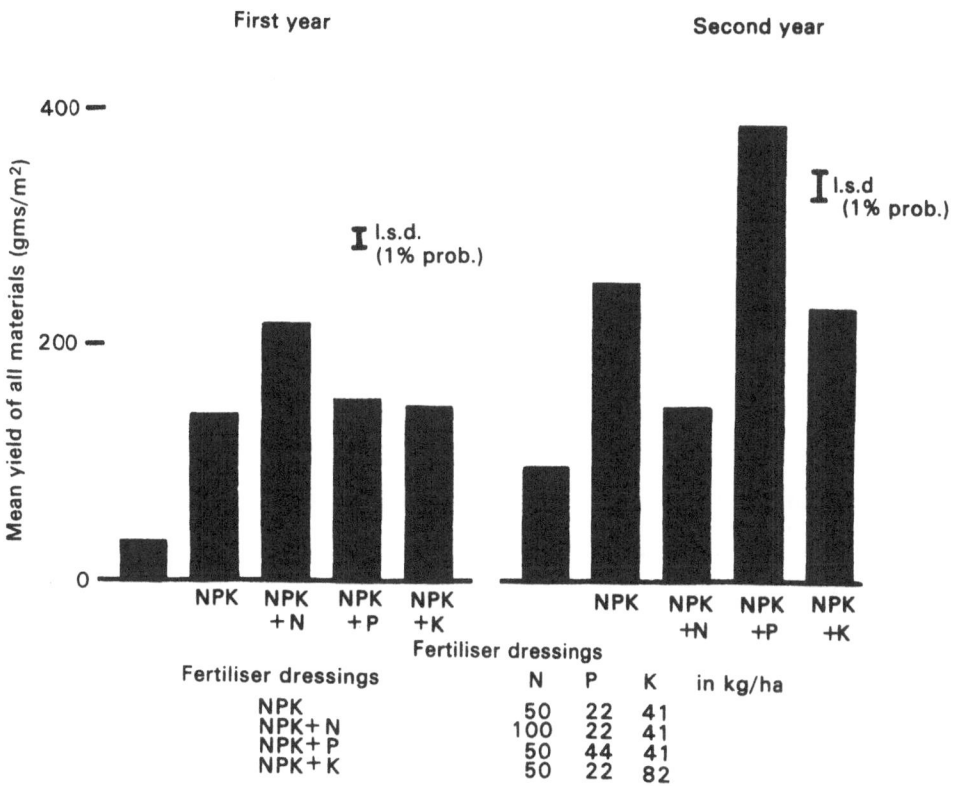

First year   Second year

Fertiliser dressings

Fertiliser dressings	N	P	K	in kg/ha
NPK	50	22	41	
NPK+N	100	22	41	
NPK+P	50	44	41	
NPK+K	50	22	82	

Fig. 1. The effects of different fertiliser treatments on the growth of a grass/clover mixture sown on wastes produced by kaolin extraction (Bradshaw, Dancer, Handley and Sheldon 1975)

Subsequent management requires that the clover be maintained
in an active form. This depends on an adequate supply of lime and
phosphorus, about 1,000 kg/ha^{-1} of CaCO$_3$ and 25-50 kg/ha^{-1} of P
(Figure 2), applied every two years.

But grazing is also necessary. In some sites particularly in
open areas such as Bodmin Moor sheep can come in from the surrounding moorland. This they do and a reclaimed site can carry an
estimated 10 sheep/ha^{-1} during the summer.

But in others the mining operations make sheep grazing
difficult. In these there is deterioration of the sward, and loss
of clover. If these areas cannot be grazed an alternative reclamation treatment must be found. This is being investigated.
It will have to be based on trees and shrubs because any other
vegetation will naturally become colonised by woody species in the
absence of grazing, and it therefore seems sensible to encourage
the development of woodland from the beginning. However, it will
be necessary to ensure a nitrogen supply to the developing woodland ecosystem, perhaps through leguminous shrubs such as Tree
lupin and gorse, although care must be taken to ensure that these
do not become a fire hazard.

FLUORSPAR TAILINGS IN DERBYSHIRE

Fluorspar mining is a small but very significant industry in
Derbyshire supplying important materials for steel making. The
deposits are in veins which may be up to 30 m wide, worked by open
cast and deep mining methods. These areas are restored by application of topsoil. The major environmental problem is set by the
waste material, the tailings, produced by the concentrator plant.
These are deposited in a series of lagoons, each about 5 ha in
extent, which because of the angular shape are a considerable
intrusion on the landscape. At the same time since the tailings
are very fine they are liable to erosion by wind.

The whole industry is within the Peak District National Park,
one of the most used and enjoyed of all the British National Parks.
There is therefore a very considerable need, developed from public
pressure and planning requirements, to provide a landscape solution
which will allow the tailings lagoons to merge into the surrounding
landscape which like Cornwall is small grassland farms.

The waste is fine and calcareous, containing appreciable
quantities of residual fluorspar. But it also contains significant
amounts of lead and zinc which occur in the original ore material
and are extracted during the concentration process, although not
completely. There are also deficiencies of major plant nutrients

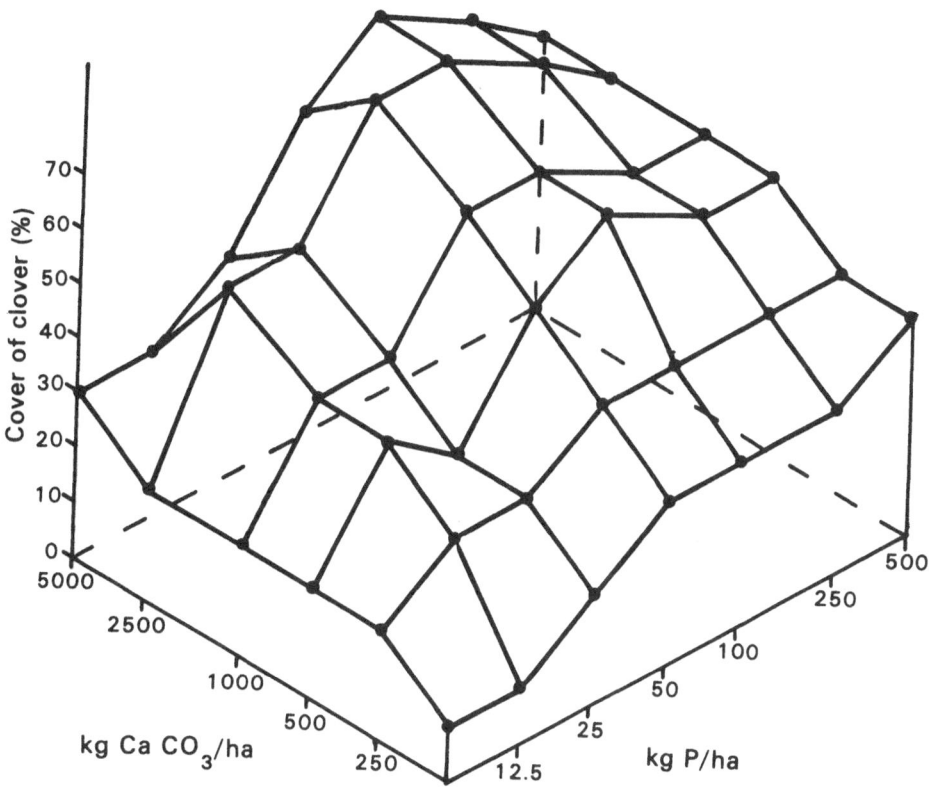

<u>Fig. 2.</u> The response of clover to additions of lime and phosphorus to a two year old grass/clover sward established on sand waste: estimate of cover lyege (three observers) two years after application.

(Table 5). Such a combination of adverse nutrient and toxicity
factors could conspire to restrict plant establishment. However
there does not appear to be any substantial constraint because
various plant species do appear on the newly deposited tailings.
Even the older tailings produced by mining for lead and zinc and
containing substantial quantities of metal (about 1% of lead and
of zinc) are colonised, but by a restricted range of species which
have evolved metal tolerant populations.

A combination of greenhouse and field experiments confirms
that there are no real restrictions to plant growth if complete
fertiliser containing N P and K is applied (Johnson, Bradshaw and
Handley, 1976). There appears to be little toxicity from the lead
or zinc in the recent tailings, for material which is not tolerant
to heavy metals grows as well as tolerant material. But on the
older tailings which contain substantial quantities of heavy metals
the situation is quite different, non tolerant material being
extremely unsuccessful (Figure 3). It is necessary to ensure an
adequate continuous release of nutrients, and for this purpose
sewage sludge has proved excellent. There is the possibility
that the sludge is also improving the physical characteristics of
the material and complexing any plant-available metals that may
occur.

The surface of the tailings begins to erode by wind action as
soon as it is dry, and so there is need to establish a cover
quickly. Since the underlying material will still be wet and
thixotropic, ordinary machinery cannot be used. So again hydraulic
methods are essential, but in this case sewage sludge as well as
fertiliser is included in the mixture. The material runs down
into the cracks of the drying surface, and seedling establishment
is very rapid. The seed mixture is a broad based agricultural
mix analogous to that used on kaolin wastes but without acid
tolerant species. An excellent sward develops within one season.

There is reasonable retention of nutrients in the waste but
a substantial input of nitrogen is necessary as in other materials.
So clover must be maintained. Under the alkaline conditions phosphate tends to become unavailable and must therefore be re-applied.
It would seen obvious that the good growth of grass would make
excellent fodder for grazing animals. However, analysis shows that
the herbage contains levels of fluoride, lead and zinc well above
those recommended for dry feed for livestock (Table 6). As a
result the grazing cannot be by animals kept permanently on the
area; but grazing by animals kept on the area for periods of about
4 weeks at one time is possible.

The problems of grazing suggest that trees and shrubs would
be a valuable alternative, especially since these would break the

Table 5.   Characteristics of fluorspar tailings.

PHYSICAL	
<u>texture</u>	fine uniform silty clay material, very soft and thixotropic beneath.
<u>stability</u>	liable to wind blow whenever surface dries.
<u>moisture</u>	very wet below 10 cms but surface can dry out.
NUTRITIONAL	
<u>nitrogen</u>	very low indeed.
<u>other macronutrients</u>	high Ca, low P.
<u>micronutrients</u>	satisfactory.
TOXICITY	
<u>pH</u>	satisfactory
<u>heavy metals</u>	considerable in old material, substantial in new material but apparently not toxic to plants.
<u>salinity</u>	none

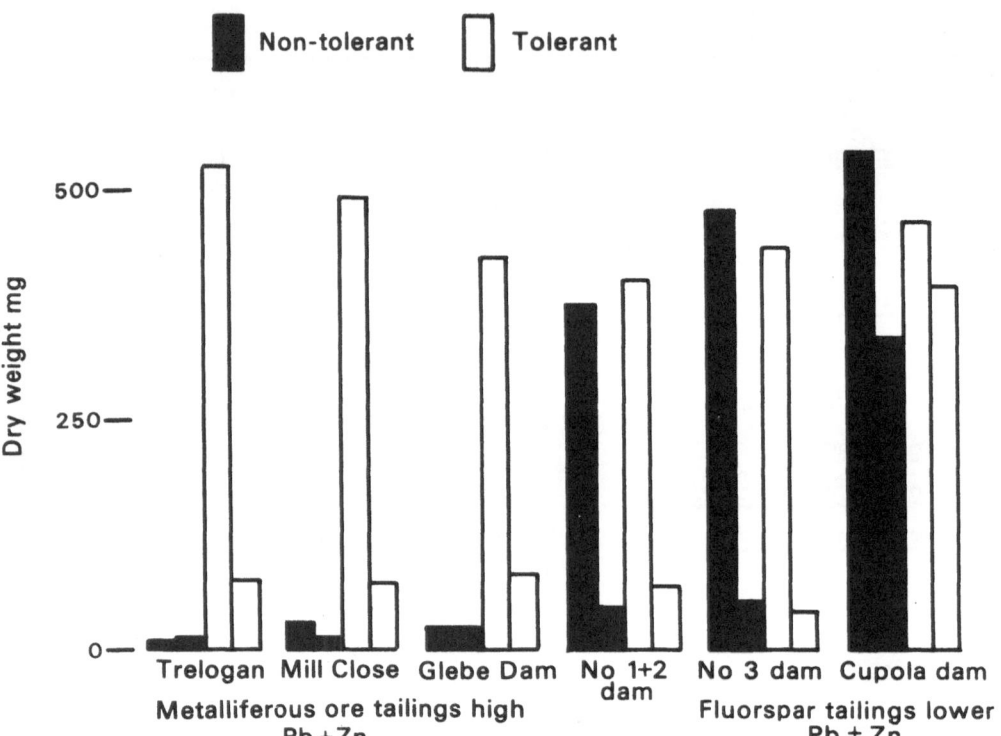

Fig. 4. The effects of nitrogen and phosphorus on the growth of
Festuca rubra on limestone fines.

Table 6.   Concentrations of lead, zinc and fluoride in shoots of
           Festuca rubra grown on fluorspar tailings and metalli-
           ferous mine waste (μg/g of oven-dried tissue).

SITE	POPULATION	LEAD	ZINC	FLUORIDE
Lead zinc wastes				
Trelogan	Commercial	1200	5000	6
	Metal-tolerant	440	2520	4
Mill Close	Commercial	660	1985	13
	Metal-tolerant	395	730	4
Glebe dam	Commercial	623	803	920
	Metal-tolerant	500	455	900
Fluorspar wastes				
No. 1 and 2 dam	Commercial	282	220	650
	Metal-tolerant	113	88	890
No. 3 dam	Commercial	98	74	600
	Metal-tolerant	61	67	495
Cupola dam	Commercial	460	238	1010
	Metal-tolerant	495	104	790

rigid contours of the lagoons. The wet calcareous tailings are not
suitable for all species and the sites are extremely exposed. But
sycamore (Acer pseudoplatanus), hazel (Corylus avellana), Mountain
ash (Sorbus aucuparia) and Violet willow (Salix daphnoides) grow
well and will form attractive woodland copses on the lagoon surface.
The surface dries out and consolidates well under the vegetation
and rapidly forms an attractive amenity area.

LIMESTONE QUARRIES IN DERBYSHIRE

Derbyshire is also a major area for the production of limestone for
the chemical industry and for aggregates. Much of the quarrying
is in an area excluded from the National Park but enclosed by it.
But some quarrying occurs in the Park and more could extend into
it when the present quarries are exhausted. The quarrying process
leads to complete destruction of topography and soils, producing
bare rock faces and quarry floors, a loss of land use and an un-
attractive intrusion on the landscape.

In some areas disused quarries can be used for the disposal
of industrial and other wastes such as fly ash, or for the siting
of light industry. However, in an area of outstanding beauty it
is preferable to restore a vegetation cover. Since the topography
has been drastically altered and the thin over-burden lost it is
not possible to restore the original agricultural grazing land use.
However, the new form provides the opportunity to substitute a
new use, and create a wilderness area with trees and open grassland in keeping with the surrounding deeply cut limestone dale
landscape.

In establishing a vegetation cover there are considerable
problems. The quarry faces are almost vertical walls of pure
limestone. However, they are cracked by the blasting and contain
occasional natural fissures, often filled with clay material.
Physical and chemical factors would appear to be rather extreme
(Table 7). Neverless natural colonisation of rock faces does
occur slowly as seedlings, mainly of trees and shrubs, establish
and root into the cracks and ledges where there is accumulation
of fine material. The quarry floor is flat and compacted, but
possesses some fine material and is therefore colonised more
rapidly. But the poverty of nutrients in the material maintains
a very open vegetation cover in which rare and local species
such as the Bee orchid (Ophrys apifera) and mezereon (Daphne
mezereum) may find refuge. As a result several limestone quarries
are now nature reserves (Ratcliffe, 1974).

Since the main restrictions to plant growth appear to be
lack of appropriate seed parents, low nutrient status particularly
nitrogen and phosphorus (Figure 4), and lack of a moisture retaining

Table 7.   Characteristics of limestone quarry faces and floors.

PHYSICAL	
texture	face — large rock surfaces with crevices containing fine material. floor — heterogeneous mixture of fine and coarse material, very compacted.
stability	satisfactory.
moisture	free drainage and surface drought but fine material retains water below.
NUTRITIONAL	
nitrogen	very low.
other macronutrients	high Ca, low P, K and Mg.
micronutrients	satisfactory
TOXICITY	none

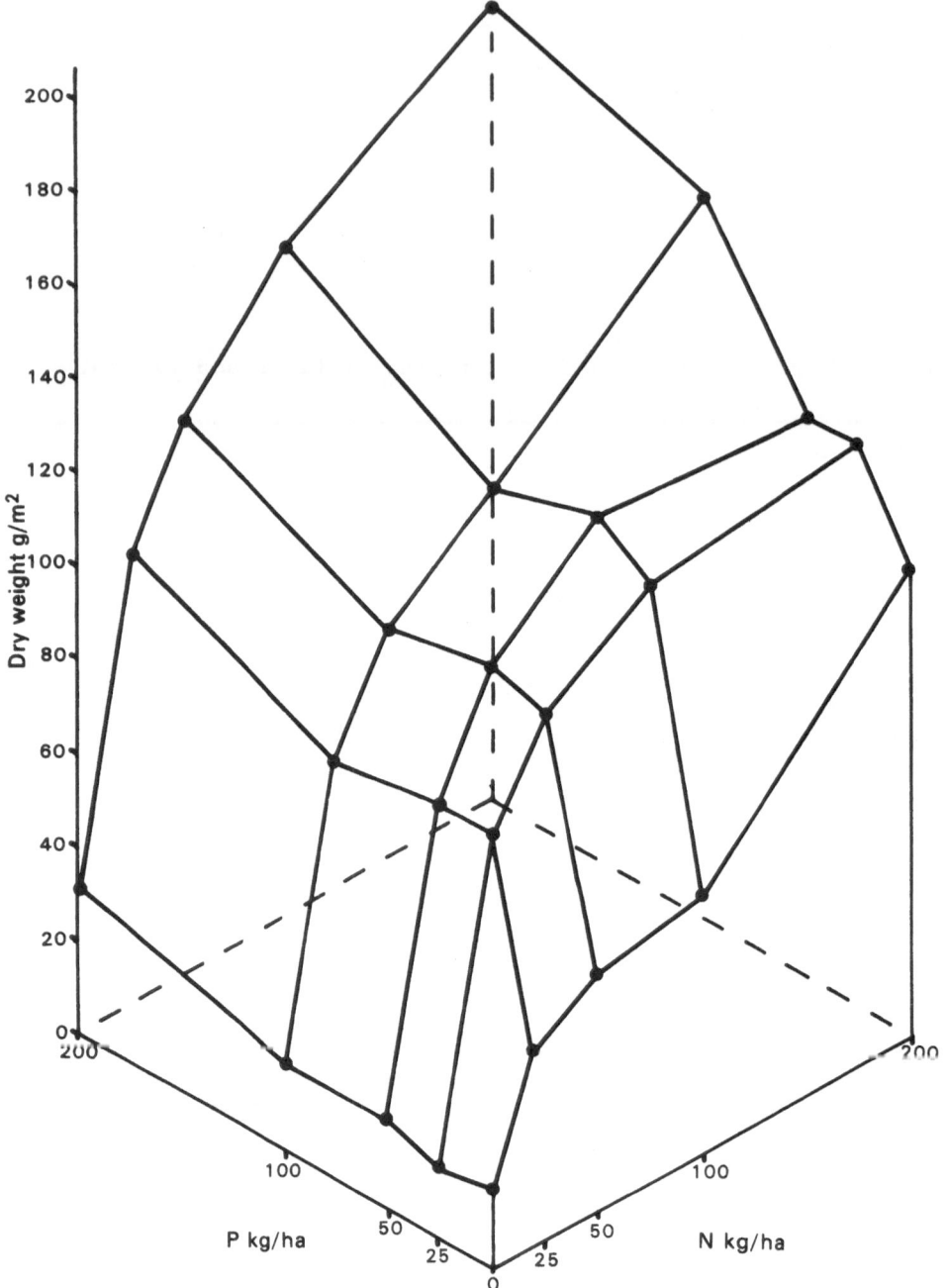

<u>Fig. 5.</u>   The accumulation of nitrogen after reclamation in sand wastes
produced by kaolin mining, compared with naturally colonised
areas and local pastures (Bradshaw, Dancer, Handley and
Sheldon 1975).

substrate, the most effective ways in which these could be overcome have been investigated. On a face that was selectively blasted to produce ledges and screes which resemble the natural topography, various semi liquid mulches were poured down. These contained seeds of grasses, legumes and trees native to the area. The mulches ran into cracks taking the seeds with them. Despite two extremely dry summers the establishment has been remarkably good (Table 8). The mulch richest in plant nutrients, the spent mushroom compost, gave extremely good grass growth but has tended to prevent tree seedling establishment by excessive competition.

There is no doubt that this is a remarkably effective way of introducing species into an inhospitable habitat. However, a sub-sidiary trial in which seed and fertiliser alone were scattered on the quarry face worked well where there was sufficient fine material on ledges and in crevices, suggesting that physical factors may not be as important as supply of seed and nutrients. On the quarry floor which is consolidated and smooth, physical factors such as the availability of protected microsites for germination appear very important (Harper, Williams and Sagar, 1965), as well as the presence of material which retains moisture and can be penetrated by plant roots (Table 9). The seed mixture contained readily available agricultural, herb and tree species. A wider range of species characteristic of limestone and other calcareous rocks could be included. This is presently being investigated.

Since the ultimate object is a naturalistic vegetation cover little subsequent management should be necessary. However, by analogy with other areas some attention will need to be paid to major plant nutrients. Legumes are included in the mixture and should persist because of the openness of the habitat: but it might have been better if a wider variety of legumes could have been included. But regardless of species, legume growth will be aided by some further addition of phosphorus if they are to contri-bute the nitrogen needed to get reasonable growth rates of the woody and herbacious species. However, excessive application of nutrients particularly phosphorus will lead to excessive eutro-phication and the elimination of species adapted to open habitats, and consequently loss of wild life interest (Green, 1972). This can be seen from the data in Table 8.

CONCLUSIONS

In each of these areas there are substantial constraints which mean that natural colonisation does not occur at all rapidly. Yet if appropriate action is taken based on a proper appreciation of the problems, a vegetation cover can be readily established, without the expense of moving or finding topsoil. It happens

Table 8.   Establishment of plants on limestone quarry face (shelf area) after treatment with 3 different seed-containing mulches one year previously.

	seed number/m^2 in original seeds mixture	SEWAGE-SLUDGE	SOIL	MUSHROOM COMPOST
TOTAL VEGETATION COVER		55	40	70
BRYOPHYTE COVER		15	50	-
HERBAGE HEIGHT (cm)		5-30	5-16	5-40
SPECIES COVER				
Festuca rubra	2800	10	5	5
F. ovina				
Lolium perenne	300	20	15	60
Dactylis glomerata	1150	20	5	10
Poa pratensis	50	+	+	
Agrostis tenuis	250			+
Phleum pratense	50		+	+
Cynosurus cristatus	50		+	+
Trifolium pratense	100	1	5	(+)
T. repens	150	1	10	+
Medicago sativa	50	+	+	
Vicia sativa	50	(+)	(+)	(+)
Onobrychis viciifolia	50	+	+	
Lotus corniculatus	100		+	
Quercus robur	1	(+)	(+)	
Rosa canina	40	+	+	(+)
Fraxinus excelsior	19	+	+	+
Cratagus monogyna	6	+	+	
Acer campestre	19	+	+	+
Viburnum opulus	14			
Sorbus aucuparia	1			

+   present in small numbers

(+) present in first summer but not subsequently.

Table 9. The effect of various treatments on seedling germination and survival of ryegrass (Lolium perenne) on a limestone quarry floor.

Number of seeds sown/500 cm	500
Number of seedlings established	
un-amended	18
with 4cm layer 1st fines	60
with 4cm layer soil	186
with 4cm layer compost	254
scarified but otherwise untreated.	142

that in the examples chosen considerable use has been made of waste materials such as sewage sludge and novel methods such as hydraulic seeding. In the appropriate place these are valuable, but other more normal methods are just as important. The criteria must be to use methods which are cheap and effective.

There are many factors to be considered which must be brought back to normality before ecosystem development can take place properly, but in our experience the over-riding problem common to all situations is the provision of adequate nitrogen. Natural communities have a total of at least 700 kg N/ha^{-1} in the soil and vegetation. Raw wastes of all sorts, no matter what their origin, usually have less than 10 kg N/ha^{-1}. Nitrogen in soils is mainly in an organic form, and an adequate supply of mineral nitrogen depends on the size of this organic store. For this reason 700 kg N/ha^{-1} seems the minimum size of the store, although the rate of microbial decomposition will affect the precise value (Dancer, Handley and Bradshaw, 1976a).

Attention must therefore be directed towards the accumulation of this amount of total nitrogen. Natural aerial fall-out of N is only about 10 kg N/ha^{-1} yr^{-1}. Only a proportion of nitrogen applied as fertiliser is retained, and anyway nitrogenous fertilisers are now expensive and energy consuming. It is essential therefore to arrange for nitrogen build up through the use of leguminous plants and other species with nitrogen fixing micro-

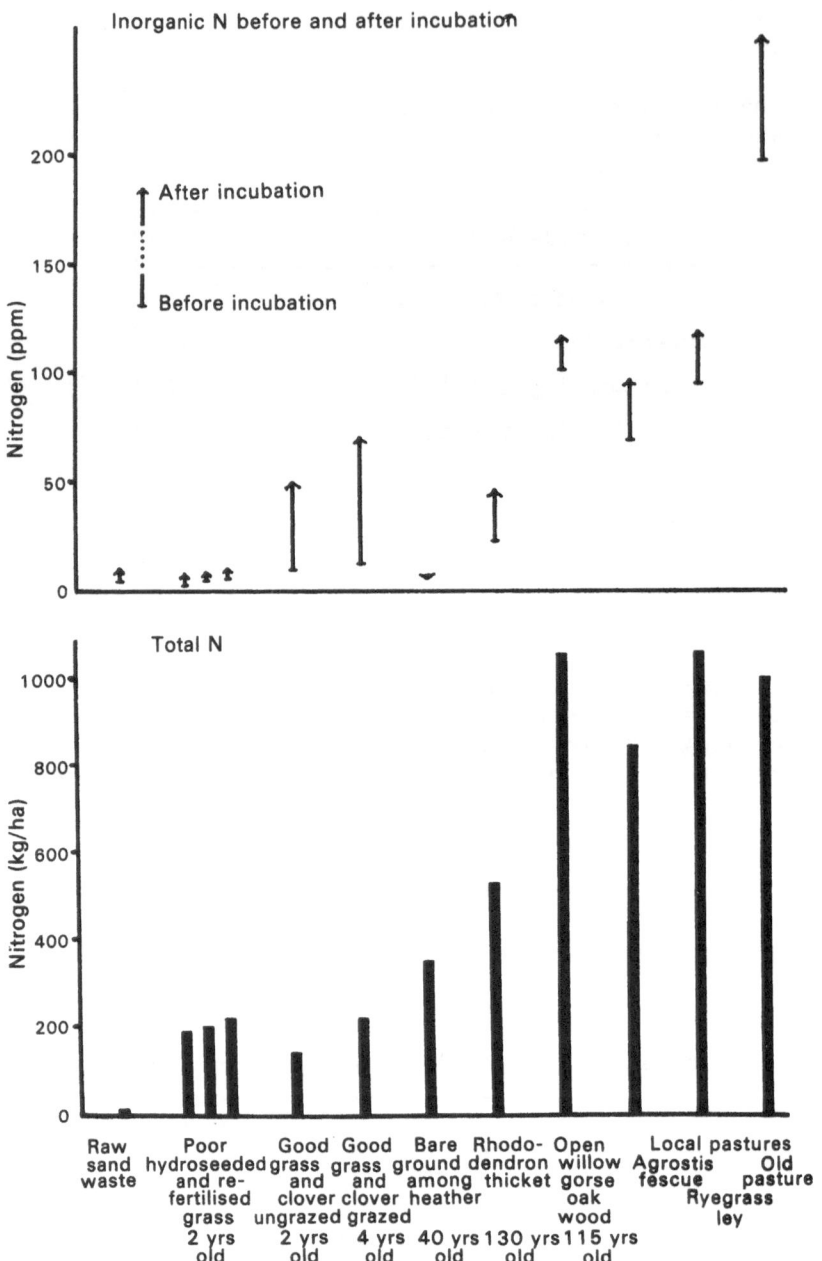

Fig. 3. The growth of lead/zinc tolerant and non-tolerant populations of _Festuca rubra_ on fluorspar tailings and older metalliferous tailings containing higher levels of metal (Johnson, Bradshaw, and Handley 1976).

organisms. But it must be remembered that the process of accumulation is slow, with consequent effects on the amount of available
mineral nitrogen (Figure 5) (Bradshaw, Dancer, Handley and Sheldon,
1975). Nevertheless considerable progress can be made.

In each of the sites it would have been possible to carry out
a complete restoration to give ecosystems which, as nearly as
possible, resembled the original. But since all three sites are
in subclimax man-made ecosystems it is rather difficult to know
what these should be, since they could either be the more recent
sub-climax ecosystems or the original climax of woodland. If we
assume that man was part of the earlier ecosystem then we would
presume that restoration of the original ecosystem would involve
reinstatement of a man-influenced sub-climax. But this could be
(i) heath or rough grassland if we assume only a moderate influence
by man or (ii) agricultural grassland if we assume a heavy
influence by man. Amongst the woodland end points we can not only
consider natural woodland but also woodland heavily managed for
forestry. Arable farming is another possible end point.

All this puts us in a dilemma because there are different end
points depending upon our view of what restoration of the original
ecosystem means. This suggests that rather than get into an arid
discussion as to which represents the proper end point from an
ecological point of view, it is better to tabulate all the possible
end points (Table 10).

From this it can be seen that there are a wide variety of end
points. But some of them can only be achieved with considerable
difficulty, by the importation of top soil, or major land reshaping:
these have been indicated in the Table. End points resembling the
ecosystems present immediately before the disturbance may be in
this category. The end points that have been chosen in the present
investigation do not therefore necessarily restore the orginal
ecosystems. They have been chosen because they represent the
best compromise of ease of reclamation and conformity to the landscape, agricultural and amenity requirements of the areas. Although
they may be new and different this does not mean that they are any
less valuable.

Indeed the new ecosystems may on the combined criteria of
landscape, agriculture and amenity be better than the ones they
replaced. The reclaimed kaolin sand waste is certainly much
better for agriculture than the poor moorland it replaced. The
limestone quarry, when the new ecosystem is fully developed, could
provide an outstanding addition to the amenity and wild life interest
of the National Park which would more than compensate for the loss
of agricultural land. The fluorspar tailings, perhaps, are one

Table 10. Possible end points for the reclamation of three contrasting degraded areas.

AREA	ARABLE	PASTORAL HIGH INTENSITY	PASTORAL LOW INTENSITY	WILDERNESS	FORESTRY
Kaolin sand waste heaps	mixed arable	grassland*	heather*	oak woodland	pine + larch forest
fluorspar tailings dam	mixed* arable	grassland*	rough grassland and scrub	mixed woodland	ash forest
limestone quarry	mixed* arable	grassland*	rough grassland and scrub	ash woodland	ash forest
KEY		grassland*			

most difficult ←——————————→ most easy

* ecosystems in existence before degradation

case where, because of the problems of toxicity of the herbage, the final product does represent some loss: but if the woodland develops well, the gain in amenity may compensate. In other areas where metalliferous wastes are to be rehabilitated in areas of intensive agriculture, it is necessary to explore ways of covering up the wastes completely with an inert material such as colliery shale which is itself a waste product (Johnson, McNeilly and Putwain, 1976).

From the point of view of nature conservation, rehabilitation should not necessarily mean restoration of the original often un-interesting ecosystem. Farming is exerting great pressures on our native flora and is the direct cause of most of the losses of rare species (Perring, 1974). These man-made habitats can be manipulated to become the refuges for wild species which are being oppressed elsewhere. The limestone quarry habitat offers outstanding opportunities for such creative conservation; there are at least 30 rare or local species that could be given a home in such a habitat (Humphries, in Bradshaw, 1976). Many other habitats also offer important opportunities if a grazing land use is not required.

The possibilities for rehabilitation for ecosystems are there-fore considerable, even when these have been damaged to the extent of total destruction. The normal processes of ecological succession make this clear. But what we should realise is that with proper understanding of the biological problems involved the rehabiliation process can be rapid and complete.

ACKNOWLEDGEMENTS

We would like to acknowledge the support of the Science Research Council for research studentships (R. N. Humphries and M. S. Johnson), the Natural Environment Research Council for a research grant (A. D. Bradshaw and R. D. Roberts), and the support of our three collaborating companies, English China Clays, Laporte Industries and Imperial Chemical Industries, who have been very generous in encouragement and facilities.

REFERENCES

Bradshaw, A. D. 1976. Conservation problems of the future, Proc. R. Soc. Lond. A. (in press).

Bradshaw, A.D., Dancer, W. S., Handley, J. F. and Sheldon, J. C. 1975. The biology of land revegetation and the reclamation of the china clay wastes in Cornwall. In: The ecology of resource degradation and renewal, edited by M.J. Chadwick and G.T. Goodman, 363-384. (Brit. Ecol. Socl Symp. 15) London.

Coaldrake, J. E. 1973. Conservation problems of coastal sand and
    open case mining. In Nature conservation in the Pacific,
    edited by A.B. Costin and R.H. Groves, 299-314, Canberra:
    Australian National University Press.

Dancer, W. S. 1975. Leaching losses of ammonium and nitrate in
    the reclamation of sand spoils in Cornwall. J.environ. Qual.,
    4. 499-504.

Dancer, W. S., Handley, J. F. and Bradshaw, A.D. 1976a. Nitrogen
    accumulation in kaolin mining wastes in Cornwall. I. Natural
    communities, plant and soil (in press)

Dancer, W. S., Handley, J. F. and Bradshaw, A. D. 1976b. Nitrogen
    accumulation in kaolin mining wastes in Cornwall. II. Forage
    legumes, plant and soil (in press).

Goodman, G.T. and Bray, S. (eds.). 1975. Ecological aspects of the
    reclamation of derelict and disturbed land: an annotated
    bibliography. Norwich: Geo Abstracts.

Green, B. H. 1972. The relevance of seral eutrophication and
    plant competition to the management of successional
    communities. Biol. Conserv.,4, 378-384.

Harper, J. L., Williams, J. T. and Sagar, G. R. 1965. The behaviour
    of seeds in the soil. I. The heterogeneity of soil surfaces
    and its role in determining the establishment of plants from
    seed. J. Ecol.,53, 273-286.

Johnson, M. S., Bradshaw, A. D. and Handley, J. F. 1976. Re-
    vegetation of metalliferous fluorspar mine tailings. Trans.
    Instn. Min. Metall. A, 85, 32-37.

Johnson, M. S., McNeilly, T. and Putwain, P. D. 1976. Revegetation
    of metalliferous mine waste contaminated by lead and zinc.
    Environ. Pollut. (in press).

National Coal Board, 1974. Open cast Operations,I. London:
    National Coal Board, Open Cast Executive.

Perring, F. 1974. The last seventy years. In: The flora of a
    changing Britain, edited by F.H. Perring, 128-135. Hampton,
    Middx: Classey.

Ratcliffe, D. 1974. Ecological effects of mineral exploitation
    in the United Kingdom and their significance to nature
    conservation. Proc. R. Soc. Lond. A, 339, 355-372.

DISCUSSION: PAPER 14

E. M. NICHOLSON   Professor Bradshaw's title was in my view
unsatisfactory, for the topics he dealt with
were much more important than just the restoration of vegetation on degraded lands. He
dealt with the much wider question of the
management of stocks of land temporarily under
use for mineral extraction. On any civilised
basis such land must of course be restored
for something: the process is simply a part
of the rational management of the total land
stock. It is clearly important that the
future use of the land is decided before it
is taken for extraction since this, in turn,
will determine the way in which the land is
treated and will constrain the actual operation
of exploitation. The techniques Professor
Bradshaw has described are, in fact, a step
towards strategic land use, minimising the area
left in bad condition and ensuring that in future
the land meets a new need: his paper is thus
about how we move from one form of land use to
another. Moreover, it is possible to experiment
with landscapes for example creating multiple
use patterns and combining mining, agriculture
and conservation. There is, in addition, a
scientific spin-off because these kinds of
practice are natural experiments from which we
can derive fundamental ecological knowledge.

A. D. BRADSHAW   Mr Nicholson has raised an important point
about what we mean when we speak of "restoration".
The ultimate option in many temperate environments is to re-establish a woodland wilderness -
but that would often be nonsensical as a practical
objective.   The kind of worked-out limestone
quarry we have studied, equally, could not easily
be "restored" to agriculture. But it can be
covered with a mosaic of rough grasslands and
scrub woodlands with a high potential recreational value and considerable attraction to
botanists and this could be invaluable in an
area like the Peak District where the adjacent
agricultural land is coming under very heavy
pressure from tourists.

M. W. HOLDGATE

Professor Bradshaw repeated a commonly
made statement that "prevention is better than
cure", but is this not an emotive view? It
may be cheaper to overcrop a resource if we have
time to let it recuperate, and if it will do so.
For example, the population of fur seal in the
southern hemisphere was devastated by the sealing
boom of 1780 - 1820 but is now substantially
recovered. Like shifting cultivations in
tropical zones, if the population or system is
not pushed out beyond certain limits and there
is time and space to allow it to be set aside
for recovery, it might be cheaper to extract
intensively for a short period and then allow
a natural curative process with or without
human assistance to take place.

May I suggest as a simple rule of thumb,
that prevention is better than cure if:
(1) the costs of prevention are low (i.e. in
economists' language, the control costs are
small and probably exceeded by the damage costs)

(2) the value of the site is high (i.e. the
damage costs would be high, thereby justifying
substantial control costs).

On the other hand, cure may be better
than prevention if:

(1) the damage costs are low (i.e. the land
is not valued as a resource) while the control
costs are high

(2) the control costs will be paid by some-
one else - including nature in the latter
category! If nature will remedy the situation
on an acceptable time scale this may be a
perfectly rational choice.

Finally, we should remember that
affluent societies tend to bid up the damage
costs by attaching higher values to amenity
while poor countries do the opposite: as a
result the same actual situation may lead to
a different answer in different economies and
social circumstances.

A. D. BRADSHAW   I agree that the primary need is to cost
what is being done and relate it to need.
Whether any remedial action at all is taken
must depend on the situation: for example
in Australia I have seen an extremely
intractable site high in pyrite, carbonate
and salinity and with an exceedingly low rainfall — since this site was five hundred miles
from anywhere it seemed reasonable to leave
it untreated. On the other hand, the quarry
site that I described in the Peak District
where the whole area is coming under extreme
recreational pressure is clearly a candidate
for restoration if the costs are low and the
result is to relieve that pressure.

E. M. NICHOLSON   In 1840 there was a landslide at Lyme
Regis on the English coast and it left an area
that looked like a bare quarry for some 60 years.
Today an ashwood, highly valued for its wild
life conservation value covers the slope: an
excellent example of a thoroughly satisfactory
cure wrought by nature. We must, however, be
cautious: we need to know more and the value
of these mineral sites for research must be
exploited. If we leave the recovery process
to nature, we make little use of this potential
value.

G. VAN DYNE   In his paper Dr Köpp spoke of converting
agricultural lands to forest and reversing this
process if the land subsequently was needed to
produce food. I would like to ask Professor
Bradshaw which treatment in his opinion restores
fertility most rapidly and gives the most
flexibility in terms of future options.

A. D. BRADSHAW   I think the important thing is to inject
nitrogen into the system at the highest possible
rate. On the least toxic of the three kinds of
site I described, the kaolin sand, where there
was no toxicity and less physical problems than
in a quarry site the fertility was limited by
the extremely low level of nitrogen. The best
way of accelerating nitrogen fixation is to
establish plants with root nodules. This seems
to be best carried out by herbaceous plants
but bushes like Ulex europaeus and for the
matter of that Lupinus carboreus could be

effective as well as attractive. Either would
be flexible in terms of future options. The
important point is to establish a vegetation
which ultimately produces a satisfactory store
of nutrients in the soil.

B. ULRICH

If the nutrient store in the soil is too
low, for example because of low levels of
organic matter, the nitrogen levels will also
be low and the cycle will not function. It is
hard to start such a cycle by simply injecting
inorganic nitrogen because it persists in the
soil only for a short period. It is thus best
to inject organic matter and raise the nitrogen
to a more constant level and this is best done
with a leguminous crop. The organic matter
should be built up as fast as possible and
this will cause a rapid increase in the rate
of the nitrogen and phosphorous cycles.

H. REGIER

Dr Holdgate's rule of thumb elaboration
of cost benefit analysis is insufficient for
the kind of decisions we have to take. When
information is quantified in money terms it
tends to drive out of circulation other highly
relevant data, as B. Gross, J. K. Galbraith and
other economists have noted. Cost benefit
analysis, because it selects information that
can conveniently be stated financially, tends
to cancel out long term ecological and social
values. It is useful to a limited extent where
a broader analysis is undertaken, especially
through a participatory process. As ecologists
we must stress that if these economic techniques are to be used they must neither be oversimplified nor over-extended or they will give
an unsatisfactory answer.

15. FORESTRY IN THE FEDERAL REPUBLIC OF GERMANY, WITH PARTICULAR
REFERENCE TO THE REHABILITATION OF FOREST ECOSYSTEMS

H. Köpp

Institut für Forstpolitik

Busgenweg 5, Federal Republic of Germany

INTRODUCTION

Forestry in Germany is a subject which affords a course of at least
4 years for the academically trained forester and some practical
and administrative under- and post-graduate training. Even then
the graduate forester is usually only engaged in either a rather
broad professional range of work in the field or some specialised
disciplines in teaching and research or administration within the
forest services. It is, therefore, almost impossible to write
about forestry in Germany without limiting the scope. Particular
reference will thus be made to the rehabilitation of forest ecosystems. Rehabilitation of forest ecosystems once was both
"mother and father" of forestry in Germany (Kopp, 1975a).

Before details are given of various reasons for and different
ways of forest rehabilitation work, however, an outline on the
historical development of forestry in Germany and a description
of the present situation seems to be essential. There is hardly
any other discipline in the field of applied natural sciences which
really needs the understanding of its historical origin and development in order to objectively understand its situation. Present
criticism usually lacks this knowledge and is thus only of limited
use (Köpp, 1973).

HISTORICAL REVIEW

Under natural conditions more than 95 per cent of the total area
of Germany would be under forest cover. Exceptions would include
high mountain regions in the south, the coastlines and some islands
in the north, some moorlands, steep slopes and the inland waters.
Following human population growth, the forest area decreased
steadily, with some minor oscillating to much less than the present
percentage of 29 per cent. Historical records and well-known
paintings give a terrifying impression of how the country looked
2 centuries ago (Mitscherlich, 1974).

Overuse of timber, overgrazing of livestock and other usages
must have created the fear of a severe wood shortage similar to,
or even greater, than the recent threat of a worldwide energy shortage. Forest ecosystems were destroyed and depleted, in many
cases beyond their capacity to regenerate themselves.

This situation led to the development of forest legislation.
The first acts of forest legislation can be traced back to the
second half of the 15th century. Between 1550 and 1850 the then
roughly 350 territorial states in Germany passed almost countless
forest laws and regulations. They often regulated even the
smallest details of forest management and also of the timber market and wood consumption. Plochmann (1974) cites a few examples:
the time for felling trees, for instance, was limited to short
winter periods with a crescent moon, since that phase of the moon
was believed to enhance the durability of wood; the stump height
in felling operations had to be kept as low as 10 cm. in order to
save timber; and details of wood consumption were regulated even
to such matters as the width of stove openings in order to save
firewood.

This early development of forest legislation, frequently
amended, contained 3 elements which continuously remained in
effect:

(a)   the obligation to replant after cutting,

(b)   the right of the state to authorise any clearing of forest
      land, and

(c)   the right of the state to impose special rules for the
      treatment of protection forests.

The period of centrally planned and organised forest management came to an end with the beginning of liberalism and the
capitalistic theory of the economy. It was not until the middle
of the 19th century that those new ideas began to influence forest

legislation and management in Germany. A typical example of a
liberal forest law, as Plochmann said, resulting in liberal
management, was the Bavarian Forest Law of 1852, which remained
in force for 113 years. Main aims were to relieve forest owners
as far as possible of limitations and rules laid down during the
previous 400 years and to protect private forests against encroachments from outside. The free decision of the forest owner in
managing his estate was only restricted by the obligation to re-
forest after cutting. State permission for clearing of forests
and fixed rules connected with protection forests were the only
remaining limitations.

These liberal forest laws worked well and satisfactorily as
long as the social functions of the forests were merely considered
to be by-products of timber production and their value was not
particularly appreciated (Plochmann, 1974). Up to very recently
some of those liberal ideas were still apparent in the forest
legislation particularly in the northern states. One of the results
of this period was a large scale conversion of mixed hardwoods into
coniferous stands, the usage of fast growing species including
several exotics.

PRESENT SITUATION

The present situation, to quote Plochmann (1974) again, can be
characterised by:

1. different regional changes in the proportion of forest
    cover,

2. rapidly growing public demand to use forest land and
    its products.

3. rapidly declining rate of net income in forestry.

1. Regional Changes - 7.3 m ha of forests in the Federal Republic
of Germany cover 29% of the total land area with extreme regional
differences. There are counties which have almost 80% under
forests, for instance in the Black Forest or in the Bavarian
Forest. There are others with only 4% or less, for example near
the North Sea and the Baltic coasts.

Although the daily losses of open country (agriculture and
forestry) to building and road construction, to industry and
other developments exceeded 100 ha in recent years, the forest
area has grown slowly but steadily during the past 150 years, and
particularly recently. The problem, however, is that in areas of

high population concentration where forests are urgently needed
for various social reasons, there were and still are heavy losses,
sometimes up to 80 per cent within the last 100 years, with an
exponentially accelerating trend. Forest land, particularly that
in public ownership, is often still considered as a "cheap land
reserve" for development.

On the other hand there is an alarming increase of forest and
potential forest land resulting from structural changes in agriculture and agricultural policy in the Common Market countries, and
the afforestation of marginal and submarginal agricultural land.
Unfortunately this lies mainly in hilly and mountainous areas which
already have a forest cover well above average and often a weak
economic infrastructure and few people. This increases the undesirable impact on the landscape, because the afforestation of open
valleys and of pasture land would basically change the character
of the countryside formed and developed by more than 1,000 years
of land use, and it would obviously lower the scenic and aesthetic
values. It is sometimes difficult to estimate and evaluate the
ecological consequences of the changes in the relationship of open
country and forestry. There are, therefore, various small and
large-scale research projects monitoring this development and
trying to give advice both to public authorities as well as private
owners.

In 1968 the then High Commissioner for Agriculture of the
European Communities in Brussels, Sicco Mansholt, estimated the
total area of abandoned agricultural land as some 6 m ha of which
about 80 per cent were to be afforested in the member states,
which meant a large scale rehabilitation of forest ecosystems, often
after many centuries of agricultural land use (Speer, 1955). At
that time the figure was believed to be far too high, but it is
now clear that Mansholt's data were not over- but under-estimates.
This trend can still be seen in Germany and other European countries.

2. Public Demands - The majority of forest lands in Germany serve
a multi-purpose function. They are managed to give local priority
either to one or two functions or to a compromise, establishing a
"well-balanced equilibrium". Conflicts are often unavoidable.

The total population is about sixty million inhabitants, which
means about 0.12 ha of woodland per capita. There are only about
5 ha per 100 inhabitants in the North and North West, but 24 ha
in Bavaria. The federal average with 12 to 13 ha per 100 inhabitants
is significantly lower than the equivalent European figure of 32 ha
and is closer to the position in the USA (Köpp, 1971).

If one considers day and week-end recreation to cover about
50 km around urban centres, and if one draws circles with this
radius around cities with 25,000 inhabitants or more, 94 per cent
of the area of the country would be covered. If one assumes that
a new Autobahn opens up 30 km of countryside along its route on
both sides, and if one draws lines alongside those 30 km fringes,
relatively little of the country would still be free from this
influence. In fact there are figures indicating that forests close
to urban centres are used for leisure and recreation by an average
of up to 1,000 persons per ha on peak summer week-end days. This
intensity requires specialised recreation management. This situation is also the main reason for the "unhappiness" amongst German
foresters about new concepts of urban forestry in North America.
Very little of German forestry would not be covered by that concept.

It has clearly been shown that forest recreation occupies a
predominant and continuously increasing role in the recreational
behaviour of the German population. Many research projects are
based on the increasing public pressure in the field of recreation and leisure and its consequences (Tocher and Köpp, 1973).

Another example of fast growing demand may be water, closely
related to forestry in Europe, with a present average consumption
of 120:1 per head of population in Western Europe, a figure which
is estimated will rise to 200:1 in less than 25 years time, an
increase of 80 per cent. A large amount of this water in Germany
comes from reservoirs in forest areas (Köpp, 1975c).

There are many other examples of rapidly increasing demands
towards forestry, for instance the protection against noise and
other emissions. Forestry is continuously trying to meet those
by restrictions in the choice of tree species, by not using
existing technical means of timber harvest, transportation etc.
Speidel (1975) has made it clear that these have an influence on
both the quality and quantity of timber production and thus on
costs of forest operations. He strongly rejects the common belief,
particularly on behalf of private forest enterprise, that forestry
can automatically meet all those demands without receiving
compensation.

3. Declining Income - Forestry in Germany was never a "big
business", although the state of Prussia once received its major
revenue through marketing its timber. Unlike Scandinavia and
North America, the combination of land ownership with forestry
industry did not develop for historical reasons. German forestry,
with few exceptions, does not share in the processing of its raw
material. Moreover, it is exposed to heavy competition on an
internationally free and open timber market. Mechanisation is

handicapped by a high diversification of forest sites, stand
composition and ownership structure. There is also a growing
public pressure against everything that normally goes with man-
made forests, such as monocultures, clear fellings, use of pesticides, and use of heavy machinery.

Timber prices have not risen much beyond the 1955 level, but
labour costs have. At the beginning of 1973 they were about eight
times, and total management costs about three times as high as
1955. This explains why many private forest owners do not make
profit. All State Forest Services are also losing money. Many
and in the first instance the small forest owners, have discontinued
management operations. Many more just harvest mature stands, without investing for silvicultural management of young stands, road
maintenance, erosion control or other operations. In spite of
this critical situation relatively few try to sell their land, one
of the results of monetary inflation but even more of attitude.

As a result of various regional and global studies, for
example by FAO, the main features of the world forestry situation as
it is likely to be between now and 1985 are fairly well defined.
Kalkinnen (1974), has produced one of the more recent studies on
the future aspects of timber production and consumption in Europe.
The dynamism which has characterised the demand for processed
forest products over the past decade or so is likely to continue.
The share of world consumption of processed forest products by
the developing countries will rise appreciably. For this and
other reasons Henry Kissinger has just repeated the demand for new
forests to be planted in the Sahel Zone of Africa.

However, the great bulk of the increase in world consumption
will continue to take place in the more advanced countries. Supply
problems which are already limiting production in those countries
will become even more severe. From this point of view there is no
doubt that timber production in Germany, in Europe and abroad
should be increased both through the afforestation of marginal
agricultural land and through the intensified utilisation of
existing forests (Hummel, 1974).

The regional studies have also clearly shown certain areas
where a concentration of efforts could yield worthwhile results,
e.g. increased utilisation, establishment of domestic processing
facilities, harmonisation of forest and forest industry development etc. According to a recent joint ECE/FAO-study made by
Eckmullner (1975), it is unlikely, for a least a medium term
period that one will face a crucial shortage of timber and wood
products, as others have predicted, but there are and will be
bottlenecks. The problem of supplying, distributing, transporting
and marketing the resource and its various products, however, at

what one calls "a reasonable price" may become a serious one.

4. Ownership - Ownership has been touched upon briefly. Some
more data are necessary, because ownership is indeed a major problem
of forest policy. It shows a wide diversity in the 9 member states
of the European Communities and also within Germany. Private
owners account for about 60 per cent of the total forest area in
Europe, municipalities and similar public bodies, such as the
churches, for 25 and the state for only 15 per cent. Private
forests comprise more than half of the forests in all member
states except in Germany where the figure is 44 and in Ireland
where it is only 12 per cent. The percentages for the publicly
owned forests vary widely too, not only from state to state but
also within some countries. In Ireland nearly all forests belong
to the state, in Italy virtually none; Britain and Ireland have
almost no forests owned by local communities, while France, Italy
and Germany have each about 2 m ha.

The private woodlands are divided among some 3.5 m owners in
Europe, less than 50,000 of whom own more than 50 ha; in Germany
more than 75 per cent of 700,000 owners own less than 20 ha,
many of those less than 2 ha (Kopp, 1975b). The average holding
is only about 4 ha, and those are often divided into separate lots.
More than 250,000 lots are smaller than 0.4 ha. The disadvantages
of this fragmentation are self-evident, especially for intensive
silviculture, mechanised harvesting, and profitable marketing.
Attempts at improving this situation, dating back some 30 years,
were not very successful (Plochmann, 1974).

Hummel (1975) has summarised this combination of circumstances,
which makes efficient woodland management difficult, facilitates
integration of farming and forestry, contributes to a varied landscape, and leads to a resistance to change, because to most owners,
the woodlands are more than just future logs and money, more than
a well established landscape including vital forest ecosystems;
they are part of a way of life.

REHABILITATION OF FOREST ECOSYSTEMS

Efforts at rehabilitation of forest ecosystems, based on the background information given above, can thus be divided into a long
term historical development, following structural changes in land
use, changing and growing demands for certain resources and social
needs, and a short term one following the same criteria and the
need for improving the environment, particularly in densely
populated areas. In principle, there are no differences between
these two. They are based on economic and on social terms.

Practically, however, the impact and scale of modern interferences
are much larger. Much more knowledge is available of what can
be done in technical terms, and, of course, schemes of rehabilitation of forest ecosystems as an integral part of regional
planning and landscape management are planned ahead of the removal
and destruction of existing forests (Federal Republic of Germany,
1967, 1972; Erz, 1970).

Unlike the situation in the U.S.A. the long term changes from
forestry to agricultural land use, went on slowly in Germany
throughout its historical development, but accelerated during the
last war and immediate post-war period. Food production became the
vital aim. However, during the emergency situations the need for
timber and other forest products at the same time generally led to
a stricter safeguarding of existing forests. Vital food production
without sacrificing forest resources became a first priority. It
is interesting to learn, that even during the later years of world
war II the need for the intangible values of forests and woodlands
and their multipurpose functions were clearly integrated into the
planning and management concepts of the forest service, especially
in the Berlin area (Hasel, 1975).

Under present conditions of an expanding total forest area
and taking into consideration a regional and world wide growing
demand for timber, the question which arises in establishing new
forests is to find out which species, mixtures, compositions etc.
will lead to forest ecosystems that can meet all the different
demands both in the economic and the social sphere, and which
are more resistant against diseases, pests and other threats. The
planning process in establishing those forests has become a very
complicated task.

To a much lesser but regionally and locally not minor degree
one has to maintain certain minimum areas of open country (Zundel
1973). They may differ in size mainly depending on unwanted changes
of the countryside, for instance in the field of farming, climate,
leisure and recreation, or the visual amenities. Contrary to former
times, the public shows interest and participates in this planning
sector.

Most of the present forest operations are in fact rehabilitation efforts, that is reafforestation of areas which naturally were
once forests or replacing man-made forests by other ones after a
thorough investigation of soil, site, climate, economic factors
and the multitude of present demands and those likely to occur in
the foreseeable future.

1. Rehabilitation following Excavation - Specific rehabilitation
schemes are connected with small and large scale excavation work,
mining, and the sand and gravel industry. In this particular field
Germany has a rather long experience. Classical examples can be
found in the Ruhr industrial region, around Cologne and Aachen in
Northrhine-Westphalia with its huge open cast mining and to a
smaller extent in the area between Brunswick and Helmstedt in the
state of Lower Saxony.

Many international conferences, meetings and symposia have
been concerned with specific problems and technical details in
this field (e.g. the Ruhr Workshop, 1972). Sooner or later the
situation may arise in rehabilitation schemes where there is the
background knowledge, modern techniques, and an abundance of
experience, both negative and positive to show that a project is
feasible and funds are available, but should it be done? What are
the total costs, are there limits for investments, what are the
returns for society? The answers are not easy, particularly in
times of economic recession. They may differ from one example to
the other, but they will have to be stated. - The question of
whether marginal agricultural land should be converted into
forest land in Germany surely received a different answer 10 or 20
years ago.

2. Rehabilitation following Emmissions - Another sector of
rehabilitation of forest ecosystems, particularly in industrial
regions, is the work against pollution, i.e. emmissions into the
air. This normally leads to a replacement of coniferous trees
by hardwoods and here again to a further limitation in the choice
of species. Information in this field is also available. Specific
problems, as for instance compensation to private owners within
one state or even beyond its boundaries, are very difficult and
important, but there is no room to discuss them in this paper.
Recent legislation tried to tackle this.

3. Rehabilitation following Changes in the Ground Water Level -
Forest ecosystems suffer further damage through the lowering of
the ground water level. Rehabilitation schemes can, to a limited
extent by changing species, reach and try to overcome those
difficulties. Normally the consequences not only lead to quite
different ecosystems, but also result in forest stands of lower
economic value. Classical examples were those of the upper Rhine
valley, following major engineering projects in Germany and France.
But there are others almost everywhere throughout the country.

4.   Rehabilitation following Road construction and related
Engineering Projects – New roads, railroads and motorways, airports,
canals and many other infrastructural measures not only require
forest land, but influence the remaining forest stands quite considerably. To a limited extent again, rehabilitation schemes can
try at least to lessen the impact.

5.   Rehabilitation following Recreational Pressures – Another
specific case is over-use from leisure and recreation. There
are only a few examples in Germany as yet, where forest ecosystems
suffer badly from recreational over-use. Normally a forest area
can absorb more pressure without sustained disturbances than open
country. However, certain limits do exist. It is very difficult
indeed to determine those limits in advance. Little has been done
so far in Germany to actually restrict visitor numbers in forests,
although certain reserves are not open to the public. More
research work in connection with carrying capacity is going on in
North America. It is obviously essential to watch these trends
carefully, particularly because modern forestry and conservation
legislation opens up private and public forest areas to the general
public for recreation. It is too early to comment on the consequences.

     Bijkerk (1975) has recently pointed out the necessity of a
classification of forests according to their recreational potential,
which – as he said – will prove to be of great benefit to planning
the possibility of increased recreational use which is very likely
in the future.

6.   Rehabilitation following Wildlife Damage – Intensive wildlife
damage to trees and shrubs in forest stands, particularly from
Red deer, locally from Roe deer, rabbits or grey squirrels in the
U.K. should be mentioned without going into details. There are
countries where wildlife damage is of major concern and difficult
to tackle; for instance in New Zealand. Concern is growing in
Germany.

7.   Rehabilitation following Windblow – There are further major
disturbances of forest ecosystems which Germany has experienced
recently. Severe windblow in forest stands happened in the winter
of 1967, particularly in Bavaria, and in November 1972 hitting the
northern parts of Germany, and also Holland, Denmark, and GDR and
Poland.

     Even with the help of modern heavy machinery mainly from
Scandinavia and Canada, and with extra labour forces from Sweden,

Yugoslavia and Austria, it will take about 10 years to re-establish
proper forests in areas damaged by catastrophes like the latter one,
which destroyed more than 100,000 ha of forest land and blew down
almost 17 m m^3 within a few hours (Köpp, 1973).   Another storm in
January 1976 has accelerated the review of afforestation plans,
the choice of species etc.

8.   Rehabilitation following Forest Fires - Last not least, major
forest fires occurred in Northern Germany in August 1975.   In
terms of U.S. forest fires, or those in Southern France, Australia
and some other countries, 8,000 hectares of destroyed pine forest
is not unusual.   However, in German terms, it was the biggest
forest fire ever recorded in this part of the country including
war and post-war years.   Again reafforestation will take many
more years and several of the ecological consequences are as yet
not or hardly known in Germany.

Despite major achievements in national and even international
research programmes like the Solling-Project, the main German
contribution to the International Biological Programme (IBP), and
similar forest ecosystem research projects in many other countries
in Europe and throughout the world, including UNESCO's Man and the
Biosphere Programme (MAB), and the WWF/IUCN efforts towards the
tropical forests, there are still many open questions (Ellenberg,
1971; Pesson, 1974; UNESCO, 1971).

CONCLUDING REMARKS

"An astronaut orbiting the earth a century ago would have looked
down upon a globe encircled by a broad belt of green broken only
by the oceans.   This belt was the product of over 50 million years
of evolution.   Photographs taken by astronauts today show that the
belt is now broken, fragmented, and greatly reduced in area.   The
last of the world's great ecosystems is being torn apart by man
as his numbers and demands for new products increase and he seeks
to feed and otherwise support the population" (WWF, 1975).

This recent statement made by the International Union for
Conservation of Nature and Natural Resources (IUCN) in the Conservation Programme 1975-76 of the World Wildlife Fund (WWF), introduces and reflects the growing efforts to save the tropical rain
forest.   This statement can also be applied to certain regions of
the temperate zones, although there are obviously major differences
in time and scale and also in man's present efforts to re-establish
forest ecosystems (Köpp, 1975b).

An astronaut or normal traveller within Europe some 100 years
ago would have looked down upon the British Isles, West Germany
and many other countries on the continent dominated by green pastures,
fields with individual trees or groups of trees, forests and woodlands; and even in our day he would still get the impression of
looking down upon a countryside which does not show major disturbances, despite some concentrated local ones. This is still true
for at least 85 per cent of the Federal Republic or even Europe's
total area, despite built-up areas, urban and industrial conglom-
erations and all their infrastructure.

Almost certainly the present traveller in his comfortable
city-jet does not recognise that even within the green colour he
sees some 10,000m below, there are major environmental disturbances.
Perhaps he also does not know that more than 50 per cent of the
developed areas were taken from the countryside, i.e. from agriculture and forestry within a single generation of his own life-
time. And almost certainly he does not know the average figure of
daily losses to urbanisation and development. Should he then happen
to read a newspaper or the proceedings of this conference which
refers to the present overall, and regionally sometimes extremely
undesirable increase of forest land in Germany, he would not only
misunderstand our worries - against the background of a raw
material shortage - but be confused.

Rehabilitation work in forest ecosystems has a lot to offer
both for the economic and social and also for the cultural development. Sometimes there seem to be unlimited techniques. However,
it is not the forest scientist or expert in the field who sets
the limit, but the treasuries, as we all know.

REFERENCES

Bijkerk, C. 1975. Recreation values of forests and parks. -
    Phil. Trans. R. Soc. B, 271, 187-198.

Eckmüllner, O. 1975. Cost and price statistics relating to
    forest products. - ECE/FAO study (joint working party on
    forest products statistics, 10th session).

Ellenberg, H. 1971. Integrated experimental ecology: methods and
    results of ecosystem research in the German Solling Project,
    Berlin, New York: Springer.

Erz, W. 1970. Nature conservation and landscape management in
    the Federal Republic of Germany (English translation by
    H. Köpp). Bonn. 1970.

Federal Republic of Germany, 1967. Forestry and wood economy of
    the Federal Republic of Germany. 5th Edition. Bonn.

Federal Republic of Germany, 1972. Report of the Federal Republic
    of Germany on the human environment (prepared for the U.N.
    Conference, Stockholm 1972). Bonn.

Hasel, K. 1975. Die Berliner Wälder (Gutachten im Auftrage
    des Berliner Senats). Limited distribution. Berlin.

Hummel, F. C. 1974. Probleme der Holzversorgung, Holzverwertung
    und Aufforstung in der EWG. - A. G. Deutscher Waldbesitzerver-
    bande e. V., Bonn.

Hummel, F. C. 1975. Forest policy at Common Market level, -
    Brussels.

Kalkinnen, E, 1974. Future aspects of timber production and
    consumption in Europe. Geneva: FAO.

Köpp, H. 1971. Forest recreation and conservation in Germany
    (paper presented to the 7th International Seminar on National
    Parks). Limited distribution. U.S. National Park Service,
    Washington, D.C.

Köpp, H. 1973a. Documentation of the 1972 storm damage in Germany.
    Unser Wald, 1, 4-7, 15-16.

Köpp, H, 1973. Forestry, wildlife conservation and amenity
    (paper presented to the Council of Europe's international course
    on ecology applied to land use). Strasbourg, London: Council
    of Europe.

Köpp, H, 1975. Energy resources and societal needs: future stra-
    tegies and alternative futures - resource management in forestry
    (paper presented to Session 162 of the Salzburg Seminar in
    American Studies). Limited distribution. Salzburg.

Köpp, H. 1975a. Trees, woodlands and countryside - a contribution
    towards Europe's architectural heritage year. Notes from
    Europe, 3, Hamburg 1975.

Köpp, H. 1975b. Forestry policy in a European perspective
    (paper presented to the Conference on Forest Management of the
    Royal Society of Arts). RSA Report 76-88. London.

Mitscherlich, G. 1974. Vom Nutzen des Waldes in Vergangenheit
    unds Gegenwart. - Biol. unserer Zeit, 4, (2).

Pesson, P. 1974. Ecologie forestiere. Paris, Bruxelles, Montreal.

Plochmann, R. 1974. Trends and aspects of forest legislation in
    Germany. - J. For. 4, 202-207.

Rheinische Braunkohlenwerke. 1975. Wo neue Wälder wachsen.
    Forstliche Rekultivierung der Rheinischen Braunkohlenwerke
    A.G. 5th Edition. Cologne.

Ruhr Workshop 1974. Green colliery spoil banks in the Rhur
    (proceedings of an International Special Congress on Mine
    spoilheaps in the Ruhr and their integration in the landscape,
    1972). Essen: Ruhr kohlen bezirk.

Speer, J. 1955. Wiederaufforstung von Kohlflachen in England
    AID, 80. 28 pages. Bad Godesberg.

Speidel, G. 1975. Forstpolitische Leitlinie 1975. Forstarchiv,
    46, 133-137.

Tocher,R and Köpp, H. 1973. People and forests - the challenge
    of forest recreation. Oxford University For. Soc. Journal, 7,
    (3), 5-17.

UNESCO 1971. Productivity of forest ecosystems: proceedings of the
    Brussels symposium organised by UNESCO and the International
    Biological Programme, 1969, edited by P. Duvigneaud. Paris:
    UNESCO (Ecology and Conservation, 4.)

W.W.F. 1975. WWF Programme 1975-76. Morges: World Wildlife Fund.

Zundel, R. 1973. Wald, Mensch, Umwelt. Mitt. bad. forstl. VersAnst.,
    52.

DISCUSSION: PAPER 15

O. BOERSET   Dr. Köpp mentioned clear fellings and
                    storm damage: was there any problem over re-
                    afforestation? Are there any political
                    objections to the re-afforestation of
                    agricultural land. In Norway such removals of
                    land from cultivation are opposed. Finally,
                    Dr. Köpp stated that mixed plantings were
                    replacing pure spruce - is this a trend in
                    Germany?

H. KÖPP

The clear fellings themselves cause no re-afforestation problems. The soil on the site was not disturbed and the ground vegetation remained. The whole process of replanting after World War II was quite rapid.

So far the areas damaged by storm have not completely been reforested. Labour did present a problem: much of the heavy machinery and also labour force came from Scandinavia after the 1972 storm.

As a result of increasing knowledge of soil and site potential more and more areas that formerly only supported coniferous trees are being replanted with mixtures including hard-woods. This is especially so in state forests, it is less easy to persuade private forest owners to do so because of costs involved.

B. ULRICH

All Government forests have been mapped and their site and soil potential assessed so that we can establish forest types that use the real potential of the environment. In the north of the country, where pine was planted extensively in the last century, Douglas fir and oak may now be used, Scots pine being restricted to areas of poor soil. Our aim is to create mixed forests and mosaics that correspond to the natural variability of the environment.

H. KÖPP

Dr. Boerset asked a question about the extension of forests onto agricultural land. Mansholt estimated in the late sixties that there would be some six million hectares of land in Europe that had been or would be abandoned by agriculture and transferred to other uses. At the time this figure was thought by many to be a sweeping over-estimate: it now appears an underestimate and the true figure may be as much as 15 million hectares. Much of this land - may be up to 80% regionally, can be expected to be transferred to forests.

Economically it is desirable to establish more forests in the European Economic Community as all member countries rely on imported timber. Some areas are, however, already 80%-90% forested and there are doubts about planting the remaining patches of agricultural land here. From many aspects it is important to retain the landscape pattern and diversity. Public attitudes are highly emotional and inevitably colour such a debate.

May I make one final comment: mixed age and mixed species forests are a general goal so far as the German forest services are concerned.

J. N. R. JEFFERS

I have two questions: first how much of the forest land in Germany is in private hands rather than state ownership, and second how far it is possible to sustain land managed for hard wood production despite the economic models that suggest that this is a bad form of land use?

Forty per cent. of the land is in private hands: the remainder is either state owned (about 30%) or owned by communities (for example, towns, villages, churches and ancient co-operatives). The pattern is thus very mixed.

H. KÖPP

The German approach is not based only on economic models. I showed a slide of a famous oak stand which is a tremendous investment and which many economists would argue we can no longer afford to retain unharvested - but oak of this quality, suitable for veneer, has a very high value. Even now our aim is to plant more oak harvestable only in some 250 years despite the high cost of this form of forestry. We cannot forecast market conditions two centuries ahead, but partly for this reason we consider that one of our aims should be to sustain a potential diversity in forest yield.

G. VAN DYNE

The problem of converting agricultural land to forest raises a general point. It is confidently predicted that world population

will double and food needs increase by 4 –
yet the EEC is already a food importer.
Surely, the situation is far more likely to
demand the transfer of land from forests to
agriculture within 30 years or so rather
than the other way about?

H. KÖPP

This is an important point and has been
much debated. The fact is that at present
the EEC has certain food surpluses rather
than shortages. A second point is that
experiment shows it is quite easy to convert
forests into productive agriculture. If,
therefore, by twenty or thirty years hence
a new need for food production arises – and
it may not in Europe – that need could be met
fairly quickly. Meantime, it is surely better
to have the land in forest than lying derelict.

J. BALFOUR

Increased agricultural production is
related to the size and structure of land
holdings. For example the average size of a
West German agricultural holding is about 4
hectares in contrast to the United Kingdom
average of 110 hectares. Would it not follow
that there was a greater potential for
agricultural efficiency in Germany if the
pattern of land holding was changed?

H. KÖPP

The problem is being partly solved
because small farmers are leaving their holdings. On areas of good land, rationalisation
is easy because the adjacent farmers will
gladly take on the land thus released. There
are also co-operatives for farmers i.e. for
marketing their products. However, it is
correct that there is still a great difference
between the U.K. and Germany in the average
size of holdings.

N. POLUNIN

Once again in this conference, as in so
many others, we have heard a projection of
population trends spoken of as a confident
prediction that world population will double
within the next 25 years or so. This
projection may be dangerous. On the other
hand, confident assertion that these mounting
human numbers can be sustained may lull people

into false security while on the other the
projection may raise fears of some form of
population crash. Many of us believe that
this projected doubling will not happen or
even be approached — and that nature herself
will intervene if man does not. Let us
therefore be very cautious in what we say
about such projections.

H. KÖPP   The population in the Federal Republic
of Germany, as in many other European
countries, is actually stabilising, if not
declining. As a result there are unused capacities in our environment, for instance in
the social field. This may not be the
pattern in other parts of the world: I agree
with you that there is a need to plan for
realistic forecasts. I believe that the
projections of a world population doubling by
about 2,000 A.D. are in part based on
the present age structure of the world
population and even if the present generation
under 15 years of age reproduce at no more
than replacement rate a substantial increase
is inevitable.

E. M. NICHOLSON   I would simply like to point out that in
the United Kingdom the most recent figures
indicate a net reduction in population for the
first time in many years.

16. PLANT GENETIC RESOURCE CONSERVATION AND ECOSYSTEM

REHABILITATION

J. Heslop-Harrison and G. Lucas

Royal Botanic Gardens

Kew, England

RELEVANCE OF PREVIOUS EXPERIENCE IN THE UK

Despite its beauty, the rich rolling countryside of southern England
with its patchwork of fields and woods is in fact a rehabilitated,
once severely damaged forest landscape. The present acceptable and
beautiful scene is synthetic and some of its most striking components were created in the minds of a small number of eighteenth
century landscape architects like "Capability" Brown who, when
asked what he was doing planting so many trees and then surrounding
them with oak palisades, instructed the questioner to come back and
see in a hundred years. Today, 200 years later, the overall pattern
remains to please the eye.

From this experience, two important lessons can be learned.
One is the time scale - unlike governments who work to a three,
five, or ten year plan the Conservationist has to think in terms of
centuries. It is the long-term maintenance of genetic resources
that is of vital importance to man's future. The second lesson is
the need for diversity. Many species of trees and shrubs from many
sources were tried out by British 'landscape improvers' for whom the
various landscapes were living long-term experiments and when species
proved unsuited to a habitat they were replaced by others. For
example, selections were made for fastigiate or weeping habit depending on the position they were to fulfill in the landscape pattern,
so that diversity was maintained almost unconsciously.

Only a small portion of Britain (or any other country) has been
reconstructed deliberately for its beauty. In most areas new
ecosystems developed in place of the forests, under the influence

of progressively evolving agriculture. In some areas early
industrialisation caused degradation and when the degrading
influence was removed natural rehabilitation took place slowly but
surely.

The Severn valley provides an interesting example of an area
once under industrialisation which, for now approaching a century,
has been allowed to recover naturally following the transfer of
most of the operations elsewhere. The earliest coke-fired iron
smelting processes began in the valley with the opening of Abraham
Darby's first blast furnace in 1630, and prints from the late 18th
century and early 19th century show that much of the woodland, even
on the steepest valley sides, was devastated, with heavy fall-out
of pollution from kilns and furnaces, all of which were coal or
coke fired. The recovery here has been almost entirely a natural
process, with the seeding in of native species from neighbouring
woodlands which were not affected by the industrial operations
in the river valley. Although there has been no systematic study
as yet of the present day ecology of the old industrial sites, it
is clear that there has been strong selection in favour of certain
rapid colonists among the arboreal flora. The regenerating woodland has a species composition both smaller in number and different
in kind from the neighbouring residues of the old mixed deciduous
forest characteristic of this part of England.

The scale of this early devastation was such that plant colonists could heal the damage by invasion from the edges of the devastated region. It is because of today's massive scale of activities
that we have to provide generous support to rehabilitate some of
our major industrialised sites. The planned rehabilitation of
industrial wastelands in the UK is a comparatively recent activity.
Examples include efforts to utilise mine spoil heaps, power station
ash dumps and the wastes from quarrying operations. Much of the
work has involved re-grading, flattening and dispersing the wastes
and modifying the hydrography, all basically engineering activities.
Biological studies have included attempts to identify suitable
species for establishing pioneer soil binding, water-retaining and
humus-creating communities. In several instances success has been
achieved using races of native species seeded as pioneers with no
subsequent introductions, the succession thereafter depending on
what was available from natural seedings from natural communities
in the vicinity. Considerable success can be claimed in providing
semi-natural environments in the vicinity of industrial cities and
it may be that this should always be an aim in heavily populated
countries like the UK. But beyond this, encouragingly enough,
relatively long-term experience of some sites in Britain shows that
self-maintaining communities can be restarted readily enough and
can quickly become a refuge for the native fauna and flora of the
area.

THE REHABILITATION OF VEGETATION
AND THE CONSERVATION REQUIREMENT

Now notwithstanding the modest success that can be shown it is
clear that much more can be done to improve the rate of environmental restoration and diversify the flora and fauna of the restored
areas. It must be emphasised that the last aim is not necessarily
linked with the provision of amenity. For the average user of
recovered industrial lands in England the diversity of the flora
and fauna is never likely to be more than a minor factor, and it
may well be that landscape planners, local authorities and governmental agencies will take a considerable amount of convincing that
this conservation aspect of land rehabilitation is an important one.
If it is established as one objective of rehabilitation, then a
range of problems, both practical and scientific, arises immediately.
Decisions are needed as to whether the aim should be to attempt to
simulate the natural plant associations and their linked fauna
known to have existed in the area prior to the devastation, or whether the recovered areas should be looked upon as repositories for
the native biota in a more general sense. It is possible to take a
purist line here and argue that re-synthesis should not involve the
transfer of species not previously known in the area; but it is
evident now that this line is largely indefensible since natural
migrations have constantly brought about re-distribution of species
throughout the period of human occupation in a country like Britain.
A classic example is provided by the Norfolk Broads, created as
mediaeval peat-diggings and now supporting a rich aquatic flora for
which there are few natural habitats in the neighbourhood. It would
seem perfectly acceptable to create further refuges for the diminishing wetland flora and fauna of Britain by using the opportunities
available in areas like clay and gravel pits, even though this might
involve the introduction of species not now present in, or even
previously known from the particular area. The reasonable limit
might be the point at which introductions from outside of the country
came into question.

      In a country like the U.K., landscape restoration and the
rehabilitation of industrial wastelands might therefore well be used
as an opportunity for creating conservation centres into which
diminishing flora and fauna of the country could deliberately be
introduced. There seems no real ethical or other reason why this
should not be done, provided that appropriate records are maintained.
The alternative could be complete loss and it is surely illogical
to argue that this should be accepted because of some principle that
the distributions of flora and fauna within the country should not
be tinkered with deliberately when those distributions are in fact
very substantially the outcome of man's interference with the
habitat. Such a policy need not involve introductions to 'natural
areas' such as sites of Special Scientific Interest whose present

biological patterns have their own value. Such areas are often
habitats for the rarer plants and animals, and a main aim should
be to safeguard these species by appropriate habitat management and
if necessary special measures like artificial breeding to increase
their populations.

GENETIC RESOURCES AND REHABILITATION

If it be accepted that rehabilitation areas are to be treated as
refuges for diminishing biota, then the problem of the sources to
be used for introductions comes into sharp focus. Again, taking
the U.K. as a model, one would normally suppose that, wherever
feasible, genotypes from the British Isles should be used, and that
from within this area the maximum possible diversity should be
sought. This certainly defines a task which has not yet been taken
seriously, and up to now there is no basis to evaluate the scale of
effort that might be required.

In the last few years there have been attempts to ensure that
planting programmes for recovered areas in the country make use of
native species - for example in the planting of motorway verges and
recovered areas such as Cannock Chase. Some seedsmen and nurserymen in the U.K. already offer natural species for this kind of use -
thus native seed mixtures are available for road verge plantings
but the full implications have not been studied, and it is already
apparent that there are considerable risks in what has been done.
Nurserymen offering woody species native to Britain frequently do
not distinguish provenances sufficiently well, so that stocks
originating outside the British Isles are being freely introduced.
Moreover, the propagation policy of many nurserymen is likely to
produce an undesirable degree of genetic uniformity, since the same
individual is frequently used as the source for mass propagation.
Commercial considerations will frequently demand that parents should
be selected for ease of propagation and early establishment, and the
hazards of this are self-evident. One need only quote the consequence
of elm planting programmes of last century.

It may be necessary to mount a substantial educational programme to ensure that local authorities and others responsible for
planting understand thoroughly the need for genetic diversity in
what is planted and accept the additional initial costs that this
may entail. Nurserymen for their part must ultimately come to
accept the need for heterogeneity in their own sources. Agricultural
and horticultural monoculture techniques have shown by the rapid
destruction of whole crops through disease and pest attacks, the
problems that can arise if genetic variability is not maintained
and options kept open.

The importance of reserves of natural or semi-natural vegetation cannot be over-estimated in relation to the foregoing. Reserves should themselves be treated as genetic resource centres and should be exploited as such, subject only to the controls necessary to retain their own stability. Obviously the national agencies and private conservation organisations have an important part to play by making existing reserves freely accessible as sources of species to be introduced into other areas undergoing rehabilitation. This should include proper opportunities for the use of reserves as resource centres by commercial organisations. Nurserymen, for example, should be encouraged to acquire, hold and trade in species derived from reserves in the trade itself, which, experience has already shown, will always be limited and may be of uncertain origin.

The facilities of the commercial trade are unlikely, however, to be able to match up to the needs, particularly where long-term maintenance of stocks is required. Economic pressures are already causing many companies in the U.K. to reduce their holdings and to act more as agents. These trends point to the need for back-up from national conservation bodies and from agencies maintaining arboreta and botanic gardens. Again there seems no reason why the commercial side should be considered in isolation: nurserymen and others skilled in marketing and distribution should be freely used as agencies through which stocks maintained by national bodies are distributed. In the U.K. prototypes for such organisation already exist in the handling of crops, where governmental organisations may be responsible for breeding and multiplication while the trade undertakes distribution and marketing.

TECHNOLOGY OF PLANT GENETIC RESOURCE CONSERVATION

Hitherto attention has been paid mainly to the conservation of crop varieties and wild relatives, and little effort has been devoted to the problems of conserving natural source material. For example, there are few Seed Banks in the world beyond that being developed at Kew, dealing exclusively with wild species. The basic technology can in large measure be developed from the experience gained with banks for economic species, but it is already apparent that more research will be required to deal with a different set of problems which will inevitably arise. For the most part, economic species have been selected because of the readiness with which their seeds can be stored, and for ease of germination. Many wild species which would be desirable for use in rehabilitation programmes have quite different seed biology. Some have short-lived seeds for which adequate storage methods have not yet been found, and others have difficult germination requirements. What is true of seeds is also true for vegetative propagation. Special units like those being developed in Hawaii and at Kew may have to be set up to deal with these problems.

Research into breeding biology must be accelerated to go hand in hand with increased holding capacity. The ignorance about so many aspects of whole plant biology from taxonomy to biochemistry becomes evermore worrying and financial support for research must be forthcoming.

CONCLUSIONS

In the foregoing, experience from the U.K., a relatively small country although one with diverse topography and ecology, has mainly been called upon. It is perhaps possible to define what should be the principal aim in a country of this kind with a long history of human occupation culminating in intense industrial development through the last three centuries. They are largely to recover as much as possible of the land devastated by human activities and to use that part of it which cannot readily be restored to agricultural production first to provide amenity for a large, mostly city dwelling population, and second to produce refuges for the residues of natural fauna and flora. To set up some hypothetical target of restoring a natural ecosystem seems somewhat unrealistic in the British context, and one might well argue that what should really be done is to develop a national plan aimed at ensuring the maximum possible diversity of native wild life whether the product resembles any natural ecosystem or not.

The position is likely to be quite different in other parts of the world, where the target might be set at restoring something which is broadly comparable with the natural biome existing before the intervention of man. This is clearly one of the matters to the consideration of which the present Conference might well contribute. But it may be noted that whatever the aims of rehabilitation are, the same problems arise in relation to genetic resources. If ecosystem rehabilitation becomes accepted policy for major parts of the earth's surface, there will be a need for the genetic resources which will make it possible. It is not too early to consider how these should be collected and conserved against this possible future requirement.

REFERENCES

Heslop-Harrison, J. 1973. The plant kingdom: an exhuastible resource? Trans. bot. Sco. Edinb., 42, 1-15.

Heslop-Harrison, J. February 1974. Genetic resource conservation: the end and the means. Jl. r. Soc. Arts, 122, 157-169.

Lucas , G. Ll. 1976. In: Conservation of threatened plants, edited by J. C. Simmons et al. New York, London: Plenum.

Perring, F. H. 1970. The last seventy years. In: The flora of a changing Britain, edited by F. H. Perring, 128-135. Hampton, Middx: Classey, for the Botanical Society of the British Isles.

Thompson, P. A. 1974. The use of seed banks for conservation of populations of species and ecotypes. Biol. Conserv. 6, 15-19.

Thompson, P. A. 1975. The collection, maintenance and environmental importance of the genetic resources of wild plants. Environ. Conserv. 2, 223-228.

DISCUSSION - PAPER 16

J. CHRISTIE
In fish, it is very doubtful whether exotic species should be introduced in preference to native ones. Exotics have often been imported in ignorance of the niche they will occupy. Native species, in contrast, are a known quantity. Selective breeding from native stocks may have a future in fish - but under carefully determined circumstances.

G. L. LUCAS
In my paper, when I spoke of introducing plants to habitats I was especially referring to moving native species around within a country. I would hate to import French orchids, for example, to re-stock a British habitat: there are too many horror stories of exotics that have run rampant for that!

H. SUKOPP
Fifty per cent of the urban flora of Berlin is not native, but imported by man. The habitats in urban areas are often distinctive: they are warmer, the ground water table is often lowered, and the soils show many changes. Non-native species may be needed: Buddleia, for example, can be useful in rehabilitation.

May I ask how many British native species are held in the seed bank at Kew?

G. L. LUCAS   Perhaps 60 or 70. This is not many
                      compared with the Spanish seed bank maintained
                      by Professor Gomes-Campo. The Botanical
                      Society of the British Isles is about to
                      launch a campaign with the NCC and Kew to put
                      as many as possible of the British species into
                      the seed bank.

                      The Royal Botanic Gardens, Kew, trains
                      people in the diploma course, who very often
                      join local authorities as parks superintendents
                      and the like. One part of this training is to
                      instill the idea that some parts of even an
                      urban park should be left wild, and for native
                      species, even if the bulk of the area is
                      managed to support exotics.

L. M. TALBOT   This stress on soil, water, and plants
                      needs broadening: for me these components of
                      the system exist to support herbivores for
                      my predators to eat! Seriously, the basic
                      message is surely that we must retain diversity
                      and preserve options for a future we cannot
                      foretell. In considering the conservation of
                      habitats and ecosystems, this retention of
                      diversity must be a main thrust.

G. L. LUCAS   There are many examples of the value of
                      preserving species and strains - for example,
                      grasses with nitrogen-fixing nodules are
                      proving valuable in Arizona, and a member of
                      the Celastraceae which is yielding an anti-
                      cancer drug has recently been found, confined
                      to a nature reserve established for Sable
                      antelope in East Africa.

A. D. BRADSHAW   It is important to stress the value of
                      disturbed and degraded areas as places where
                      wild life can find refuge. We should not
                      restore all limestone quarries or flooded
                      gravel pits to intensive management: areas
                      should be left open for wild species to
                      colonise and hide in.

G. L. LUCAS   Some reserves may need continuous dist-
                      urbance - for example Teucrium needs rotovation
                      to maintain open soil to succeed, and poppies
                      need similar open ground.

O. BOERSET   Propagation in commercial seed orchards
                      can reduce genetic diversity, just as clonal
                      propagation from cuttings does.

G. L. LUCAS   If material of equal age and equal
                      performance is required, then such methods of
                      propagation may well be necessary. But we
                      must also set aside areas we deliberately
                      manage to retain diversity. A few years ago,
                      only two strains of peas were grown on any
                      scale in the United States, and when disease
                      broke out there was an intensive search for
                      older resistant varieties being cultivated
                      by a few obstinate old-fashioned farmers!
                      Reserves of genetic stock must be kept along-
                      side efficient propagation of high-yielding
                      crop strains.

J. N. R. JEFFERS   Today the trends in taxonomy and in teach-
                      ing surely run counter to the study of variation
                      and diversity?

G. L. LUCAS   In Europe the "lumpers" have it in
                      taxonomy, whereas in the United States and
                      U.S.S.R. the "splitters" are in the ascendency
                      - but the plants go on regardless! There are
                      serious taxonomic inadequacies today - espec-
                      ially in the tropics: our conference here is
                      about a well-studied area.

E. VAN DER MAAREL   May I endorse what has been said about
                      the maintenance of genetic diversity and plead
                      for means of preserving species by safeguarding
                      their habitat, since this is generally both
                      cheaper and best. It is also clear that when
                      we aim at restoration it is not a natural
                      system but a man-made approximation to it,
                      requiring continuing management effort, that
                      we produce. Resources of skill are needed,
                      and some old-fashioned agricultural skills
                      that are highly relevant are being lost at
                      present.

R. W. J. KEAY   If we agree that the preservation of
                      genetic diversity is important and that money
                      is limited, what are the priorities in Europe?

G. L. LUCAS   First we should find out what is endang-
                        ered on a European scale, picking out the key
                        species in each country. But we should
                        remember that Europe has a well-known flora:
                        there are other parts of the world with far
                        greater reserves of species, far less well-
                        known and scarcely protected at all.

C. The Role of Predators in Terrestrial Ecosystems

17. THE ROLE OF PREDATORS IN ECOSYSTEM MANAGEMENT

L. M. TALBOT

President's Council on Environmental Quality

722 Jackson Place, N.W., Washington, D.C. U.S.A.

INTRODUCTION

Predators rarely receive priority attention in considerations of
maintenance or restoration of ecosystems, yet both functionally
and symbolically, predators are central to these considerations.
Predators play a key role in maintaining ecosystem integrity in
terms of species and genetic composition, ecosystem functions, and
long term stability.

Historically, predators have been one of the first and most
completely exploited elements of ecosystems. Predators have been
perceived as a direct threat or competitor to man and throughout
history strenuous efforts have been made to exterminate or "control"
them. One result is that larger predators comprise over half of the
recorded exterminations of continental mammals (Talbot, 1964). The
image of predators as man's enemies is deeply rooted in folk lore
and "common wisdom". Versions of the belief that "the only good
wold is a dead wold" are still widely held, even among many wildlife and wild land managers, and this attitude still constitutes
a real impediment to rational management of ecosystems.

Public attitudes are changing, however, and in North America
and Europe there is increasing interest in predators and concern
with their status. This change brings policy implications of greater
tolerance toward maintaining predators - rather than removing them.
At the same time scientific understanding of their role in the
ecosystem is also changing, with attendant implications for the
role that predator management may take as part of ecosystem
restoration and management.

The purpose of this paper is to consider the role of predators
in temperate zone ecosystems and the consequent importance of maintaining predators as an essential component of ecosystems which we
wish to safeguard or restore. Maintenance of predators has certain
implications to ecosystem management and these will also be briefly
considered.

The discussion will focus on mammalian predators: however the
basic principles also apply to some degree to avian and invertebrate
predators.

First, however, it is important to define the objective of ecosystem management. The types of ecosystem management to which this
paper is directed are those where the objective of management at
least explicitly or implicitly requires predators. Thus the objective can be to restore an "original" ecosystem – or at least one which
existed, presumably in a stable condition, before recent human
disturbance. It can also be to create or restore one which supports
some exploitation, such as grazing, lumbering, or some cultivation,
where the role of predators in controlling potential "pest" herbivores and in contributing to long term stability will be significant.
Lastly, the objective can be to maintain or restore any ecosystem
which has a significant herbivore component.

THE FUNCTIONAL ROLE OF PREDATORS

Predators play a significant although still not fully understood
role in the ecosystem. They are an important link in nutrient
cycling. They may affect the genetic composition, morphology,
physiology and behaviour of prey species. They may mediate interspecific competition, and they may play a variable role in regulating
or controlling numbers and distribution of prey species. It is
this latter point on which I shall focus most attention.

Role in Food Chain: Predators are at the top of the food chain and
consequently they are an important link in the nutrient cycle within
any ecosystem. In the absence of predators, the secondary producers
would still die and, in effect, be recycled. But the process would
be different and to that extent, the ecosystem would not maintain
the integrity it had with the predator component.

Selective Impact on Prey: Predators act as a powerful selective
force on the prey species. Predation on larger prey often entails
considerable risk to the predator, and always involves a considerable
investment of energy. Accordingly predators select for the most
accessible prey, sick, weak, young, or otherwise abnormal individuals
whenever they are available.

One example of the impact of this selection is the precocity of young of ungulates which are born in open situations as opposed to those born in thick cover. In East Africa, where the greatest spectrum of ungulates available today survives, the young of wildebeests and gazelles may be on their feet and running beside their mother within five minutes of birth (Talbot and Talbot, 1963). In these species birth usually occurs on open short grass plains, where there is no shelter or cover from the abundant predators. In contrast, the young of impals may not gain their feet for over an hour, but the thick vegetation in which these young are usually born provides effective cover and protection from predation.

It could be argued that such selective influences occurred in the past, and that we are concerned with maintenance of ecosystems in the present. However, this overlooks the fact that ecosystems are dynamic and that selectivity is a continuing process. Further, the selection by predators of ill or abnormal individuals reduces the chances of the spread of disease or deleterious genetic abnormality to the population as a whole. This consideration becomes especially significant when the ecosystem involved is a relict one, and the prey population sufficiently limited that extermination would be a threat.

Mediation of Interspecific Competition: Where this function occurs it would stem from predation pressure on several potentially competing prey species, with the predators maintaining the prey numbers somewhat lower than would be expected in the predators absence, thus reducing their interspecific competition and thereby enhancing community diversity and stability. This function has clear significance to management aimed at maintenance of ecosystems, most particularly relict or isolated ones.

Regulation or Control: For purposes of this discussion I distinguish between population regulation and control along the lines proposed by Keith (1974) as follows: Population control is the maintenance of a population and may involve both density dependent and density independent processes; Population regulation is the dampening of numerical fluctuations by density dependent processes.

It has long been an accepted tenet of ecology and of wildlife management that predators regulate or control the numbers of their prey. This has been basic to the thinking of wildlife and other land managers, and historically removal or reduction of predators has been a primary method used to maintain or increase numbers of game species.

Recently, however, the accepted dogma of the role of predators has been seriously questioned. Paul Errington's detailed work on muskrats and Bob-white quail showed that, at least for systems with smaller predators and prey, predation per se had little if any effect on those prey populations (Errington, 1945, 1946). There was predation, but most of it involved killing animals which were surplus or "doomed" - old, sick or others in excess of the carrying capacity of the habitat. In effect predators were seen as scavengers consuming the prey before the fact of inevitable death rather than after it. Other studies of small mammals or birds further confirmed or extended these findings, but the belief has persisted that in the case of larger prey, mostly ungulates, the situation is somehow different; that predators constitute the primary limiting factor for populations of prey.

The "case history" of the Kaibab deer herd has become established dogma used to prove or reinforce this concept (Leopold, 1943). The Kaibab Plateau is on the northern rim of the Grand Canyon, Arizona. When the area was given protected status the larger predators were systematically and energetically removed in an attempt to rebuild the depleted deer population. Subsequently, there was a substantial increase in the number of deer. The actual magnitude of the increase is now subject to considerable question. It was assumed for many years, however, that it was truly dramatic, with deer numbers rising from a few thousand to over 100,000. The increased numbers of deer predictably destroyed their forage base and the population crashed. The assumption was that this constituted proof of the role of the predators in holding down the deer population, that once the predators, i.e., presumed limiting factor, were removed the population increased dramatically far beyond the capability of the habitat to sustain them and subsequently crashed through starvation.

In the first place, the magnitude of the increase is subject to real question. There were no good census figures for deer present in the park in the early years. There are only assorted estimates. Most estimates placed the population under 5,000 about 1908, in the neighbourhood of 30,000 at the "peak" around 1924, with a drop to below 20,000 in the early 1930s. One estimate (the basis for the dramatic boom and bust story) placed the 1924 peak at 100,000, and the post peak level at 10,000 (Colinvaux 1973: 399 -401). However, regardless of the actual figures, there was clearly some increase followed by a decrease. Far more important is the question of why.

The text book reason is that "the predators were removed." However, there were a series of other concurrent factors. Prior to its declaration as a park, the area had been intensively grazed, probably overgrazed, by a recorded 200,000 sheep, 20,000 cattle, and numerous horses. Presumably, the ranchers had also used fire in modifying the vegetation of the area, and doubtless had exerted considerable hunting pressure on the deer (Rasmussen, 1941). Clearly, the removal of this hunting pressure, competition, and other vegetation modifying factors would be expected to result in a substantial increase in the deer population with or without predators. In

short, the major "proof" historically cited for the role of
predators in controlling prey populations does not stand up to
critical examination (Caughley, 1970). Predation may have been a
factor, but the evidence does not support the idea that it was the
factor.

My own work in East Africa provides striking evidence that in
some cases predators play no significant role at all in population
regulation or control of prey ungulates. In the roughly 15,000
square mile savannah ecosystem in the Serengeti-Mara region of
East Africa, there were in 1961 roughly 240,000 wildebeest, 172,000
zebras, 22,000 buffalos, 6-9,000 elephants and 480 to 800,000 Thompson
gazelles (Talbot and Stewart, 1964). These were among the most
abundant of the 30-odd species of ungulates common in the area. The
predators were diverse, including lions, leopards, cheetah, hunting
dog, hyena, and three types of jackal, but they were relatively low
in numbers. Lions apparently made up the largest single component
of the predator biomass, and they were the primary predator of
wildebeest in the open plains area, being responsible for over 90
per cent of the total predation on these antelopes (Talbot and Talbot,
1963). There were an estimated 700 lions in the region, and their
calculated maximum predation toll on the wildebeest - assuming they
all only ate wildebeest - was between 12,000 and 18,000 per year.
This would have amounted to between 5 and 7.5 per cent of the total
population which is insignificant in terms of any population regulation or control. Roughly half of the wildebeests were female and
virtually all of the adult females produced one calf a year. Thus
there was a potential recruitment of nearly 50 per cent per year.
Yet, during three years of study, the total population remained
roughly stable, clearly indicating that factors other than predation
were at work.

The primary regulatory mechanisms were partly density dependent
and involved the carrying capacity of the environment. They were
activated by weather and therefore range condition. During dry years,
(as 1959, 60 and 61 had been) at the time of calving the wildebeest
were aggregated into great herds which were concentrated on the few
areas where rain had fallen and therefore green grass was available.
Under such conditions we found that roughly 40 per cent of the year's
calf crop was lost during the first several weeks of life when, as
a result of crowding, the calves got separated from their mothers.
In contrast, in good, i.e. wet years, the animals were dispersed
over vast areas with individual concentrations rarely of more than
25 animals. Under such conditions, calf survival during the first
few weeks of life was nearly 100 per cent.

Several months later, during the dry season, in a dry year the
animals were constantly on the move seeking areas where rain had
produced green grass. Weakened by nutritional stress, the young
animals were particularly vulnerable to disease which in such years

killed roughly half of the surviving calves. In contrast again, during this period in wet years the losses from disease were insignificant.

To further emphasize the insignificant role of predation in regulation of wildebeest numbers, following the drought period which ended in 1961, there were a series of "wet" years which resulted in a nearly 4-fold increase in the total wildebeest population. During the same period, the lion population approximately doubled.

While the details of the regulatory mechanisms operating with the other ungulate species differ from the wildebeest in some degrees, it is clear that in most cases predation played no role in control of their numbers either. There were simply too few predators. We postulated that the only cases where predation could exert a significant impact over prey numbers were situations involving a small, discrete population of a non-migratory ungulate in the presence of a relatively large population of predators.

In the temperate zone there is conflicting evidence about the role of predators in controlling ungulate populations. Huffaker (1970) reviewed a number of early studies, showing cases where predators apparently did control population levels and other cases where they did not. Keith (1974) analysed these studies and more recent ones involving moose, Dall and Bighorn sheep, caribou and deer, and concluded that under "pristine" or natural conditions, predators did regulate prey ungulate populations, but that where the environment had been significantly modified by man, and particularly where there was habitat change providing additional food, along with reduction of the predator population, the regulating or controlling function of predation was overridden.

This phenomena is best exemplified in the case of ungulate irruptions. The earlier view was that the reduction or removal of large predators removes the limiting factor initiating a period of uncontrolled expansion of the prey population (Leopold, 1943). More recent thinking holds that the factor involved is a major discrepancy between the ungulate population and the carrying capacity of the environment (Caughley, 1970; Riney, 1964). However, both of these conditions have been involved in the irruptions that have been typical of North American ungulates during the past century. While good data are the exception, rather than the rule, Keith, (1974) has cited several case histories which he believes "strongly implicate a lack of predation as the chief cause of these irruptions." These include caribou in Newfoundland, moose in various parts of North America, Dall sheep in Mount McKinley National Park, Alaska, and Bighorn sheep in the Rocky Mountains.

In my view, the attempt to find a single answer may tend to
mask the real significance of predators in the ecosystem. There
are few simple answers in ecology. Regardless of whether or not
predation or its removal was the triggering factor in a certain
situation, predation is clearly one of the central factors operating
in all these situations.

THE EXAMPLE OF ISLE ROYALE

It is difficult to find good data on situations where ecosystems
containing large predators have been reconstructed or rehabilitated.
While not strictly an intentional rehabilitation of the ecosystem,
Isle Royale probably provides the best case history data available
in a northern temperate zone ecosystem.

Isle Royale, a 210 square mile island, roughly 45 miles long by
9 miles at its widest, lies within Lake Superior in the northern
U.N. near the Canadian border. The island is protected as part
of the U.S. National Park system. The island can be considered a
discrete ecosystem bounded by the waters of the lake, although the
waters do not form an absolute barrier. Animals which are strong
swimmers, such as moose, occasionally swim to and from the mainland,
and in unusually severe winters wolves have crossed to and from the
mainland over the ice. However, these occurrences are relatively
rare.

The dominant vertebrates at present are moose and wolves. Prior
to around 1900, the island fauna apparently included caribou, coyotes,
and various other mammals, but not moose or wolves. The earlier
fauna disappeared from the island probably because of human activity,
and in about 1908 moose replaced them, doubtless swimming from the
mainland (Murie, 1932). Precise figures from the early years are
not available, but the moose numbers increased rapidly until they
severely over-utilised and damaged the forage, and then crashed,
first in the mid-1930's and again in the mid-1940's. Wolves arrived
on Isle Royale in the late 1940's and the wolf population quickly
grew to roughly 20. From then into the late 1960's, the two populations remained roughly stable, with a population of about 24 wolves
and a herd of roughly 600 moose, along with about 2,000 beavers which
provided summer food for the wolves. Studies initiated in 1958
(Mech, 1966; Allen, 1974) indicated that the wolves were indeed
controlling the numbers of moose on the island.

This appeared to be a clear example of the role of predators
in controlling the numbers of the prey species and providing stability in a system in which, prior to introduction of the predators,
there had been gross instability.

Subsequent research, however, has shown that the situation is not that simple. The excellent wolf studies on Isle Royale have been continued to the present, and have shown that after its apparent stability up to the mid 1960s, the moose population (by midwinter census) increased to approximately 1,000 in the early 1970s. Unusually heavy winter snowfall, coupled with limited available forage due to continued regrowth of the earlier burned area, have again changed the picture in the past several years. Calf production in the moose population has been lower since 1970 than in any of the earlier years. At the same time, the wolf population more than doubled to over 40 animals, comprising three packs and several individual animals or families. Winter predation by wolves has increased both because of the increased numbers of wolves and the increased vulnerability of moose in deep snow. Wolf predation during the recent years has been estimated to have taken 10-15 per cent of the moose population each winter, whereas there has only been a recruitment of the adults to the population of 5-10 per cent per year. As a result, it is estimated that there has been an approximate 40 per cent reduction in the moose population over the past six winters (Allen, D.L. 1974, 1976; Peterson, 1976).

There also has been a change in the pattern of wolf predation. Whereas through the early 1960s, it was entirely young and very old animals that wolves took, wolves are now observed also taking prime age animals. One hypothesis links the unusually heavy winters with this apparent change in the food habits of the wolves. During "normal" winters moose calves have adequate nutrition and those which reach adulthood have the strength and vigor to avoid or fend off wolves. However, during unusually severe winters with deep snow conditions, the moose population has inadequate nutrition. Adult moose have died of malnutrition, and the young which survived clearly suffered nutritional stress. It is postulated that when some of these animals have been sufficiently weakened by this experience, they are vulnerable to wolf predation on reaching adulthood (Allen, D. 1974; Peterson, 1976).

In any event, it is clear from the 19 years of studies on Isle Royale that under these circumstances predators play a major role in controlling or regulating the numbers of their prey, but this effect is intimately interrelated with climatic conditions and other factors affecting the carrying capacity of the habitat for the prey In this case, unusually severe winter snowfall has been a major factor, coupled with changes in moose carrying capacity of the habitat associated with forest regeneration following forest fires.

It is also clear that predator numbers are associated with those of prey, or of available prey. An increase in the prey population coupled with increased accessibility due to deep snows has been followed by more than a doubling in the predator population. With

this has been a significant change in the social behaviour of the
predators. During the period of relative stability of predator
and prey populations, up to the early 1960s, there was one single
wolf pack and the entire island constituted its territory. The
wolf population averaged about 24 and it was believed that this
constituted the maximum wolf density possible (Allen, 1974; Mech.
1970). Subsequently the area was first divided among two packs and
now among three, each with specific territories which are vigorously
defended and the total wolf population is now 44 (Allen, D.L. 1976;
Peterson, 1976).

        It would appear both from these studies and from research
currently in progress in Alaska (Cowan, 1976) that once wolf numbers
reach a high enough level relative to moose numbers, and in the
presence of winters with heavy snowfall, that wolves can significantly reduce the moose population. These studies will have to be
continued further before we know what the ultimate result will be.
Clearly there will be a reduction in availability of prey sooner
or later, either through significant reduction in the prey
population or through reduction in the severity of the winter snowfall which reduces the preys vulnerability - and hence availability
to the wolves - or both. When this happens and if there are no
alternative sources of prey, it seems probable that the predator
population will be reduced through lowered reproductive success and
possibly through starvation, which may contribute to another increase
in the moose population, if other habitat factors are favourable.

                STABILITY - INSTABILITY: MICROTINES AND LAGOMORPHS

While there is still uncertainty as to whether or not predators
control the numbers of a prey population - particularly ungulates -
in the sense of holding it to a certain level over a long period
of time, there is a growing body of data on another form of population regulation exerted by predators which might be termed "destabilisation". This occurs when the population levels of the prey
species follow periodic cycles, usually 3-4 years in microtines and
8-11 years in hares.

        In such cases the population of the prey species, usually
microtines or lagomorphs, increases rapidly, reaches a peak and
crashes. The predator populations increase in response to the
increased prey, and are still increasing when the prey crash; they
do not cause the crash. However, the high predator population exerts
heavy pressure on the declining prey, hastening the decline and
lowering the prey population at the bottom of the cycle below the
level that the habitat would support, keeping it there until the
lack of food causes a reduction in the predator population. Removal
of predation, coupled with excess food, allows the process to repeat.

The changes in microtine and lagomorph population levels are
affected by a series of factors, both density dependent and density
independent. Predation in this situation has the effect of increasing the magnitude of the fluctuation at its lower end - but
not apparently at its peak - and of increasing the interval between
peaks through extending the duration of the low periods, to some
degree synchronising the different cycles throughout the area
accessible to the predators. Data supporting these phenomenon
have been provided by studies of a variety of species under widely
differing conditions, among them Black-tailed jackrabbits in Utah,
Snowshoe hares in a Boreal forest, California, voles in grassland,
and Brown lemmings in the Arctic tundra (Keith, 1974; Lidicher, 1973;
Gross et. al., 1974).

The probable significance to the ecosystem of this form of
predation is to foster long term stability. Lagomorphs and microtines can exert an enormous impact on vegetation. By depressing the
herbivore populations below the carrying capacities during cyclic
lows, and extending the low periods, predators provide the vegetation
a "rest" period. Without such periods during which recovery is
possible it is probable that long term vegetation changes could
occur.

While the evidence for this "destabilisation" comes from
rodents, it is interesting to speculate on its application to ungulates also. The developing situation on Isle Royale might prove
to be at least partially a phenomena of this sort. The wolves may
drive the already declining moose population well below the carrying
capacity of the environment; reduced food could lead to a reduction
in the wolves, allowing an irruption of sorts of the moose population.
There are obvious differences between the rodent and ungulate
situations, but the basic ecological principles may be similar, and
in any case, the significance of the predators to the exosystem is
profound.

IMPLICATIONS FOR MANAGEMENT

Achieving the functional presence of predators within a managed ecosystem has certain implications for management. Chief among these
is the size of the area which is required. Large and wide ranging
predators such as Mountain lions and Polar bears require substantially larger areas than do smaller for more resident ones such as
fox and lynx. The key factor is the adequacy of the prey population.
The Isle Royale case has shown that in the presence of adequate food,
wolves could adapt behaviourally and socially to a density of twice
what had formerly been considered limiting, i.e. one per five square
miles rather than one per ten. An attendant consideration is the
compatibility of predators with surrounding activities; the like-

lihood of encounters between the predators and livestock or humans
in the neighbouring areas.

Particular care should be exercised to anticipate – and possibly
control – damaging irruptions of herbivore in situations where an
ecosystem is being rehabilitated. Rehabilitation implies disturbed
conditions, which, particularly in the absence of adequate numbers
of predators, may trigger herbivore population irruptions functionally identical to those where herbivores have been introduced into
new environments (Riney, 1964; Caughley, 1970).

A further important consideration is that where rehabilitation
is the objective, the ecosystem will be damaged, and as noted above,
the predators will almost certainly be gone or greatly reduced.
Consequently, there is a need for initial determinations of which
predators were a part of the ecosystem one wishes to rehabilitate –
and what prey they depended upon. The re-establishment of those
species and their relative "balance" with each other and with the
environment, is essential to successful rehabilitation.

Long term research and continued monitoring are a necessity.
The current state of knowledge of the role of predators is still
fragmented and conflicting. Again the Isle Royale case illustrates
the importance of long term studies to provide the evolving data
base for continued management.

CONCLUSIONS

Where the objective of management is to maintain or rehabilitate
a given "original" ecosystem, or any one which includes herbivores,
it is clearly necessary to include predators. Where predators are
not present the habitat may have the appearance – for a time at
least – of the original ecosystem. But without the predator it will
not be the same, in terms of nutrient cycling; species composition;
genetic composition; morphology; physiology and behaviour of prey
species; ecosystem function and stability; and ultimately, human
welfare.

REFERENCES

Allen, Durward L. 1974. Of fire, moose and wolves. Audubon Mag.
    76 (6), 38-49.

Allen, Durward, L. 1976. Personal communication.

Colinvaux, Paul A. 1973. Introduction to ecology. Chichester:
    Wiley.

Cowan, I. M. 1976. Personal communcation.

Caughley, G. 1970. Eruption of ungulate populations with emphasis on Himalayan Thar in New Zealand. Ecology, 51, 53-72.

Errington, P. L. 1945. Some contributions of a fifteen-year local study of the Northern Bobwhite to a knowledge of population phenomena. Ecol. Monogr. 15, 1-34.

Errington, P. L. 1946. Predation and vertebrate populations. Q. Rev. Biol., 21, 144-177, 221-245.

Gross, J. E., Stoddart, L. D. and Wagner, F. H., 1974. Demographic Analysis of a Northern Utah Jackrabbit Population. Wildlife Monogr., Chestertown, 40.

Huffaker, C. B. 1970. The phenomenon of predation and its roles in nature. In: Dynamics of populations, edited by P.J. den Boer and G. H. Gradwell, 327-343. Proc. Advanced Study Institute of Dynamics of Numbers in Populations, Oosterbeek, 1970.

Keith, Lloyd, B., 1974. Some features of population dynamics in mammals. Transactions of the International Congress of Game Biologists, 11th, 17-58.

Leopold, A. 1943. Deer irruptions. Publs. Wis. Conserv. Dep. No. 321, 3-11.

Lidicher, W. Z. 1973. Regulation of numbers in an island population of the California vole, a problem in community dynamics. Ecological Monogr. 43, 271-302.

Mech, L. D. 1966. The wolves of Isle Royale. Fauna natn. Pks. U.S., Fauna Series 7.

Mech, L. D. 1970. The wolf. Garden City: Natural History Press.

Murie, A. 1932. The moose of Isle Royale. Misc. Publs Mus. Zool. Univ. Mich., no 25.

Peterson, R. O. 1976. The role of wolf predation in moose population decline. ms. for First conference on Scientific Research in the National Parks. American Institute of Biological Sciences and National Park Service, Washington, D.C.

Rasmussen, D. I. 1941. Biotic communities of Kaibab Plateau, Arizona. Ecol. Monogr. 11, 229-275.

Riney, T. 1964. The impact of introduction of large herbivores on the tropical environment. Tech. Meet. int. Un. Conserv. Nat. nat. Resour., N.S. 4, 261-273.

Talbot, Lee M. 1964. The international role of parks in preserving endangered species. In: World Conference on National Parks, 1st Seattle, 1962, 295-304.

Talbot, L. M. and Stewart, D.R.M. 1964. First wildlife census of the entire Serengeti-Mara region, East Africa, J. Wildl. Mgmt., 28, 815-827.

Talbot, Lee M. and Talbot, Martha H, 1963. The wildebeest in Western Masailand, East Africa. Wildl. Monogr. 12.

DISCUSSION: PAPER 17

V. GEIST   There is another possible function of carnivores. The disease of rabies appears to be most prevalent in disturbed systems where the large predators have been removed and the distribution of large herbivores also altered. The small carnivores increase to fill the gaps in the system. The evidence is that all large carnivores are cannibals on conspecifics which display unusual behaviour. If rabies appears in skunks for example the affected animals are quickly killed by wolves or bears which, because they are skilful hunters normally escape injury. If a wolf is bitten its behaviour becomes aberrant and it is quickly killed by other wolves. There is thus advantage in the complexity of the system. May I add a point about the way of life of bears which have a particularly interesting strategy. As is well known they eat vegetation and can, therefore, fill the herbivore niche where ungulates are absent as occurs in parts of Kantchatka and Alaska. These are areas where ungulates cannot pass the winter because of the heavy snow but bears can because they hibernate and emerge to consume the spring flush of vegetation. Densities can be quite high: up to one bear per two square miles.

L. M. TALBOT   In some parts of the United States bear densities can reach one per square mile.

W. HINDS   One aim of ecology is to give a good
                       explanation of events in natural landscapes.
                       In areas disturbed by man "weedy" species are
                       a problem: when a former strip mine region
                       is being reclaimed weedy cats, dogs, foxes and
                       even coyotes need to be handled. Are these
                       undesirable and how should they be dealt with?

L. M. TALBOT   I have recently had a battle with sheep
                       farmers over predator control, especially
                       coyotes. On the other hand we do have in some
                       parts of the United States a problem with feral
                       dogs which are not a natural part of the eco-
                       system. So far as coyotes are concerned the
                       unrestricted use of poisons has been banned on
                       publicly owned land, and non-selective poisoning
                       has been prohibited everywhere. The results
                       have been to demonstrate that coyote abundance
                       is due to the abundance of jack rabbits rather
                       sheep: in the absence of poisoning some pop-
                       ulations went down in 1975 while others went
                       up. The disturbance by strip mining is surely
                       confined to small chunks of the landscape,
                       and the real impacts in such areas are from
                       the influx of people with their consequent
                       demands on sewage, police, education etc?

W. HINDS   My concern is that as population increases
                       in an unpopulated area the feral house cats
                       and dogs will increase and my question is
                       whether they will fill a function like the
                       original species?

L. M. TALBOT   I doubt it. Feral cats in the United
                       States are never really populous in arid areas
                       that get cold in winter. Dogs can survive in
                       such areas, but they are often not truly feral,
                       spending the day in houses and ranging and
                       killing sheep at night. They do, however, tend
                       to be kept down by hunters.

W. J. WOLFF   Dr. Köpp said that German forest deer
                       were too numerous. Has the reintroduction of
                       wolves to Germany been considered? Would the
                       restoration of large predators generally be
                       desirable in ecological systems where they have
                       previously been eliminated by man?

L. M. TALBOT   The general question is how compatible
                    predators are with people and livestock. I
                    think the answer is that reintroduction may
                    be satisfactory if contact with humans and
                    livestock is not a problem. Predators are
                    desirable elements in the ecosystem if
                    possible. However, social values are attached
                    to hunting and this also provides some protein
                    value and the problem is to work out how the
                    resource should be shared between biped and
                    quadruped predators!

H. KÖPP   We have considered reintroducing predators
                    in Germany. The fact is that most European
                    countries have developed beyond the point where
                    predators are tolerable at a density where they
                    are in any way effective. At attempt at re-
                    introducing wolves in a strictly controlled
                    way in little frequented forests along the
                    Czech border was made recently but the experi-
                    ment failed when all the wolves escaped leading
                    to great public concern and a rapid control
                    operation.

D. The Restoration and Management of Freshwater Ecosystems

18. THE RESTORATION OF FRESHWATER ECOSYSTEMS IN SWEDEN

C. Gelin

Institute of Limnology

University of Lund, Fack, Lund, Sweden

Many lakes throughout the world are undergoing eutrophication (accelerated ageing) due to man's activities. These lakes often become an environmental nuisance, because of impaired water quality, fermentation processes during anoxic periods, prolific weed growth, excessive algal blooms, and deteriorating fisheries. International concern has stimulated research on the nature and causes of the lake-aging process including the development of various control techniques. Lake restoration has become a subject of considerable interest (Bjork, 1974a, Dunst et al, 1974) and several measures have been proposed to restore lakes. But lakes are complicated ecosystems and the ability to predict the response of a lake to various measures is as yet limited. Moreover, each lake has its own "unique personality" which means that it is impossible to establish a standardised treatment. A co-operative effort with public administrators, politicians and technicians is necessary when planning restoration of lakes.

Since the middle 1960s lake restoration projects were initiated in Sweden by limnologists at the University of Lund (Bjork. 1968, 1972). The programme included different types of lakes which had been damaged by man. In this paper three different methods for restoring freshwater ecosystems are reported.

SUCTION DREDGING OF POLLUTED LAKE SEDIMENT (LAKE TRUMMEN)

The formerly oligotrophic Lake Trummen, South Sweden, (surface area 0.8 km^2, maximum depth 2.1 m, mean depth 1.1 m) was used as a receiver for waste water from 1936 to 1958. The lake then had all

the characteristics of an over-exploited recipient, expanding
dense reeds along the shore, excessive blooms of blue-green algae
(Microcystis spp.) during summer, and oxygen deficiency and fish
kills during winter.

In spite of diversion of the waste water in 1958 the lake did
not recover during the 1960s. The 20-40 cm thick, reduced sulphide-
rich sediment layer, deposited during the pollution period, was
considered to be the main reason for the delayed recovery. The
surface sediment became heavily loaded with nutrient; phosphorus
increased at least 5 times and nitrogen about 3 times, on a dry
matter basis. The content of PO_4-P in the interstitial water of
the uppermost sediment layer was 2-3 mg P^{-1} (Bengtsson et al, 1975).
Most of the nutrients fertilizing the lake were being released from
the sediment.

Lake Trummen was restored by pumping (suction dredging) the
nutrient-rich 'cultural' sediment layer into sedimentation ponds
on land. In 1970 and 1971 ca 0.5 m sediment was removed (300,000m^3).

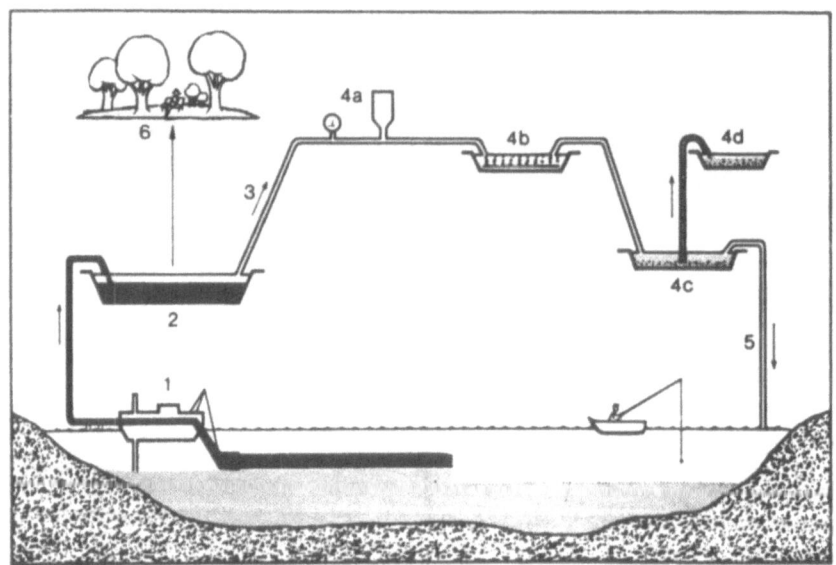

Fig. 1.   The restoration scheme as practiced for Lake Trummen.
          1. Suction dredger. 2. Settling pond. 3. Run-off water,
          4. Precipitation with aluminum sulphate (4a-automatic
          dosage, 4b-mixing through aeration, 4c-sedimentation, 4d-
          sludge pond). 5. Clarified run-off water. 6. The dried
          sediment is used as fertilizer for lawns and parks. The
          area utilized for sediment deposition will become recrea-
          tion grounds. (From Björk 1972).

The run-off water from the ponds was treated with aluminium sulphate
for precipitation of phosphorus and suspended matter (Fig. 1).   The
phosphorus reduction was 80-90%.   In 1971 all vegetation along the
shores was removed by a dragline (for further details, c.f. Bjork,
1972).

Results

During restoration, the underlying oxidised, brown and consolidated
sediment was exposed, which is indicated by the increase in dry
weight (Fig. 2).   The content of nitrogen and phosphorus of the
surface sediment decreased (Fig. 2) and are comparable with those
of oligotrophic lakes within the same area (Gelin and Ripl, in press).
The oxygen consumption of the brown sediment is much lower than that
of the black sediment.   The concentration of phosphate in the interstitial water was in 1973 200-500 times lower than in 1969 (Bengtsson
et al. 1975).   From the new surface sediment an average release of
0.23 mg P m^{-2} a day was calculated in long term experiments under
aerobic conditions (Graneli, 1975).

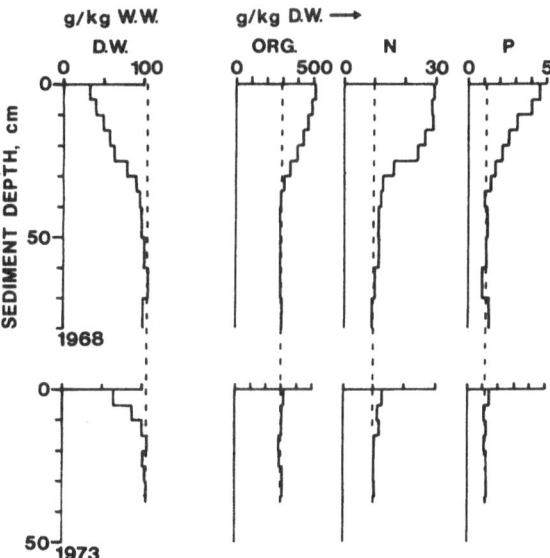

Fig. 2.   Mean values of sediment dry weight (D.W.) per wet sediment
           (W.W.); organic matter (ORG.), Kjeldahl-nitrogen (N) and
           phosphorous (P) per D.W. from the same three sampling
           stations before (1968) and after (1973) the restoration of
           Lake Trummen.   The broken veritcal lines indicate the mean
           value of the unpolluted sediment (40-80 cm in a single core
           from the main sampling station).   (From Bengtsson et al.
           in prep.).

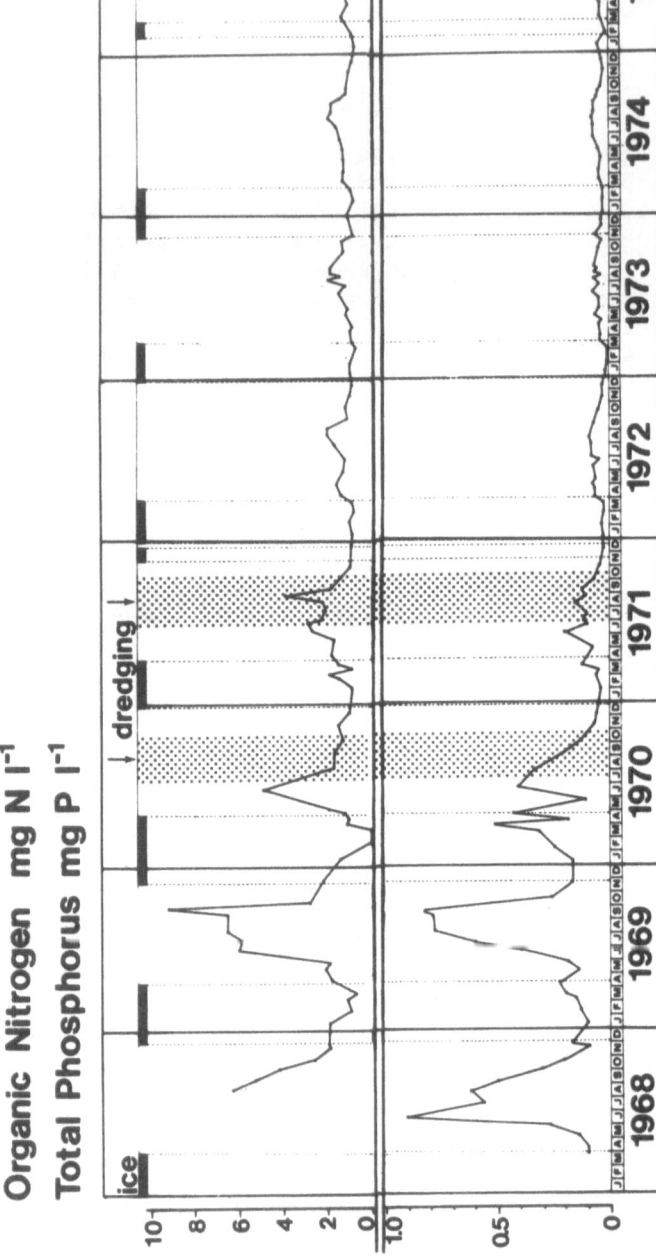

Fig. 3. Organic nitrogen and total phosphorus in Lake Trummen 1968 through 1975 (surface water).

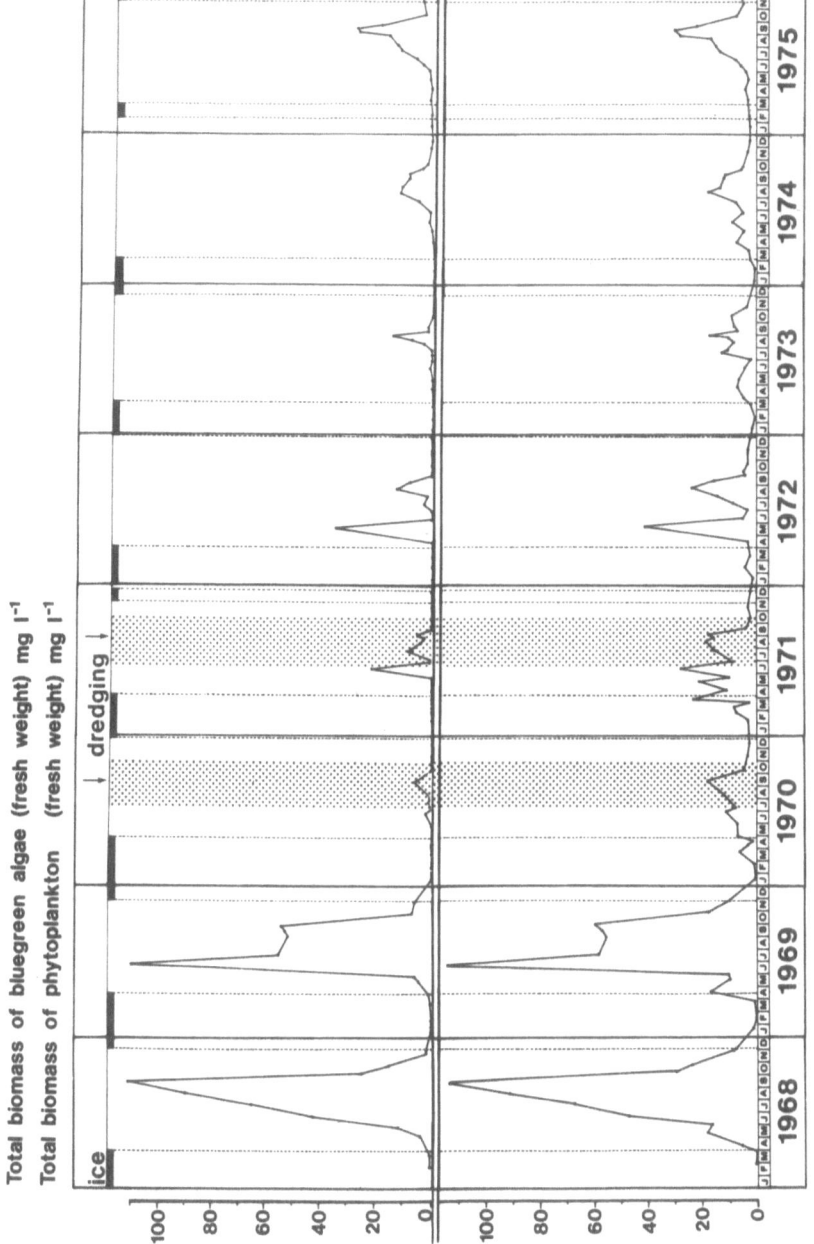

Fig. 4. Development of the biomass (fresh weight) of the bluegreen algae and that of the total phytoplankton community in Lake Trummen 1968 through 1975 (surface water).

The average concentration of PO_4-P in the water decreased from ca 200 μg P^{-1} to ca 10 μg P^{-1} during June through September 1969 and 1973, respectively. During the summer 1975 the PO_4-P concentration was still below 10 μg^{-1}. The content of organic nitrogen of the water during the summer decreased from about 6 mg N^{-1} before the restoration to about 1.5 mg^{-1} in 1973. It varied between 2.0-2.5 mg N^{-1} during August and September 1975 (Fig. 3). Total phosphorus decreased from about 800 μg P^{-1} during the summer before restoration, to 70-100 μg^{-1} in 1973. During August and September 1975 the development of phytoplankton was correlated with an increase in total phosphorus at levels between 120-150 μg P^{-1} (Fig. 3).

The ionic composition of the water and the concentration of major constituents except sulphate did not change in connection with the restoration. The use of aluminium sulphate in the treatment plant caused an increase in the sulphate concentration from 1970 to 1972, which then decreased again during 1973 and 1974.

Before the restoration, excessive algal blooms of Microcystis spp. occurred during summer (Fig. 4). The Secchi disc transparency during this period was about 20 cm but increased to about 70 cm when the abundance of phytoplankton decreased drastically during the two years after the restoration. As the biomass of phytoplankton decreased, the light conditions of Lake Trummen improved and the vertical distribution of the phytoplankton productivity was changed. The maximal productivity per unit volume decreased from about 10 g Cm^{-3} day^{-1} to 1-2 g Cm^{-3} day^{-1}. As a consequence of the lowered phytoplankton biomass, pH dropped from 9-10 and even more (at noon) during the summer before restoration, to mostly below pH 8 during the corresponding period 1972-1973. The calculated mean annual phytoplankton productivity decreased from 375 g Cm^{-2} to 225 g Cm^{-2}. After the restoration a much greater part of the total phytoplankton productivity was carried out by small algae (nanoplankton) (Gelin and Ripl, in press).

There was no submerged vegetation at all in the lake before restoration but a slight recolonisation by underwater vegetation has already taken place (e.g. Potamogeton obtusifolius, Juncus supinus and Nitella sp.). The effect of the restoration on the zooplankton community was a decline in the abundance of most of the species. During the summers (June through September) after the restoration (1972-1973), the abundance of cladocerans was only 15% of that in 1968-1969, Only half of the rotifer population was present in the lake after the restoration. The decrease in abundance of rotifers and cladocerans is directly correlated with the decrease of phytoplankton biomass. The abundance of phytoplankton has increased during 1974 and 1975 (Fig. 4). The Secchi disc transparency has

decreased again (Fig. 5) and pH exceeded 9 (at noon) during July,
August and the beginning of September, 1975. Since 1975 experiments
have been performed in Lake Trummen to examine the possibilities
for further changing the trophic status of the lake, for example
manipulating the planktivorous fish populations.

The Lake Trummen restoration project proves that a proper
suction dredging technique can bring about rapid improvements
in shallow, formerly polluted water bodies.

Costs

The total price of the restoration of Lake Trummen in 1970 and 1971
was 2.6 million Sw Cr or US $580,000. This excludes the costs for
the scientific investigations.

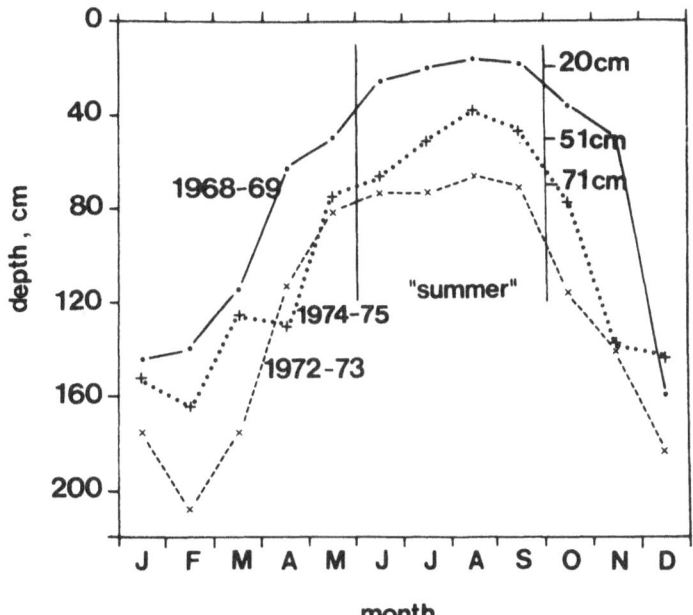

Fig. 5. Secchi disc transparency in Lake Trummen 1968-69, 1972-73,
and 1974-75 (monthly means). The figures 20, 71 and 51
(cm) indicate the mean during each summer period.

BIOCHEMICAL OXIDATION OF POLLUTED LAKE SEDIMENT WITH NITRATE
(LAKE LILLESJON)

A new method has been developed for restoration of lakes which
have been polluted by sewage (Ripl, 1976). Instead of pumping
(suction dredging) the sediment to settling ponds on land, which
is both expensive and requires land areas close to the lake, this
method is based on the in situ oxidation of the sediment by nitrate,
through the process of denitrification. Three compounds, iron
chloride, slaked lime and calcium nitrate, are injected into the
sediments.

This combination of compounds improves the sediments' binding
capacity for phosphorus and reduces at the same time the amounts of
easily degradable organic matter. In this way, nutrient recycling
as well as the biochemical oxygen demand (BOD) at the interface
between sediment and water are reduced.

This technique was applied in the restoration of Lake Lillesjon,
(surface area 0.04 km^2, maximum depth 4.2 m, mean depth 2 m) situated
close to the town of Varnamo in the southern part of Sweden. Formerly this lake was oligotrophic but was used as a receiver for
sewage for several years. Though the sewage was diverted in 1971,
the ecosystem had been severely damaged: H_2S was produced in the
sediment and hypolimnion. The surface water was on several occasions
during summer almost free of oxygen. When the lake was stratified
during summer the phosphorus concentration in the hypolimnion
increased to ca 3 mg P^{-1} and that of NH_4-N to ca 20 mg N^{-1}.

Treatment of the Sediment

The polluted sediment was treated within an area of 0.012 km^2 in
the centre of the lake in the following way:

a)   A solution of trivalent iron-chloride with an iron content
of 175 g Fe^{-1} was mixed in a container and diluted ten-fold
with lake water. This solution was applied to the sediment
by a special prototype device developed by Atlas Copco's Central
Laboratories, Stockholm, Sweden. This device blows compressed
air into the sediment through a large number of small nozzles.
The upper layers of the sediment are thus lifted towards the
surface. The device is continuously drawn forward and at the
rear the chemicals are distributed through special tubes. The
chemicals are thus thoroughly mixed with the resettling sediment.

b) After the introduction of the iron-chloride, the sediment pH decreased. As denitrification processes are retarded in acidic environments lime in the form of $Ca(OH)_2$ was applied (180 g Ca^{-2}) in the same way as the iron-chloride. Thereby the pH of the sediment was adjusted to between 7.0 and 7.5.

c) After the lime treatment the oxidising agent Ca $(NO_3)_2$ was distributed into the sediment (141 g NO_3-Nm^{-2}).

Results

There was a rapid oxidation of the upper sediment layer (5-8 cm). Sulphides disappeared and the sediment turned brown. The bio-chemical oxygen demand of the sediment was reduced by about 50% (Fig. 6). The phosphorus concentration of the bottom near water decreased to about 40 ug P^{-1} and the oxygen conditions of the water near the bottom were improved. The recycling of phosphorus from the sediment was highly reduced (Ripl op. cit.).

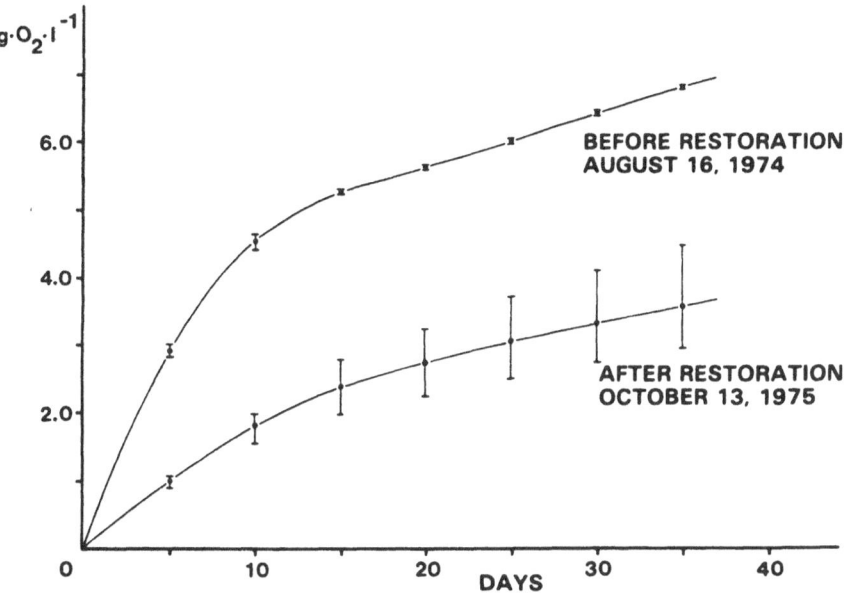

Fig. 6. Comparison of biochemical oxygen demand (BOD) curves of Lake Lillesjön sediments before and after sediment treatment. The maximum and minimum values in the different series are given. (From Ripl 1976).

Two weeks after the treatment the Secchi disc transparency
increased from 2.3 m to 4.2 m, the maximum depth of the lake.
A vigorous gas production of especially N_2 took place after this
period, stirring up the sediment. Within seven weeks after the
treatment all nitrate disappeared from the water column as well
as from the interstitial water. Besides the treatment of the
sediment 83 tons of macrophyte biomass (fresh weight) was removed
within an area of 0.018 km^2. Only a small part of the vegetation
was left to serve as a nesting place for waterfowl.

Costs

As the restoration of Lake Lillesjon was an experimental test of
a new technology it is difficult to state the relevant costs.
Ripl (op. cit.) has made a rough estimate of the costs to a total
of about 250,000 Sw Cr (1 US $ = ca. 4.45 Sw Cr).

REMOVAL OF EMERGENT VEGETATION AND PREPARATION OF THE BOTTOM
(STUBBLE MATS AND ROOT FELTS) IN CONNECTION WITH RAISING OF
THE WATER LEVEL (LAKE HORNBORGASJON)

Most of the lakes in the southern part of Sweden are shallow.
During the latter half of the 19th and the first decades of this
century many lakes were drained in attempts to obtain arable land.

Lake Hornborgasjon, Southwest Sweden, (area 30 km^2, maximum
depth 3 m before drainage) has been lowered five times since 1802.
The last two lowerings, 1904-1911 and 1932-1933, have caused an
almost complete overgrowth of emergent macrophyte vegetation (Fig.
7). The lowering of 1932-1933 has allowed the bottom to dry
completely each summer. Since 1933 the lake was completely canal-
ised, the inflowing water being led directly to the outlet. Since
the middle 1950s only a few small areas were free from emergent
vegetation and the maximum water depth in a diked-in portion during
the summer was 80 cm.

Before Lake Hornborgasjon was drained it was famous as a
nesting site and resting place for waterfowl. The Swedish government investigated the value of Lake Hornborgasjon with respect to
nature conservation and found that a restoration of the lake was
one of the most urgent conservation projects at present in our
country. The National Swedish Environment Protection Board organised
a broad study of the problems connected with an eventual restoration
of the lake. Limnological studies for a year showed that it was
possible to restore the lake (Bjork, 1972). A close co-operation
was established between the National Environment Protection Board,
the National Labour Market Board, the National Board of Forestry,
Seiga Harvester Co (manufacturer of amphibious machines) and the

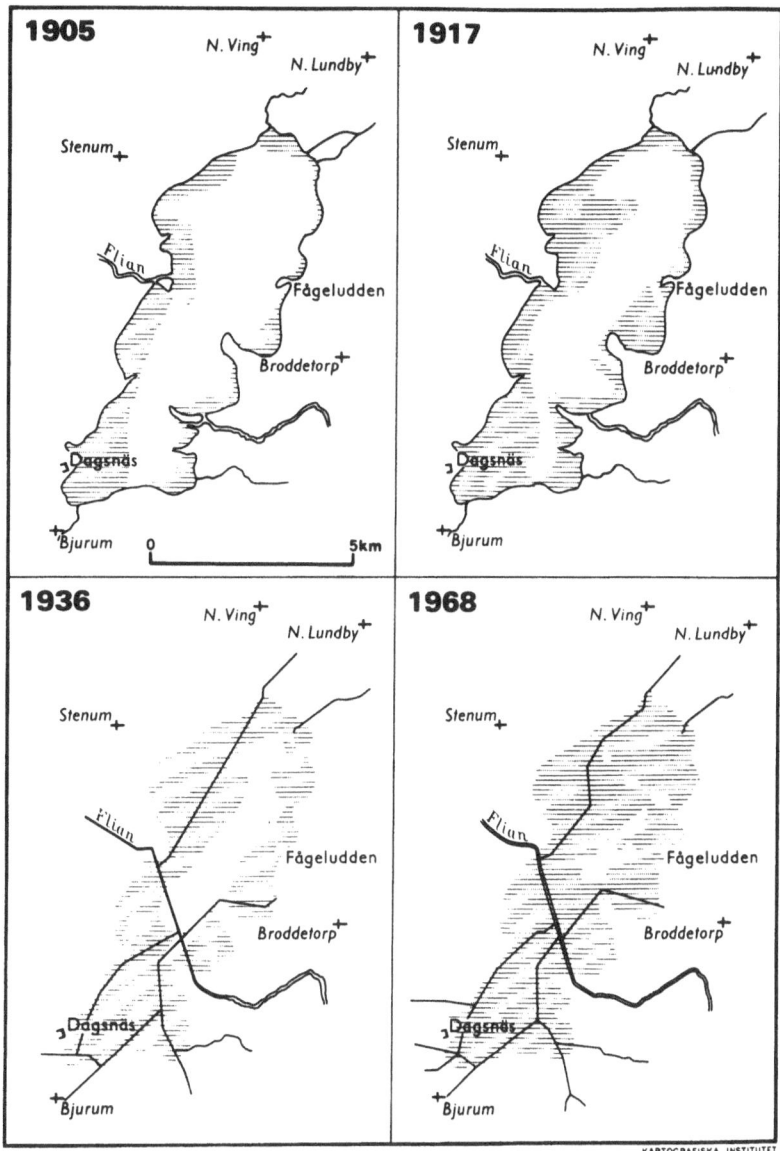

Fig. 7. Lake Hornborga. The last two water level lowerings (1904–1911 and 1932–1933) have caused an almost complete overgrowth by emergent macrophyte vegetation. At the beginning of the restoration investigation (1968) the emergent vegetation comprised stands of primarily common reed (about 12 km^2) and sedge together with mixed stands with willow bushes. The small open waters were choked with charophytes. (From Björk 1974b).

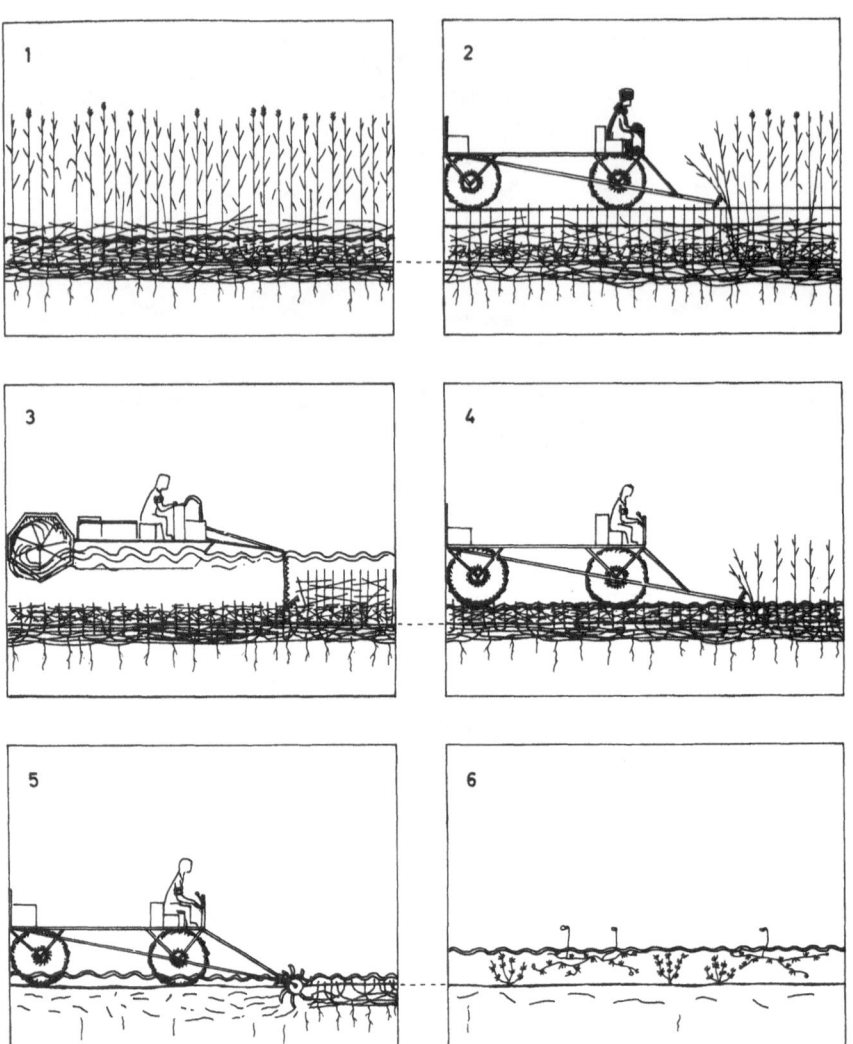

Fig. 8.   The Lake Hornborga restoration project. 1. Initial state.
The area overgrown by common reed, the consolidated mud
covered by coarse detritus and a dense root felt developed
in the top layer of the mud. 2. The work starts with cutting during the winter. Capacity 2 hectars per hour. The
reed material is being burnt. 3. During the spring high
water period, pontoon-equipped mowing machines are used for
shortening the stubble and clearing the bottom from the layer
of horizontal reed stems. 4. At low water in the summer
the green shoots are cut. 5. Final preparation of the bottom by a root cultivator. 6. Emergent vegetation is replaced by submerged plants, and bottom fauna communities
rich in species and individuals are developed. (From
Björk 1974b).

Institute of Limnology at Lund, and large-scale field experiments
began in 1968.

Since the last lowering at the beginning of the 1930s emergent
vegetation (Phragmites communis and Carex acuta) had invaded the
lake. Common reed was dominant in the northern part, while sedge
covered the main area of the southern part. The goal of the
restoration is to convert the reed area to open water (ca. 12 km^2)
according to Fig. 8, thereby changing the production of emergent
vegetation to production of submerged vegetation. The reed root
felt could be cut by amphibious rotor cultivator but the sedge root
felt caused a problem. It is impossible to remove. After preparation of the reed covered bottom the water level will be raised
(2.4 m maximum depth). As the catchment area of Lake Hornborgasjon
is large (616 km^2) the lake could be filled in one spring.

Within that area covered by sedge the root felt will float
to the water surface when the water level has been raised. This is
caused by gas production in and under the root felt through which
gas bubbles cannot penetrate. Within this area, however, it is
possible to create biotopes attractive to birds, with amphibious
excavators.

During the experimental restoration period when 1 km^2 was fully
prepared, the waterfowl fauna increased: Black-headed gull (Larus
ridibundus) from 5,000 to 8,500 couples, the Great crested grebe
(Podiceps cristatus) from 5 to 50 and pochard (Aythya ferina) from
20 to 110 couples (Bjork op. cit.). The bottom fauna was also re-
established within the experimental area.

The Lake Hornborgasjon restoration project will be used as a
case study for illustrating the possibilities to restore degraded
wetlands.

Costs

The cost of the investigations on the possibilities of restoring
Lake Hornborgasjon amounted by 1974 to about $600,000 US. Within
one year the Swedish government will probably decide on the final
restoration. The cost of the total project is in the order of
magnitude of $6,000,000 US.

ACKNOWLEDGEMENTS

In this paper, unpublished data from colleagues of the Lake
Restoration Researchers Team (University of Lund), is acknowledged
with thanks.

REFERENCES

Andersson, G. 1975. Results and experiences from the restoration
    of Lake Trummen. Proceedings from the Nordic Symposium on
    Water Research, 10th., Nordforsk. Helsinki, Finland, 495-505.

Andersson, G. Berggren, H. and Harrin, S. 1975. Lake Trummen
    restoration project. III. Zooplankton, macrobenthos and
    fish. Verh. Internat. Verein. Limnol. 19. 1097-1106.

Bengtsson, L., Fleischer, S., Lindmar, G. and Ripl, W. 1975. Lake
    Trummen restoration project. I. Water and sediment chemistry.
    Verh. int. Verein. theor. angew. Limnol., 19.

Bjork, S. 1968. Methods and research problems in connection with
    lake restoration. Vatten/Water, 24, 57-71 (in Swedish).

Bjork, S. 1972. Swedish lake restoration program gets results.
    Ambio, 1, 153-165.

Bjork, S. 1974a. European lake rehabilitation activities: Plenary
    lecture at the Conference on Lake Protection and Management,
    Madison, Wisconsin, USA, October 21-23, 1974 (Institute of
    Limnology, University of Lund, Sweden). Mimeographed.

Bjork, S. 1974b. The degradation and restoration of Lake
    Hornborgasjon. Institute of Limnology, University of Lund,
    Lund, Sweden. Mimeographed.

Cronberg, G., Gelin, C. and Larsson, K. 1975. The Lake Trummen
    restoration project. II. Bacteria phytoplankton and phyto-
    plankton productivity. Verh.int.Verein.theor.angew.Limnol.,19.

Dunst. R. C. et al. 1974. Survey of lake rehabilitation techniques
    and experiences. Tech. Bull. Dep. Nat. Res., Madison,
    Wisconsin, no. 75, 179pp.

Gelin, C. and Ripl, W. Nutrient decrease and response of various
    phytoplankton size fractions following the restoration of
    Lake Trummen, Sweden. Arch. Hydrobiol. (in press).

Graneli, W. 1975. Phosphorus dynamics - the impact of the sediments.
    Proc. Nordic Symp. on Water Research, Nordforsk, 10th.
    Helsinki, Finland, 213-222. (in Swedish).

Ripl, W. 1976. Biochemical oxidation of polluted lake sediment
    with nitrate - a new lake restoration method. Ambio.,5,
    132-135.

DISCUSSION: PAPER 18

P. LEENTVAAR   In the Netherlands, oligotrophic waters
in areas of sandy soil have been restored by
removing the sediments. The original oligotrophic microflora did not reappear and was
replaced by a richer eutrophic microflora as
a result of increased recreation in the region.
In the lower parts of our country our plan is
to supply the Loosdrecht broads with water of
the river Rhine after removing phosphates and
other substances. These lakes are about 2
metres deep and part of the lake is to be used
for water supply to Amsterdam while other areas
will be used for recreation and as a nature
reserve. We are clear that the water must have
its phosphate load removed but we are not treating the sediment and hoping that the nutrients
will wash out from it.

C. GELIN   If all sewage is diverted from the lakes
and they are filled with Rhine water after
phosphate removal and if the chemical composition of the sediments is similar to that of
Lake Trummen, many years will probably be needed
for recovery.

H. P. BLUME   A similar removal of phosphate is proposed
from lakes near Berlin, although here a decrease
in water weeds has occurred. My question is
one about the relationship between phosphorous
and nitrogen: if phosphate is removed the
nitrogen/phosphorus ratio may rise and the
stability of the reed beds in the lake may be
decreased. This may not be a good management
system.

C. GELIN   The phosphate in freshwater is usually the
factor limiting the development of phytoplankton,
but nitrogen can have this effect. If reeds
are to be retained this must of course be considered as a part of the management plan for
the lake.

E. M. NICHOLSON   How economic was the disposal of sediment
from the lakes whose treatment was described
in the paper? Is this sale of sediment usual
or was this a special case?

C. GELIN   In Lake Trummen the choice was between
                           filling the lake or restoring it, and the local
                           authority responsible chose restoration. The
                           costs, which totalled $580,000, were split
                           between the Swedish government and the town
                           authority. Sale of the sediment (about 30,000m^3
                           up to and including 1975) in connection with
                           the restoration was forced upon the community
                           by the need to do something with the lake to
                           alleviate deteriorating conditions. Whether
                           or not sediments can be used as fertiliser
                           depends on i.e. content of heavy metals, oil
                           etc. Only a few restorations by removing
                           sediments are reported so I cannot answer if
                           this sale of sediment is usual or not.

J. CHRISTIE   Did you consider attempting to extend
                           piscivore abundance instead of direct manipu-
                           lation of planktivore density? Our experience
                           in North America is that the lake strategy is
                           problematical.

C. GELIN   We have started experiments in Lake
                           Trummen directed towards removal of plankton-
                           eating fish (above all Abramis brama and Rutilus
                           rutilis). We wish to retain Esox lucius and
                           other predators. Our aim is to reduce the
                           phytoplankton abundance by increasing the
                           number of herbivorous zooplankton. This is
                           somewhat parallel to what Professor Shapiro
                           is attempting in Minneapolis, USA.

W. J. WOLFF   In my paper I commented on the influence
                           of Rhine water on the freshwaters of the
                           Netherlands and explained that there were
                           heavy silt depositions which contained toxic
                           metals and had an adverse effect on water
                           quality. One solution would be to dredge these
                           contaminated sediments - but this would displace
                           the problem from the lake to the land! Did
                           you encounter similar difficulties?

C. GELIN   I agree that sediments that are high in
                           toxic metals can pose problems for lake
                           restoration and create dumping problems if they
                           are removed.

19. SOME ECOLOGICAL IMPLICATIONS OF FRESHWATER SYSTEMS RESTORATION

B. Rorslett

Norksinstitut for Vandforskning

Blindern, Oslo 3, Norway

INTRODUCTION

Restoration of freshwater systems is a well-established practice
in Scandinavia. By this we mean the rehabilitation of ecosystems
which have been seriously impaired by human modification of the
water system or its surroundings (cf. Björk, 1975).

How restoration should take place at a given locality has up
to now been a question for ecologists and limnologists, in co-
operation with experts on the technical problems involved.

There exists, however, a definite relationship between the
technical process of lake restoration and the public interests
connected with the use of the water system. This aspect is
seldom mentioned when restoration schemes are discussed in Scandinavia.

THE ECOSYSTEM AND ITS SURROUNDINGS

Under experimental conditions we can make use of the ideal design,
a closed ecosystem. Real ecosystems are always more or less open,
i.e. subject to material transport in and out of the system. The
state of a particular ecosystem is closely related to events in
the ambient environment. A straightforward example of this is the
eutrophicated lake, which is overfertilised by nutrient supplies
from arable land, domestic wastes and other human activities in
the catchment area.

     An understanding of the structure and function of the eco-
system involves the need for knowledge of the environment
surrounding the ecosystem. Here we must be aware of an obvious
fact: this environment can be described ecologically, but it
operates within a social and political sphere. If this environment
is to be controlled satisfactorily, according to needs, an interaction has to take place between the experts in question and
political and administrative bodies in the society. As elsewhere
in scientific circles, little emphasis is placed on the political
role of the ecologists. This political aspect should, however,
be borne in mind when restoration schemes are being considered.
Direct attention should be paid to the implications of a restoration
project and to whom will benefit by it.

     In discussions we claim that restoration should return the
ecosystem to its original state. In practice it is a question of
restarting a succession to a particular point. However, the main
concern is how far back in a successional chain the ecosystem should
be led. A eutrophicated lake is an unstable ecosystem which is
kept out of balance by the continuous external nutrient load
(Margalef, 1975). By removing this load the ecosystem can be returned to a more stable complex and diversified oligotrophic state.
However this return to the stable, oligotrophic state is seldom
the ultimate aim of current restoration schemes. The actual criteria
for restoration depend on the utilisation of the ecosystem and water
resources in question.

     In densely populated districts there is a great need for re-
creational areas. From a social planning point of view the best
form of lake restoration therefore provides new recreation grounds.
The reason for this is again political, and outside the scope of
this paper. The social planner is aware of why restoration is
necessary from a user point of view. It should, however, be stressed
that this angle does not necessarily coincide with the ecologist's
viewpoint. This hidden contradiction may become apparent when the
decision is taken as to how a restoration project is to be carried
out.

RESTORATION MEASURES

Steps taken in connection with a restoration can be divided into
two main groups: (a) internal and (b) external. The internal
measures may be of immediate help only, while external efforts will
more often have a long term purpose.

Internal Measures

Among internal measures we can mention are:

1. <u>Management and control of biological components</u>. There are
many activities in this field. A well known example is removal of
components, such as harvesting of macrophyte vegetation. Efforts
involving a greater degree of management are the deliberate change
of production chains within the ecosystem. Examples of this include
stocking of plant-feeding fish in shallow, overgrown lakes, or
plankton-feeders (char) in regulated water courses and hydro-
electric basins. This approach may involve addition to already
existing production channels, or a complete re-establishment of
the food chains.

A characteristic of many restoration schemes is their destructive treatment of biological components in the ecosystems. This
applies in particular to the macrophyte vegetation, which is often
regarded as undesirable in a restored water system.

2. <u>Change of abiotic components in the ecosystem</u>. A number of
measures under this heading are of a "first aid" character. They
include oxygen supply to the hypolimnion by different types of
aeration, enhanced supply of oxygen-rich water, and so on. Such
efforts are often more effective in advertising concern for the
environment than in improving the situation in nature!

We have experience with such methods, namely the "Limnox"
aerator in Kolbotnvatn, a lake located outside Oslo. This lake
had reached a state of severe oxygen deficiency. Heavy blooms
by blue-green algae occurred and there was a high incidence of
dead fish during the winter. By artificial aeration of the bottom
water layer the oxygen conditions have been improved considerably.
Oxygen deficits still occur, however.

The danger of large-scale "first aid" measures is evident:
they are <u>visible</u>, lead to wide press coverage, and give politicians
a kind of <u>alibi</u>, showing that <u>they</u> are fully aware of the environment issue. In short, efforts <u>of</u> this nature can serve as a
deviation, paying more attention to the actual decay in the locality
than to the cure of the underlying causes. Experience from
Kolbotnvatn indicates that this danger does exist. Artificial
aeration of the lake has been performed since 1973, while treatment of the municipal waste water polluting the lake has been post-
poned. Since the restoration was begun, algal production and
nutrient content in the lake waters have increased, with the result
that the state of Kolbotnvatn is in fact deteriorating, despite
the restoration measures. However, fish kills, an obvious symptom
of a devastated lake, no longer occur.

Other efforts to restore a lake with a more long-term effect
include removal of the bottom sediments. The purpose of such efforts
is to stop the self-fertilising state, into which eutrophicated
lakes may enter. This method of lake restoration leads to profound
environmental transformation in the lake ecosytem. A reduced nutrient feedback from the sediments, increased depth and steeper lake
banks for example, have a negative effect on the growth conditions
of the macrophyte vegetation. As a matter of fact, this effect
is often intended in order to keep the macrophyte stands under
control. Not only macrophyte vegetation, but the wildlife in
general is adversely affected by such drastic restoration measures.

External Measures

External measures affect the ambient environment, thus indirectly
influencing the lake ecosystem. They may range from pollution
control by modern treatment plants to water management in the
adjoining watercourses. From a time aspect and financial point
of view such measures are more comprehensive than those performed
in the lake itself.

Among external restoration schemes we can include environmental preservation in the broadest sense. If the eutrophic state
of a lake is to be maintained, it may be essential to provide a
sufficient nutrient supply. This may be the case with nutrient-
rich bird lakes situated in the agricultural districts of South
Norway. The natural values of these lakes are closely associated
with the agricultural character of the surroundings.

RECREATIONAL AND NATURAL RESOURCE
INTERESTS RELATED TO LAKE RESTORATION

The potential conflict between basic viewpoints as to why a lake
restoration should be carried out and whom it is to benefit, may
be illustrated by an example taken from Norway.

Østensjøvatn is a shallow, vegetation-rich lake situated just
outside Oslo. The bird and plant life are unusually rich.
Luxuriant reed swamps, composed mainly of Typha, Phragmites,
Glyceria and Acorus, encircle the lake. The open lake itself
harbours large stands of water lilies (Nuphar and Nymphaea) and
submergent vegetation (Ceratophyllum and Elodea).

In the areas around Østensjøvatn, satellite towns have grown
up after the last world war, with a total population of approximately
60,000. This has over the last years led to severe nutrient loading
of the lake, due to the poor sewage system and lack of waste water
treatment.

The lake is now situated in the midst of a densely populated district and it is the most accessible recreational area for the local population. There are, however, significant conservation interests associated with Østensjøvatn and its surrounding wetlands. Formerly the wetlands protected the wildlife and the lake itself from public access.

The increasing pollution of Østensjøvatn has aroused much anxiety. Biologists saw the natural conditions threatened, and the municipality feared that a potential recreational area would become less valuable.

Planning experts in Oslo council put forward plans for an extensive restoration of the lake and its surrounding area. According to the original proposal (outlined in Fig. 1) Østensjøvatn was to be divided in the middle by a rock fill. The northern part of the lake was to be dredged and sand laid on the bottom, thus serving recreational purposes such as bathing, boating and fishing. The hygienic standard of this part of the lake was to be improved by chlorination.

On the other side of the rock fill, at the southern end of Østensjøvatn, the lake was to be preserved as a nature "reservation" with abundant bird life and luxuriant vegetation. To protect the hatching grounds of the birds it was suggested that channels be dug along the lake shores. The restoration of the southern area also included the removal of macrophyte vegetation in the open waters. The areas surrounding the lake were to be taken care of by laying lawns, making footpaths etc.

The ambitious plans met with some opposition, mainly from local nature preservation societies. Some of the restoration schemes were started, while the most controversial issue – division of the lake – was abandoned for the time being. Digging of channels, planting grass along the littoral zones and adjusting the water level was carried out in the mid 1960's. Harvesting of water vegetation was begun somewhat later, from 1972, mainly affecting the floating-leaved vegetation. There were two reasons for the removal of macrophyte vegetation: to remove nutrients and organic matter from the lake, and also to prevent a feared successional overgrowth.

In parallel with the efforts carried out in Østensjøvatn, the sewage system in the catchment area was improved. However, un-treated waste water is still entering the lake. The completion of a sewage system is estimated to cost 10-20 million N. Kr.

No freshwater locality in Norway is afforded so much care and attention as Østensjøvatn. It may be worthwhile looking at the

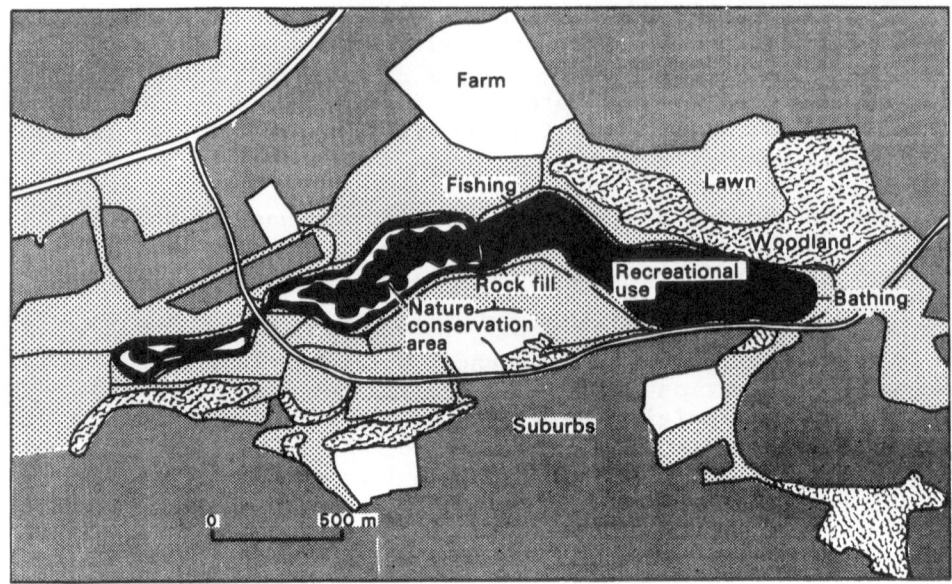

<u>Fig. 1.</u>   Proposals for Restoration of Østensjøvatn and Surrounding
            Wet Lands.

effects of the lake restoration in practice. The development in
Østensjøvatn has during the past ten years (1965 - 1975) taken the
following course:

-   the bird life has to an increasing extent been dominated by
    Blackheaded gulls (<u>Larus ridibundus</u>), now nesting in thousands
    around the lake;

-   both gulls and ducks have been attacked by botulism poisoning
    in recent years; the most severe case (1975) killed over 500
    individuals, or more than 10% of the bird population in
    Østensjøvatn. Recently, a new serious case of botulism poison-
    ing has been detected (April, 1976);

-   oxygen deficiency prevails in the winter, and fish kills occur;

-   the area covered by floating-leaved vegetation has been reduced
    from 50 to 10% since 1972;

-   blooms of blue-green algae (especially <u>Anabaena planctonica</u>)
    occur in the summer.

It is evident that the species diversity has been reduced in the Østensjøvatn ecosystem. Algal blooms and botulism poisoning indicate a serious failure in ecosystem functions. The authorities are inclined to revert to a more drastic line of action, following restoration principles laid down in Sweden. Such measures would include dredging and removal of lake sediments.

Østensjøvatn with its immediate surroundings has become an important recreation area. Further lake restoration measures can help to retain the recreational character of the environment and the lake itself, but may at the same time prevent the preservation of the natural resources associated with the lake. In order to verify this assumption, we need to take a closer look at the function of this ecosystem.

Ecological investigations of Østensjøvatn, carried out since 1960 have resulted in some self-contradictory data on the locality. Here are some examples:

- the water masses are more or less oxygen free during winter when the lake is ice-bound. The bottom sediments, however, consist of well-mineralised clay gyttja with a high density of bottom animals. The so called "culture" mud or "black" mud with iron (II) sulphide is absent from the open lake areas;

- the great supply of plant nutrients should stimulate the development of abundant macrophyte vegetation. The biomass of this vegetation can reach 400 g Cm^{-2}, but is usually below 300 g Cm^{-2}. These values can be found in many Norwegian water systems, and far higher values have also been registered. The macrophyte vegetation in Østensjøvatn is thus far less productive than expected from the nutrient load;

- the state of Østensjøvatn today, encircled by apparently luxuriant vegetation, suggests that a rapid overgrowth is taking place in the lake. Comprehensive lake restoration measures, checking the dreaded overgrowth of macrohpytes, are hence the subject of much attention both from the press and from the general public;

We have studied a representative selection of aerial photographs from Østensjøvatn taken during the years 1937-1975. These photographs show that there has at times been a negative overgrowth in parts of the lake. However, a rapid development has taken place with respect to the composition of vegetational cover: Schoenoplectus has gradually been replaced by Typha. It is interesting to note that a parallel case has been reported from the lake Lyngby Sø in Denmark (Olsen 1964). Olsen also established that the overgrowth in this lake has been negligible since 1912.

- the light conditions in Østensjøvatn are poor, with a Secchi
  depth of approximately 0.5 - 1 m. During periods with algal
  blooms the transparency can be reduced even more. However,
  dense populations of Elodea are still be be found at a depth
  2 m. These plants are spread over the bottom like a carpet and
  only attain a height of a few centimetres;

- the maximum distribution of floating-leaved vegetation was
  observed on aerial photographs from 1952. This was before the
  catchment area around Østensjøvatn was developed and urbanised.
  In Østensjøvatn, Nuphar lutea is the most common floating-leaved
  species and is otherwise to be found in heavily polluted areas.

Properties such as those mentioned above suggest that the ecosystem at Østenjøvatn in many respects functions normally, even with
the pollution load to which it is exposed. This may be due to the
short retention time of the water masses, since the theoretical
replacement time is only about one month.

The vegetation zones in and around the lake also accumulate
significant amounts of nutrients, which would otherwise have
entered the lake. The phosphorus content in the littoral vegetation
was equal to approximately 300 ug P 1^{-1} if the bound phosphorus
entered the lake (Rorslett and Skulber, 1976).

It is obviously necessary to take a critical look at the
effects of lake restoration in Østensjøvatn. As was mentioned
earlier, the main intention of the existing plans was to limit
the growth of macrophyte vegetation. Harvesting of aquatic
vegetation removes insignificant amounts of nutrients from the lake
water, but has affected the equilibrium between the macrophyte
vegetation and planktonic algae (Rorslett and Skulberg, 1976).
When the macrophyte vegetation is removed the algae will have better
light and nutrient conditions, and this is clearly evident from the
subsequent algal blooms. Investigations in Østensjøvatn indicate
that primary production of the algae on a daily basis exceeds the
amount of organic matter which the harvested water plants represent.
Removal of water plants is therefore in direct conflict with the
purpose of the lake restoration for recreational usage: the
primary production in the lake increases since some producers are
replaced by organisms with a higher turnover rate, which in turn
supply the lake with large amounts of organic matter. The protective effect of the vegetation with respect to nutrient supply will
therefore decrease with harvesting. The channels along the lake
shores are also responsible for this. By opening new littoral
zones for the pioneer phase in the reed swamp communities, production
is increased in the littoral areas as a whole. Nutrient-rich
drainage water can in addition enter the lake through the channels.

The increased production in the littoral zones supplies the shallow channels with large amounts of plant materials which are decomposed, thus causing an oxygen deficiency. It is possible that this oxygen deficiency is the cause of the high incidence of botulism poisoning among the bird population of the lake.

New and more drastic plans for lake restoration in Østensjøvatn will, it seems, make the conditions in the lake worse. This is because the restoration plans are working against the ecosystem instead of with it. The dramatic environmental transformation caused by the clearing of vegetation and removal of the bottom sediments would result in a less complex and diversified ecosystem. A simplified ecosystem of this kind cannot be expected to function satisfactorily either from a natural resources or a recreational point of view. It is therefore clear that lake restoration which preserves the natural values and at the same time keeps the existing ecosystem intact, is incompatible with the use of the lake for other purposes. The restoration procedure thus demands a choice between different values in society.

REFERENCES

Bjork, S (1975). Översikt over sjorestaureringsproblematikken. Förening för Vattenhygien, Vårmøte maj 1975,

Margalef, R. (1975). External factors and ecosystem stability. Schweiz. Z. Hydrol., 37, 102–117.

Olsen, S. (1964). Vegetationsaendringer i Lyngby Sø. Bot. Tidssk., 59, 273–300.

Rørslett, B. and Skulberg, O. M. (1976). Vegetasjonsundersøkelser i Østensjøvatn, Oslo kommune, 1974–75. Norwegian Institute for Water Research, Report O-69/72.

DISCUSSION: PAPER 19

C. GELIN   Surely it is an academic, and not very useful, point to call the measures taken in Sweden rehabilitation instead of restoration?

Did I understand you correctly when you said that you thought that Lake Østensjøvatn would recover in spite of the sewage still being discharged into it - and, if so, how could this happen?

B. RORSLETT   Yes – because the macrophytes, which contain
                       a large amount of nutrients, intercept those
                       coming in.

W. J. WOLFF   You said that the retention time in Lake
                       Østensjøvatn was one month but that the role
                       of sediments in binding nutrients was hard to
                       judge: what would you say was the ratio of
                       sediments to inflow as sources of nutrient?
                       In estuaries marine phanerogams can act as
                       "pumps" transferring nutrients from sediments
                       to water: did this happen in your case?

B. RORSLETT   The standing amount of nutrients in
                       vegetation can be several times as great as
                       in the water, but the harvesting has only
                       been of floating vegetation and this is clearly
                       insufficient.

W. J. WOLFF   Harvesting, however, removes a mechanism
                       as well as a standing stock.

B. RORSLETT   I think the short retention time is the
                       dominant factor in determining the features
                       of this lake.

R. GOSSEN   Has the micro flora of the sediments been
                       examined?

B. RORSLETT   Very little research has been done on
                       these matters in Norwegian lakes.

C. GELIN   Very little periphyton was found in one
                       example that has been studied.

J. CHRISTIE   Could you give us some other examples in
                       which low oxygen levels have been associated
                       with botulism?

B. RORSLETT   I think the Great Salt Lakes in the United
                       States was an example.

E. M. NICHOLSON   Many botulism cases in Europe have been
                       associated with hot weather, for example the
                       recent outbreak on the Coto Donana in Spain.

B. RORSLETT   In Norway Lake Østensjøvatn is the only
                       recorded botulism case and the condition
                       developed even at Easter time when the weather
                       was cold.

H. A. REGIER   Why did you object to the Swedish approach
                                to lake rehabilitation? Was it because you
                                felt that it was attacking symptoms rather
                                than causes?

B. RORSLETT   I objected because if ecosystems with
                                slow turn-over rates were desired it would
                                have been more sensible to have left the macro-
                                phytes in place rather than the algae.

A. D. BRADSHAW   If you had made the decisions what would
                                you have done with the lake described in your
                                paper?

B. RORSLETT   I would have left it alone.

A. D. BRADSHAW   Would this have retained a reasonable
                                vegetation? Other lakes left on their own
                                have not necessarily recovered.

B. RORSLETT   This lake is distinctive in its features
                                and I believe it would have recovered. It had
                                changed little since the early 1950s despite
                                the load of pollutants. This, I believe, is
                                due to the short residence time of the water
                                in the lake and the substantial macrophyte
                                vegetation.

E. M. NICHOLSON   The lake may have had a special regime
                                but your thesis is that one has to choose
                                between the preservation of the ecosystem and
                                the provision of facilities for recreation.
                                Is it your thesis that no compromise between
                                these uses is possible?

B. RORSLETT   I believe that no compromise was possible
                                in this case: there had to be a choice. Com-
                                paring my example with that given by Dr. Gelin,
                                the Swedes have chosen one way and we in
                                Norway chose another.

R. W. WEIN   You said that the plan developed had been
                                criticised locally? At what stage did local
                                people get involved? In Canada there is a
                                considerable suspicion of planners in most
                                local communities - what are the points of
                                conflict in Norway?

B. RORSLETT — Local objections were brought forward in most cases at too late a stage. Centralised planning has been a fact of Norwegian life for 40 years: many people are critical of it but are nonetheless used to living with it.

M. W. HOLDGATE — How were the plans for rehabilitation of this lake developed?

B. RORSLETT — By the Central Planning Agency. They were put before the public at a stage, and in a state, where objections from the public could not readily be incorporated. Alternative plans were not exposed for discussion.

H. A. REGIER — E. Odum has argued for a mosaic approach, zoning one area of the environment for one use and another for other functions. Multiple use is nonetheless hard to achieve. Are you in general agreeing with Odum's approach?

B. RORSLETT — Yes. I think it is better to use areas or parts of the system for distinctive purposes rather than attempt to use the whole area for several purposes.

P. MELLQUIST — I would emphasise that there are other views in Norway as well as that put forward by Mr. Rorslett. Alternative plans, moreover, are often laid before the public.

B. RORSLETT — There are, of course, different points of view on such matters.

20. FRESHWATER FAUNA AND FISH-STOCKING PROGRAMMES IN ICELAND

T. Gudjonsson

Institute of Freshwater Fisheries

Reykjavik, Iceland

RIVERS

Iceland is geologically speaking a young country, formed by volcanic eruptions. The bed-rock is basalt, whereas those of the neighbouring countries are older rock formations. This fact leads among other things to rivers and lakes with different character-istics from those of the countries both west and east of Iceland.

The rivers of Iceland, about 250 in number, differ greatly in length and volume of flow. There are three types of rivers according to their origin, i.e. spring-fed ones, direct run-off rivers and glacial rivers. In their course of flow they often be-come mixtures of these types.

These types of rivers have different characteristics. The spring-fed rivers have an even flow all the year around, their temperature is low about $4^{o}C$ at their origin all through the year, and the river beds are generally u-shaped. The direct run-off rivers vary greatly in flow, depending on the precipitation and the air-temperature. Thus their temperature is low in winter when there is some freezing. They usually have big spring floods whereas in the summer the flow varies with the precipitation. The water temperature rises in the summer and is often lower in the winter than that of the spring-fed rivers. The river beds are generally rather flat in cross-section and shift at times. The glacial rivers vary greatly in volume. Their flow is greatest during the warmest part of the year from June to September. They may change considerably in volume from day to day and even during the 24 hours of a day. The glacial rivers change course quite

351

often in flat areas. At their source at the edge of the glacier
their water temperature is 0°C all the year round. During the
warmest part of the year they gradually warm up to as much as
15-20°C, if they flow long distances, especially across low
ground. The colour of the glacial rivers is muddy, whereas that
of the other types is most often clear. There are many glacial
rivers in the country and some of them are among the longest and
carry the greatest volume of water. The glacial river Thorsa is
the longest river in Iceland, being 230 km. long.

LAKES

The lakes in Iceland are rather few compared to what could be
expected in a country which has been glaciated. It is likely that
many lakes have been covered by lava fields in prehistoric time.
There are only a little over 80 lakes 1 km^2 or more in area; about
800 are from 0.3-1.0 km^2 and about 900 less than 0.3 km^2. The total
area of all lakes in the country is about 1,200 km^2 or little more
than 1% of the area of Iceland. The largest natural lake is Lake
Thingvallavatn in South Western Iceland. This is 83 km^2 in area.
The lakes vary greatly in depth, the greatest recorded being 220 m.

The temperature of the water in lakes varies considerably. In
shallow lowland lakes the temperature may reach 20°C in summer,
whereas the deep lakes warm up slowly. In Lake Thingvallavatn,
which is a deep lake with a maximum depth of 114 m., the temperature may reach about 12°C at the surface during the summer. In
highland lakes the temperature is considerably lower. During the
greater part of the year the lakes are frozen over. Poulsen (1939)
has divided lakes in Iceland into three groups with respect to their
height above sea level and the average number of days they are free
of ice.

I	0-100 m above sea level 158 days
II	101-300 m above sea level 137 days
III	300 m above sea level 115 days

Only a little research work has been done on the freshwater
invertebrate fauna of Iceland. More attention has been paid to
the freshwater fishes including anadromous fishes. Ostenfeld
and Wesenberg-Lund (1905) studied the zooplankton of Lake Thingvallavatn and Lake Myvatn. In Lake Thingvallavatn they found
Protozoan, Rotatoria and Cladocera, and in Lake Myvatn, in North
Eastern Iceland, they found Chironomid larvae besides the other
ones mentioned. Thoroddsen (1913 and 1914) found in several highland lakes Limnea species, Pisidium, Lepidurus arcticus, Chironomid
larvae and Trichoptera larvae. Lindroth (1931) names several
species of Insecta living in fresh water, among them Simulium

species. Poulsen (1939) studied the Crustaceans. He found 47
species of this class in Icelandic lakes. Several papers on
freshwater animals have been written in the series The Zoology of
Iceland.

Important groups of food animals eaten by salmon and trout in
the river Ulfarsa, are listed by Tomasson (1975), i.e. Simulidae,
Chironomidae, Trichoptera, Gastropoda, Oligochaeta, Hydracarina
and terrestrial insects. Adalsteinsson (1975) lists the food of
char in lake Myvatn sampled in June to October 1972 to 1973:
sticklebacks, Limnaea peregra, Chironomidae, and Crustaceans such
as Lepidurus articus, Eurycercus lamellatus, Daphnia longispina
and Cyclops. The seasonal variation in food of the char was as
follows:

Spring: midges, mostly pupal stage
Summer: benthic Crustacea
Autumn: plankton
Winter: midges larvae, sticklebacks.

The fresh waters in Iceland are low in dissolved substances.
Chemical analyses of water from various part of the river system
Olfusa-Hvita in 1972 indicate the total dissolved substances lie
between 40 and 60 mg/l. pH is most often between 6.5 to 8.0.
HCO_3 is normally about 20-30 mgl^{-1}, and Ca is from 3-5μ gl^{-1}.
NO_3 runs from 20 to 170μ gl^{-1}, and PO_4 20-30μ gl^{-1}.

FRESHWATER FISHES

There are only five species of freshwater fish native to Iceland.
These are the Atlantic salmon (Salmo salar), Brown trout (Salmo
trutta), both a sea-run variety the sea-trout and a land-locked
one; Arctic char (Salvelinus alpinus) both sea-run and land-locked;
European eel (Anguilla vulgaris) and the Three-spined stickleback
(Gasterosteus aculeatus). These species have come into fresh
water from the sea. Rainbow trout (Salmo irideus) has been
introduced into two hatcheries and Pink salmon (Oncorhyncus gorb-
uscha) has entered many rivers in late summer off and on since
1960, originating from releases of pink salmon into rivers in
Northern U.S.S.R. In the summer of 1974 over 200 Pink salmon,
more than any other season, were caught in many Icelandic rivers.

The Atlantic salmon is the most valuable species of freshwater fish in Iceland. In this paper it is called a freshwater
fish in spite of its anadromous habits because fishing for salmon
in the sea is not allowed in Iceland. The Atlantic salmon spawns
in fresh water in the autumn and spends 2-5 years, most often 3

or 4 years, in fresh water before it migrates to the sea, where it
spends 1-3 years before returning to fresh water to spawn. The
Atlantic salmon frequents between 70-80 rivers in Iceland, most
of them are located in the western half of the country.

The Brown trout is found in many lakes all over the country,
and the sea-trout in many rivers especially in Southern and Western
Iceland. The char being an arctic species, is found in most lakes
and in colder streams, most frequently in North Western, Northern
and North Eastern rivers. The early freshwater life of these last
mentioned species is somewhat similar to that of the Atlantic
salmon, but differs in the marine phase, since they stay close to
the rivers of origin in the sea and spend the winter in fresh
water, whereas the Atlantic salmon stays away for one or more
years and migrates long distances in the ocean on occasions as far
as to Western Greenland and Norway from Iceland.

The European eel is found mostly in Southern and South Western
Iceland and is of little commercial value. The sticklebacks are
found in lakes except perhaps in the ones of high altitudes.

EXPLOITATION OF FRESHWATER FISHES

Fishing rights are privately owned. They go with the land that
adjoins the rivers and lakes. River rights are usually in the
hands of the farmers who own the land, since most of the fishable
reaches are in agricultural areas, whereas many of the lakes are
far away up in uninhabited regions.

For almost a thousand years the fishing gear used for catching
salmon, trout, and char was very primitive and overfishing was
generally unlikely. A little over a century ago, the fishing gear
was improved due to increase in demand for salmon for export.
There was thus a great change in the utilisation of salmon, from
being almost exclusively consumed by the farmers themselves to
become an item of commerce. This development of the salmon fishery
has continued and has recently to some extent also applied to
Brown trout and char.

The rate of exploitation of freshwater fishes through the
centuries is not known. Official catch records are on hand only
since 1897. Of these, reliable records of catches of salmon do
not extend for much more than about 30 years. Only scanty information is available on catches of Brown trout and char.

During the second and third decades of this century the
fishing pressure was intensified, causing concern that salmon was
being overfished in some rivers and predictions that much stricter

rules for salmon fishing than were in existence at the time would
be necessary. This led to the passing of a new freshwater fishery
law in 1932, with increased restrictions on salmon fishing. The
law was also extended to new fields in conservation and management
of the freshwater fisheries. Since 1932 alterations have been
made to the freshwater fisheries law. These have elaborated on
further restrictions on salmon fishing and added new items of
importance as for instance restrictions on import of live freshwater fish and their eggs.

ADMINISTRATION OF THE FRESHWATER FISHERY

One of the important new features of the freshwater fisheries law
from 1932 was the establishment of a directorate for freshwater
fisheries. This was done in 1946, under the Ministry of Agriculture, to whom the director of freshwater fisheries reports. He is
responsible for administering the freshwater fisheries matters,
collecting statistics, etc. The director is the head of the Institute of Freshwater Fisheries.   There is an advisory committee on
freshwater fisheries named the Freshwater Fisheries Council,
directly answerable to the Ministry.

At the local level fishing associations administer the
fishery. These operate according to law on each river or river
system and on lakes. All fishing right owners are obliged to be
members. The fishing associations have the authority to manage
the fishing, which in most cases is leased for each river for
angling to individuals or angling clubs. They take care of improvements, hire bailiffs, often build and run lodges for anglers, and
have stocking programmes, in which case they normally buy salmon
smolts to be released into the rivers annually. Some buy one
summer old parr and release them in barren streams above impassable waterfalls. The lakes are also stocked, where necessary.
The fishing associations also regulate water flow in rivers and
build fish passes. For the last 25 years an average of about one
fish pass has been built annually. Thus about 400 km of rivers
have been made accessible to sea-run fish.

FISH CULTURE

In 1961 work was started on building an experimental fish farm by
the Government of Iceland at the farm of Kollafjordur, which is
located only 20 km north of Reykjavik, the capital of Iceland. The
purpose of this fish farm was to experiment with fish cultural
techniques and study the survival of hatchery fish in nature, as
well as supply artificially reared fingerlings and smolt for releas-

ing into rivers and lakes. Great progress has been made in rearing of Atlantic salmon smolts. Up to 15% of the artificially
reared smolts released at the Experimental Fish Farm, have returned as mature salmon from the sea the average return being about
10% for the last four years. Isaksson (1976) has found that
keeping the one year smolts under natural light during the winter
results in good returns of mature salmon from the sea, whereas
the ones that are kept in buildings with electric lights on all
winter give practically no returns.

The Kollafjordur Experimental Fish Farm is the first rearing
station in Iceland to produce smolts in considerable numbers to
sell for releasing into rivers in various parts of the country.
During the last years it has produced 150,000 smolts annually, and
sold from 30,000 to 100,000 smolts each year to fishing associations and individuals to be released in up to 50 rivers. The rest
of the smolts have been released inside the Fish Farm. They
migrate to the sea from there and return to the ponds of the Fish
Farm. The largest return of salmon in one season (1975) numbered
almost 7,000 fish. This run to the Fish Farm amounted to 9.5%
of the total number of salmon that were caught in Iceland that
year.

At the Experimental Fish Farm work has been done on Brown
trout and char as well as salmon. Lake char has proven to be a
hardy species in rearing with the occasional exception of the fry,
although a high mortality has been experienced during the hatching
stage. Char has been raised on some scale for the table, and
the artificially reared char has been found to be a palatable food
fish.

The percentage of returns of the Atlantic salmon liberated
as smolts at the Kollafjordur Fish Farm indicated that sea-ranch-
ing could be a paying proposition, since the cost of building
release ponds and traps on small streams is not high. The salmon
smolts can be bought from a rearing station and liberated into
release ponds. The salmon returning to the traps can then be
marketed. The market price of salmon is an important factor in
how profitable this business will turn out. Iceland is specially
well suited for sea ranching since fishing for salmon in the sea
is not allowed. Thus the greater part of salmon runs will not be
caught in the sea as is the case in most other countries where
the Atlantic salmon is found.

There is a great interest in fish culture in Iceland, mostly
in salmon culture where the main emphasis is on smolt rearing.
Six rearing stations are in operation at the present besides the
Kollafjordur Experimental Fish Farm. In one instance, at Laros in
Western Iceland, salmon young are being released into an artificial

lake, 160 ha in area, made by walling off an estuary. The salmon
feed on natural food in the lake. They migrate to the sea at the
smolt stage and return as mature salmon to a trap at the outlet
of the lake, Larvatn. This undertaking has been successful.

SALMON FISHING ON THE INCREASE

The salmon fishing has decreased in some rivers due to over-
exploitation. With administrative and new managerial endeavours
after the passing of the new freshwater fisheries law in 1932, and
especially after the formal governmental agency for administration
was established in 1946, the salmon fishing has been improving
mostly since 1960. The fishing associations have played an important part in this development locally.

Figure 1 shows the catches of Atlantic salmon from 1910-1975,
based on the averages of five year periods. From 1910 to 1950
the average catches were close to 15,000 salmon. The fishing
efforts were steadily being increased in the big rivers until about
1940, showing increase in catches, whereas the catches in the
smaller rivers were declining. After about 1935 the fishing in
many of the smaller rivers changed gradually from netting to
angling, and netting in the large rivers decreased allowing a
larger number of salmon to spawn. These changes, together with
stocking of the rivers, have resulted in greatly improved fishing,
as is shown in the figure, the average catch for the last five year

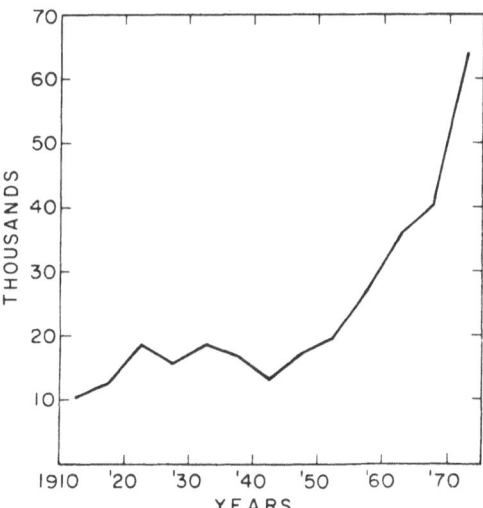

Figure 1.   Number of Atlantic Salmon caught in Iceland from 1910-1975
             based on the averages of five year periods.

period, i.e. from 1970-1975, being more than four times that of
the period from 1910-1950.

The policy of managing the Atlantic salmon in Iceland during
the last decades has been designed to ensure that sufficient salmon
escape to furnish enough ova and provide parr to fully utilise
the natural rearing capacity of the rivers, so producing optimal
number of smolts.

Due to a short growing season and rather low temperatures in
the rivers during the summer the production capacity is rather low.
In order to enlarge the salmon run far beyond what nature can
produce, fishing associations are advised to release artificially
reared salmon smolts into rivers just before the migration time in
the spring. It is estimated that up to 300,000 smolts are now
released annually in Iceland. The results in the rivers have not
been studied until recently. It is obvious that a great deal has
to be learned about release activities, choice of stock to be
released into individual rivers, and the time of release. Release
ponds for smolts built on the river banks should solve the last
mentioned problem as well as facilitate the transportation of smolts
from the rearing stations to the rivers.

The salmon runs in Icelandic rivers can be increased
considerably from what nature can produce through release of
artificially reared salmon smolts. This can be very profitable
for the fishing associations, which lease their fishing for
angling. It is difficult at this stage to say where the limit for
such activities lies. It can be expected that the demand for
angling at various times will be the regulating factor.

The lake fishery has changed in that until recently people
living in the country have netted the lake and thus kept the
Brown trout and char stocks mostly at an acceptable size. During
the last decades, the people have been moving from the country to
the towns. The people staying behind are so few in number that
they do not find time to fish the lake the way the country people
used to and the angling pressure is too low to help keeping the fish
stocks in a desirable conditions. This had led to overpopulation
especially of char, in many lakes. Experiments have been made to
reduce excessive lake populations and adjust them to the available
food supply. It is too early to talk about success in this endeavour.

Although there seem to be many cases of overpopulation of
lakes, there is also at least at the present one case of over-
exploitation of char in a lake, affecting one of the most
productive lakes in Iceland, the Lake Myvatn in the North Eastern
region. Corrective measures involving restrictions on fishing and
the release of artificially reared fingerling char, are now being
planned.

REFERENCES

Adalsteinsson, Hakon. 1975. Fiskstofnar Myvatns. Natturufroed-
    ingurinn, 45, 154-177.

Alpingistidindi, A. 1930. 6. hefti, Þingskjal 343, 810-883.

Armannsson, H., et al.
    1973. Efnarannsokn vatns. Vatnasvid Hvitar-Olfusar, einnig
    Þjorsa vid Urridafoss 1972. Reykjavik: Orkustofnun, Vatnamoe-
    lingar, Rannsoknarstofnun Idnadarins,

Einarsson, Þrleifur. 1968. Jardfroedi. Reykjavik: Heimskringla,

Fridriksson, Arni. 1940. Lax-rannsoknir 1937-1939. Atvinnudeild
    Haskolans. Rit Fiskideild. 2.

Gudjonsson, Thor. 1967. Salmon culture in Iceland, ICES, C.M.
    1967/M: 24.

Gudjonsson, Thor. 1973. Smolt rearing techniques, stocking and
    tagged adult salmon recaptures in Iceland. Spec. Publ. Ser.
    int. Atl. Salmon Found., 4, 227-235.

Isaksson, Arni. 1976. The improvement of returns of one-year
    smolts at the Kollafjordur Fish Farm 1971-73. Isl.
    Landbunadar Rannsoknir, 8, 19-26.

Isaksson, Arni. 1976. Preliminary results from the 1973 tagging
    experiments at the Killafjordur Experimental Fish Farm.
    Isl. Landbunadar Rannsoknir, 8, 14-18.

Isaksson, Arni. 1976. The results of tagging experiments at the
    Kollafjordur Experimental Fish Farm from 1970 through 1972.
    Isl. Landbunadar Rannsoknir, 8, 3-13.

Lindroth, Carl H. 1931. Die Insektenfauna Islands and ihre
    Probleme. Uppsala.

Ostenfeld, C. H. and Wesenberg-Lund, C. 1905. A regular fort-
    nightly exploration of the two Icelandic lakes, Thingvallavatn
    and Myvatn. Proc. R. Soc. Edinb., 25, 1092-1167.

Poulsen, Erik M. 1939. Freshwater crustacea. (The zoology of
    Iceland, Vol. 3, part 35). Copenhagen and Reykjavik:
    Munksgaard.

Rist, Sigurjon. 1956. Islenzk votn, 1. Reykjavik: Raforkumala-
      stjori, Vatnamoelingar.

Thoroddsen, Þ. 1913. Ferdabok, 1. Kaupmannahofn: Hid islenzka
      froedafelag.

Thoroddsen, Þ. 1914. Ferdabok, 11. Kaupmannahofn: Hid
      islenzka froedafelag.

Tomasson, Tumi. 1975. Undersokning av juvenila lax- och
      oringpopulationer i Ulfarsa, en liten islandsk alv. Manuscript.

Zoology of Iceland, 1938. Fridriksson, Gudmundsson: Copenhagen
      and Reykjavik. In progress.

DISCUSSION: PAPER 20

R. W. J. KEAY   I have two questions, both relating to
the volcanic nature of Iceland. First, are
any geothermal effects visible either because
they change the ecosystem or are useful in
managing the fishery? Secondly, has there
been any problem with the poisoning of fish
by volcanic emissions such as those of sulphur
dioxide, possibly in crater lakes?

T. GUDJONSSON   We have many thermal waters in Iceland
and some discharge into and raise the
temperature of rivers. A discharge of 200
cubic metres per second at 97°Centigrade can
heat a river a kilometre or so from the point
of emission and effectively block the passage
of salmon. In other instances the rate of
flow is so small that there is no detectable
effect. Geothermal water is used in our
experimental fish farm to warm the fish ponds.

We have little pollution in Iceland but
sulphur released from volcanos has had effects.
The 1970 eruption of Hekla spread ash across
Iceland and one river in the north of the
country had so much in suspension in its
waters that the salmon declined and took
several years to recover. The ash emitted
had relatively high fluoride levels and this,
of course, is poisonous and incidentally also
affected the bone structure of sheep grazing
on pastures covered in the ash fall.

V. GEIST

Are there differences between glacial and spring fed streams? The former must contain substantial amounts of silt and affect fish productivity.

T. GUDJONSSON

The warmer glacial rivers can be productive in spite of the silt, which is primarily present during the warmer months of the year. Tagging in the estuary of the large glacial river system of Ölfusá-Hvitá in Southern Iceland shows that most of salmon spawn in the main river which is a mixture of glacial and clear waters.

J. N. R. JEFFERS

You showed a graph which suggested that the exploitation of the fishery had doubled in ten years. Is your research adequately funded to allow you to respond to problems growing on such a time scale?

T. GUDJONSSON

No. For three years we have had a grant from the United Nations Development Programme and this has helped us to do as much research as we had previously done in ten to fifteen years, but this grant is now coming to an end and we fear that our Government may need to cut back the level of support.

J. CHRISTIE

Has tagging shown a seaward distribution of salmon? A second point, concerning overcrowding in Char: the slow growth is surely an indication that at present conditions are staying within the range of compensation?

T. GUDJONSSON

We do not know very much about where the salmon go at sea. The seas are warm off the south of Iceland, warm water flowing up the west coast and along the north while a cold southerly current comes down the east coast of the country and also down the east coast of Greenland. We believe that our salmon move south and westwards into a circular current called the Irminger Current bordering on the gulf stream on one side and the east Greenland current on the other and some go around the southern tip of Greenland. A few Icelandic tags have been recovered from west Greenland. Other fish probably go northwards in the gulf stream in the east Atlantic and

we have had three recoveries, one from Norway
and two from the Faeroes. Salmon do not stay
near the shore and they are rarely caught by
sea fishermen although a few have been taken
with herring off our north coast and some off
the west and south coasts in gill nets mostly
in early spring.

I believe that we are regulating the salmon
fishery well within the compensation range of
the population. In the lake Myvatn area where
a factory has been built the population is increasing and there has been an accompanying
increase in fishing pressure on the Char.
The U.N.D.P. grant was to allow us to measure
the size of the fish stocks especially of
salmon and trout and to design regulations
for the fishery. At present, we limit the
number of rods allowed and there is big
pressure for more.

21. THE RATIONAL MANAGEMENT OF HYDROLOGICAL SYSTEMS

H. J. Colenbrander

Toegepast-natuurwetenschappelijk Onderzoek

The Hague, The Netherlands

INTRODUCTION

From the very beginning, man on earth has used water and has also
fought it. Water is not only a primary necessity of life; from
time to time water also threatens life. This means water "shows
two faces": an amiable and a malicious one. Using water and
fighting against it, man in one or another way is influencing
the hydrological systems.

People protect their habitat against floods by building
hillocks to live upon and, later, by building sea dikes, embank-
ments and barrages. Initially these influences were of minor
importance, but gradually their number and effects increased. As
the population grew, more land was needed for the production of
food. So marshes and waste land were reclaimed, and other areas
were deforested. These measures have had a significant effect on
the water management conditions of such areas. At first drinking-
water was not a problem because the open waters were not polluted
at all.

As urbanisation and industrialisation progressed, the problem
became more serious. The surface waters have largely become
heavily polluted, and today they cannot easily be used for domestic
and industrial water supply. Large storage reservoirs have been
built and huge quantities of ground water are extracted to meet
human needs. All in all, human interference in the hydrological
cycle has considerably intensified. It has meanwhile become clear
that the various subsystems are closely inter-related. Any intervention in one of these may affect others: for instance, an

extraction of ground water often affects the surface water system.
These inter-relationships have been underestimated. Therefore,
various alternative plans have to be studied thoroughly. In a
technical and economic sense, this has already been done for
many years, but changes in the ecosystem have frequently been
neglected in early cost-benefit studies. In the last few
decades or so, the situation has noticeably altered. Nowadays, one
has to be aware of the environmental impact of each project in
the field of water management.

        Before discussing the ways and means to achieve a rational
management of the hydrological systems, it is necessary to mention
the main interests involved and to elucidate the consequences of
the various human activities.

        In this respect the following activities are of particular
importance:

- river engineering and drainage works, including the construction
  of reservoirs;

- extraction and recharge of ground-water;

- change in land use and in land treatment;

- water pollution.

        The environmental consequences of these activities will now
be described. In this respect it is often necessary to refer to
the "natural" vegetation or "natural" scenery. However, mostly
it is not clear what the "natural" condition is, nor when the
situation can be considered to be "natural".

        Another problem is that the meaning of "rational water
management" is subjective, and multi-interpretable. In this report,
"rational management" will be called the management of water after
weighing all the different interests involved. The weighing
procedure (in fact a multi-objective optimisation process) is so
complicated that it is discussed separately, in a later section.

        For a good understanding of all the processes and relation-
ships, it is necessary to describe hydrology first in general
terms and to discuss what is to be understood as the water cycle.
Some subsystems of this cycle will then be considered in more
detail.

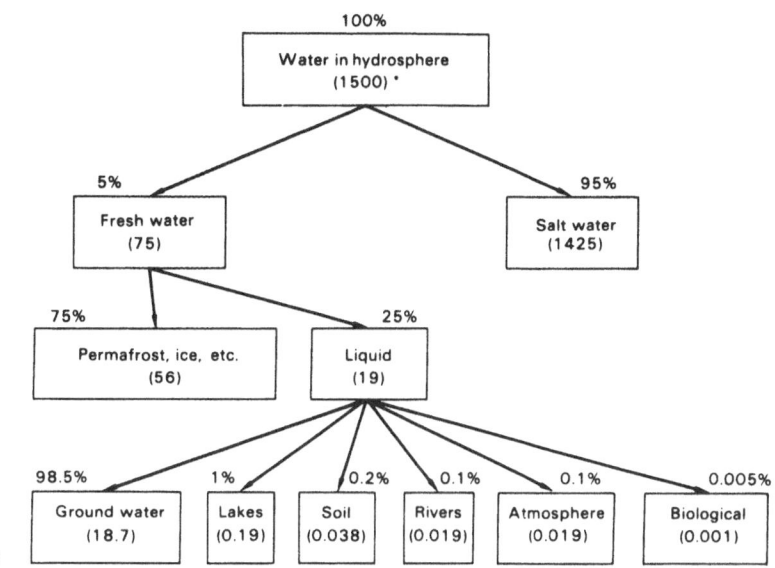

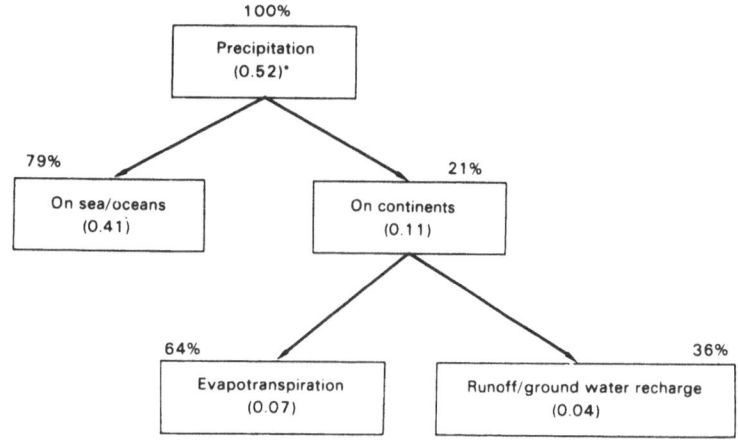

b   * Numbers in parenthesis express volume of water in million cu. km
    ** Sea, oceans, etc.   surface 140 million sq. km
       Land area   surface 560 million sq. km

FIG. 1a.   APPROXIMATE DISTRIBUTION OF WATER IN THE HYDROSPHERE
           (BASED ON DOOGE 1973).

FIG. 1b.   DISTRIBUTION OF PRECIPITATION ON EARTH.**

* Numbers in parenthesis express volume of water in million cu.km

** Sea, oceans, etc.   surface 140 million sq. km
   Land area   surface 560 million sq. km

HYDROLOGY

The total amount of water in the hydrosphere is estimated at
1,500 x 10^6 cu. kilometres. However, only part of it is readily
available for mankind (see Figure 1a).

    Although only 0.005 per cent is stored in biological species,
this water is especially important for life. Most of the water is
continuously moving; it forms a part of what is called the hydrological cycle or water cycle (Figure 2). The main processes in
this cycle are rainfall, evapotranspiration and run-off. However,
the movement of water in the unsaturated and saturated parts of
the soil is also of great importance; it covers such processes as
infiltration, percolation, capillary rise and ground water flow.

    Around 21 per cent of the total rainfall on earth is assumed
to fall on the continents (Figure 1b). As demonstrated in this
figure, this rain partly evaporates and partly recharges the ground
water, appearing ultimately as run-off. The mean annual rainfall
is about 110,000 km^3, or 0.7 per cent of the total amount of fresh
water available on the continents. This is even less than the
total amount of water stored in lakes and rivers (160,000 km^3).

    For some continents, approximate water balances are presented
in Table 1. Evidently, some considerable differences occur and
it is therefore also necessary to study water balances on a regional scale.

    In general terms, the water balance equation can be written
as follows:

$$P + U + I + \Delta S = E_a + Q \qquad\qquad (1)$$

where: P = precipitation

                    U = net subsurface flow (i.e. total underground
                                          inflow minus outflow)

                    I = imported minus exported water

                    ΔS = Change in total water storage (i.e. storage
                        at the beginning of a balance period minus
                        the storage at the end)

                    E_a= evapotranspiration

                    Q = river outflow

    Storage comprises three main components: surface water
storage; soil moisture storage and ground-water storage. In this
case, the hydrological boundary co-incides with the topographic

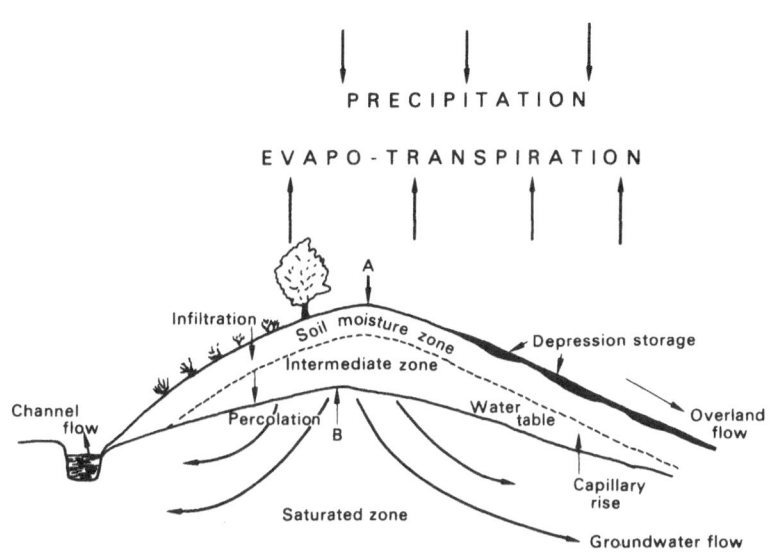

FIG. 2. SCHEMATIC DRAWING OF THE HYDROLOGICAL CYCLE OVER AN AREA.
(A = TOPOGRAPHIC WATER DIVIDE; B = PHREATIC WATER DIVIDE).

Table 1. WATER BALANCE OF CONTINENTS (Lvovitch 1971)
(All figures in mm depth of water per year)

CONTINENT	PRECIPITATION (mm)	EVAPORATION (mm)	RUNOFF (mm)
EUROPE	734	- 415	- 319
AUSTRALIA	440	- 393	- 47
AFRICA	686	- 547	- 139
S. AMERICA	1648	- 1065	- 583
AVERAGE FOR ALL CONTINENTS	760	- 480	- 280

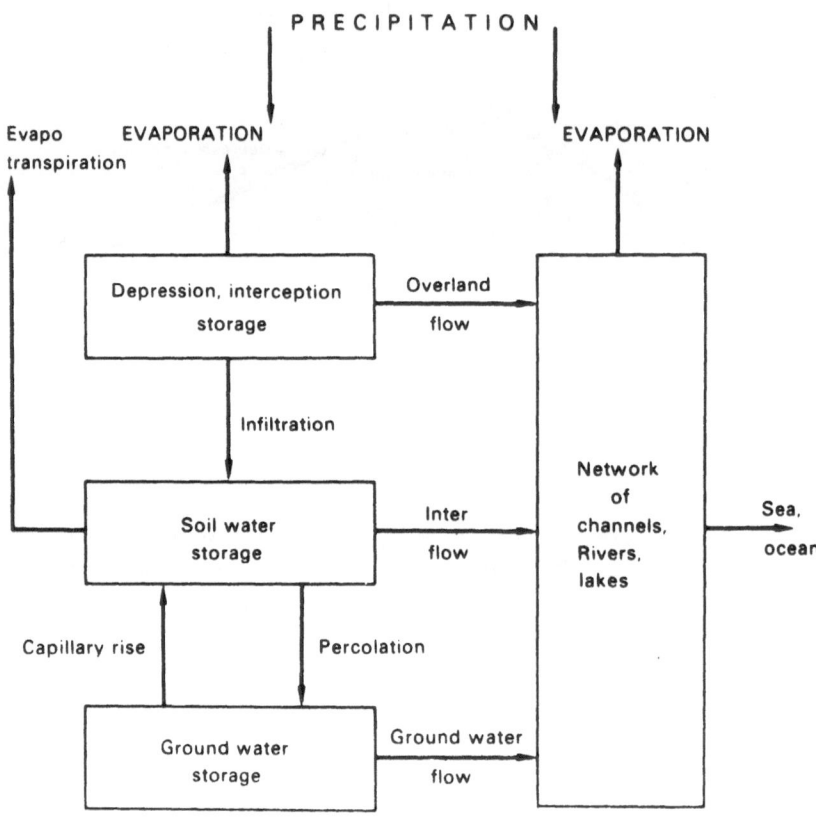

<u>FIG. 3.</u>   BLOCK DIAGRAM REPRESENTING SOME HYDROLOGICAL SUBSYSTEMS
(BASED ON DATE FROM DOOGE, 1973).

divide; no term for surface water inflow minus outflow has to be introduced. As an example of a regional water balance, equation (2) expresses the items of equation (1) in mm's depth of water for an average year:

$$810 - 45 - 40 + 0 - 440 + 285 \qquad (2)$$

This equation has been derived from the Leerinkbeek area, which lies in the eastern part of the Netherlands (Commissie, 1962). In this area (size 52 km^2), the depth of ground-water is rather small (0.4 to 4 m. minus soil surface) as is the thickness of the aquifer (1 to 40 m.). The area has a humid climate representative of the temperate zone.

For Hungary as a whole Bogardi (1973) presents the following equation for an average year (expressed in mm.):

$$P(620) = E_a(552) + Q(68) \qquad (3)$$

The evapotranspiration is increased by human activities by about 4 per cent in comparison with the original situation and run-off decreased by 23 per cent.

Living in the computer age, one has to transform the schematic drawing of Figure 2 into a more mathematical presentation (Figure 3). Within the scope of this paper, it is not possible to deal with all subsystems of the hydrological cycle: only three will be discussed in more detail:

- the surface water system, excluding oceans and estuaries;

- the ground-water system;

- the soil moisture system.

Attention will be focussed on the inter-relationships of these subsystems and the human influences on them. An example of the inter-relationship between the several subsystems and a human activity on these is given in Table 2.

HYDROLOGICAL SUBSYSTEMS

Surface water system

The surface water system consists of the network of all open water courses (= channel network), natural lakes and man-made lakes. Oceans and estuaries are excluded. The aspects related to

Table 2. COMPONENTS OF THE WATER BALANCE OF AN AREA OF 14,300 HA IN THE
EASTERN PART OF THE NETHERLANDS.
(PERIOD 1 APRIL 1973 – 8 MARCH 1974)
Given are the data without and with an artificial extraction of
groundwater (after De Laat and Van de Akker (1976))

| COMPONENT | SITUATION | | DECREASE DUE TO GROUNDWATER EXTRACTION | | DECREASE AS A PERCENTAGE OF THE TOTAL AMOUNT OF EXTRACTED GROUNDWATER |
	WITHOUT GROUNDWATER EXTRACTION (mm)*	WITH GROUNDWATER EXTRACTION (mm)*	(mm)*	(%)	
ARTIFICIAL EXTRACTION OF GROUNDWATER	—	26	- 26	–	–
PRECIPITATION	647	647	0	–	–
EVAPOTRANSPIRATION	436	432	4	0.9	15.4
RIVER OUTFLOW	162	140	22	13.6	84.6
GROUNDWATER OUTFLOW	29	29	0	–	–
INCREASE OF WATER STORAGE	20	20	0	–	–

* For this area 1 mm depth of water equals a volume of 143,000 m^3

the channel network on the one hand, and the man-made lakes on
the other, will be discussed separately.

The channel network is likely to be the first system ever
influenced by man. The building of river and sea dikes, dams and
other regulation works, was started in more or less prehistoric
times. Other regulation works are, for example, the straightening
and lining of channels, the construction of levees, weirs and other
hydraulic structures. Many of these are to reduce the water level
gradients and, consequently, the flow velocity so as to prevent
erosion. To concentrate the deposition of the remaining sediment,
sand traps have often been built. Many flood-water reduction
works have also been constructed. These are mainly reservoirs
in which water is temporarily stored during high floods.

Another very early influence of man on the flow behaviour of
a river is caused by the reclamation of waste land, especially
for agriculture. As a consequence the existing drainage system has
to be enlarged and usually intensified. The run-off from undrained
land is generally smaller than that from reclaimed land, because
under natural conditions much water is stored in depressions, pud-
dles etc., and this evaporates before reaching the river. Special
attention must be paid in this connection to wetlands (marshes, fens,
peatland, etc). The Council of Europe is to mount publicity camp-
aigns in 1976 for the conservation of this type of habitat. To
illustrate their importance, a part of the introductory article of
the Magazine Naturopa (1976) states:

"Of all Europe's natural heritages, its wetlands are
undoubtedly among the most vulnerable and most endangered
types of the environment. These biotopes, drained over the
centuries - indeed over thousands of years - and polluted
in recent decades, are currently undergoing a process of
decline and degradation whose implications are far more
serious than is usually imagined."

Other changes in land use besides the reclamation of waste
land (heath moorland and scrub) often have a considerable effect
on run-off. The most striking example is the disafforestation of
large areas. Besides a considerably increased run-off and higher
peak flows, this often causes severe erosion. Therefore, in
the last decades or so, large areas have been reforested. But
even a rather minor change in land use, e.g. from grassland into
arable land, vice versa, has a certain effect.

In certain areas, mining activities have a great influence on
the original channel network, In these areas enormous quantities
of ground-water are often extracted and drained elsewhere, changing

the run-off situation. As a consequence of mining, land subsidence
often occurs, affecting the gradient of the drainage system. Another
human activity which influences the channel system is the process
of urbanisation and industrialisation. Large areas of formerly
vegetated soil are now paved and have houses, factories, etc., built
upon them. The percentage of fast run-off consequently increases
markedly and the base-flow component decreases. Higher peak flows
occur (Figure 4). The total volume of run-off increases because,
generally speaking, the evapotranspiration of an urbanised area is
smaller than that of vegetated soil.

So far the quantitative hydrological consequences of man's
activities have chiefly been discussed. However, the effects in
a qualitative sense are frequently even more important. The waste
water production of urban areas, and industries is especially
high, and affects the surface water quality severely. Pollution
of surface water mofidies the aquatic ecosystem, for example by
eutrophication causing an overwhelming growth of algae and,
consequently, a steep decrease in oxygen content. The aquatic
animal population will change too. Other interests which are also
affected by surface water pollution include water supply for
horticulture and, sometimes, even for arable and grassland, domestic
water supply and recreation.

Chemical and biological water quality is not only influenced
by the direct disposal of waste water; certain hydraulic constructions, such as weirs, sand traps, etc. play a role too. These
mostly decrease the water level gradient, and flow velocity.
Especially in periods of low flow this will adversely affect the
oxygen content of the water and the flow velocity then becomes
rather uniform over the whole length of a river. The variety of
aquatic life consequently diminishes.

The natural vegetation of river banks is strongly influenced
by water depth and modified by depth changes. Consequently, maintenance operations usually have a substantial influence on bank
vegetation. In the past, the channels were largely cleared out
manually. Recent increased labour costs have led to the work being
mechanised. The machines need working space and many trees, bushes
and hedges have been cut on the channel banks. Consequently, the
sunlight can now easily reach into the river. This causes a
different, far denser growth of aquatic vegetation: the growth of
filamentous algae often expands appreciably, and the overall water
quality worsens.

Sometimes weirs and dams are installed to raise the water level
in certain drainage channels, and this in turn causes a rise of
the water table in the adjacent soils. This activity is often
called water conservation, because the higher water table increases

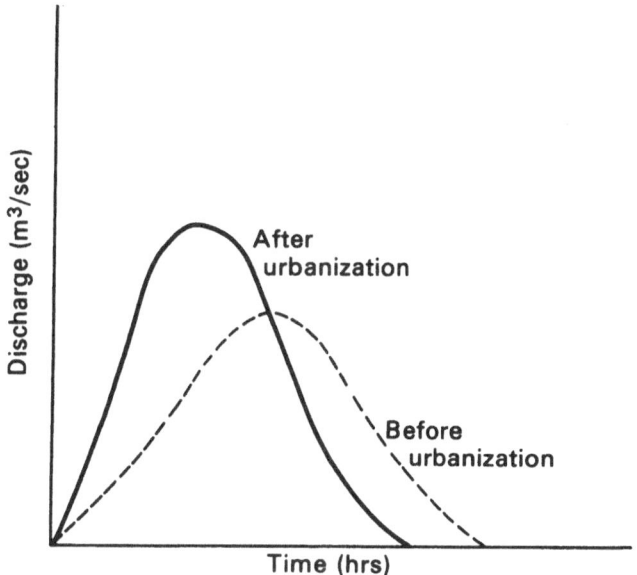

FIG. 4. SCHEMATIC HYDROGRAPHS SHOWING EFFECT OF URBANIZATION AS
REDUCING LAG TIME AND INCREASING PEAK DISCHARGE (AFTER
STRAHLER AND STRAHLER, 1973).

the amount of water available for transpiration of the vegetation
by capillary rise from the ground-water. This will generate not
only a higher production of agricultural crops, but may also change
the natural vegetation. Plants typical of dry conditions, may be
replaced by water-loving vegetation. A comparable effect occurs
where a network of irrigation canals is constructed in areas
with a very deep water table, and when infiltration of water from
the canals may alter the vegetation.

Obviously, the opposite effect may also occur. An artificial
lowering of the original water levels in the drainage channels
increases run off towards these channels and the water table
drops. The soil becomes drier and vegetation changes are again
likely.

In summary, the human activities discussed tend to alter the
following river characteristics:

- the run-off volumes;
- the height of flood peaks and the shape of the run-off hydro-
  graph;
- the frequency distributions of river flows;
- the water level gradients and flow velocities;
- the bed roughness;
- the amount of penetrating light;
- the sediment transport;
- the water quality.

Such changes in the river characteristics have a great
influence on the aquatic ecosystem. This system must be evaluated
carefully, because only then an adequate balancing of all interests
can be achieved. An evaluation of the aquatic ecosystem has been
described e.g. by Gardeniers and Tolkamp (1976).

Man-made lakes or reservoirs, are used to store water for
power generation, irrigation, domestic and industrial use, and
recreation.

In Ackermann, White and Worthington (1973), Fels and Keller,
arrive at a figure of over 400 man-made lakes now in existence,
about 315 of which have a surface area exceeding 100 km^2. The
total area of these lakes amounts to over 400,000 km^2 and their
total storage capacity is about 3,500 km^3. Lvovitch (1971)
estimates they nowadays regulate 10 per cent of the total surface
run-off in the world.

Man-made lakes superimpose an aquatic ecosystem on a
terrestrial one. The immediate result is an unstable situation
that cannot be defined easily. In the past the changes, due to
the construction of lakes, have very often not been given enough
attention in the planning stage of a project. These changes can
be subdivided into two groups: (a) on-site and (b) off-site.

The on-site changes are related to the catchment area up-
stream, the area immediately around it and the section of the river
directly below the man-made lake. Such changes include the
degradation of the river system directly below the reservoir,
because the flow distribution becomes quite different. Not only
peak flows are reduced, or at least occur less frequently, but
the total run-off volume also changes. For example, it is reported
that for an area of 2,000 ha in Brazil, due to the construction of
a number of storage reservoirs, in a dry year the run-off volume
was reduced by 25 per cent. The water level gradients in the up-
stream rivers and the ground-water regime in the area around the
reservoir also change, affecting the ecosystems in these areas.

The off-site changes concern areas far below the dam, such
as alterations of the salinity in the river's estuary, and changes
in the amount of sediment that the river carries into the sea.
Another effect of the reservoir is the transformation of the
quality of the incoming water in terms of its physical, chemical
and biological properties.

A quite different effect of man-made lakes arises because
they impose new stresses on the earth crust and may generate
seismic movements. It is likely that large reservoirs will moreover affect the weather, and the micro-climate. Up to now, however,
not much is known in detail about these effects. From what is
stated above, however, it will be clear that man-made lakes have
a strong environmental impact, and that the local conditions are
altered severely. However, much research is still needed on the
complex relationships in ecological respects.

Limnological aspects of man-made lakes are discussed in many
reports, as for example by Ackermann, White and Worthington et al.
(1973) and by Lowe-McConnell (1966).

Sub-surface water system

Sub-surface water can be divided into ground-water and soil
moisture. Ground-water is here defined as the water stored below
the phreatic level (= water table). The water in the unsaturated
soil longer above the water table will be called soil moisture.
Sometimes a third sub-surface water system is distinguished:

the water in the so-called intermediate zone (Figure 2). This
zone comprises the part of the soil between the top layer and the
phreatic level. It is assumed that from this layer no water will
be returned by capillary rise to the roots of the plants.
Because the division between the soil moisture zone and intermediate
zone is rather arbitrary, and in practice not very useful, both
zones will in this report be considered as one whole: the unsaturated zone.

As was discussed above, the ground-water and soil moisture
systems are closely linked and ground-water can easily turn into
soil moisture, and vice versa.

Ground-water system. There are a number of differences
between the surface water system and the ground-water system. The
main difference is that the depth to ground-water varies from
place to place. In a large part of the world, ground-water is
found at a hundred metres or more below land surface. In other
regions, however, like humid parts of the temperate zones, ground-
water tables are often no deeper than 1 to 50 metres below surface.
The water table is then often within reach of plant roots, and part
of the water needed for transportation will be provided by capillary
rise of ground-water.

Another important difference between surface water and ground-
water is that, generally speaking, the quality of ground-water is
better. Further, the temperature of ground-water is mostly relatively low and constant. These properties make ground-water very
attractive for domestic and industrial uses. In the Netherlands,
for instance, more than 50 per cent of domestic supplies are met
by ground-water. The relatively low and constant temperature of
ground-water also make it very suitable for cooling purposes in
industry, and in the Netherlands 10^9 m^3 are extracted for this
purpose annually.

Preferably, ground-water is extracted from (semi)-confined
aquifers, below impermeable layers so that the phreatic water table
is not, or only slightly, affected. However, due to the
considerably increased water need, extraction nowadays also takes
place in areas where the situation is less favourable. Consequently,
the phreatic water table often drops steeply and this causes a lot
of problems, especially in areas with high water tables.

Figure 5 (from De Laat and Van den Akker, (1976), shows such
a lowering of the water table due to an artificial extraction of
ground-water. This study includes the simulation of the processes
of evapotranspiration, unsaturated and saturated flow (De Laat,
Van den Akker and Van de Nes (1975)). Figure 6 and also Table 2

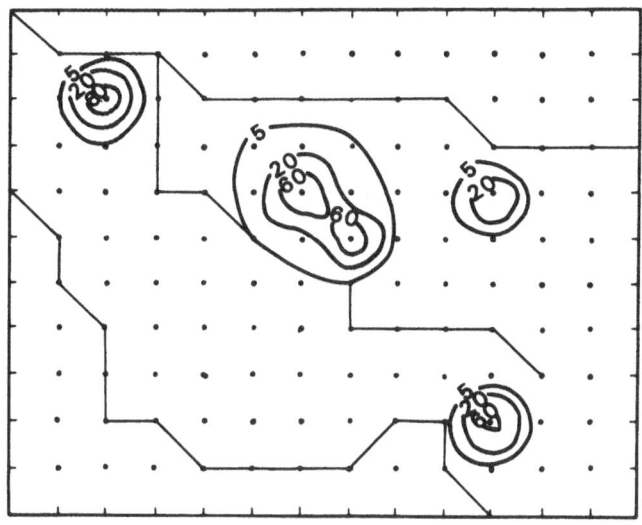

FIG. 5.   THE DIFFERENCE (IN CM) BETWEEN THE CALCULATED HEIGHTS OF
THE WATER TABLE FOR SITUATIONS WITHOUT AND WITH ARTIFICIAL
EXTRACTION OF GROUNDWATER.

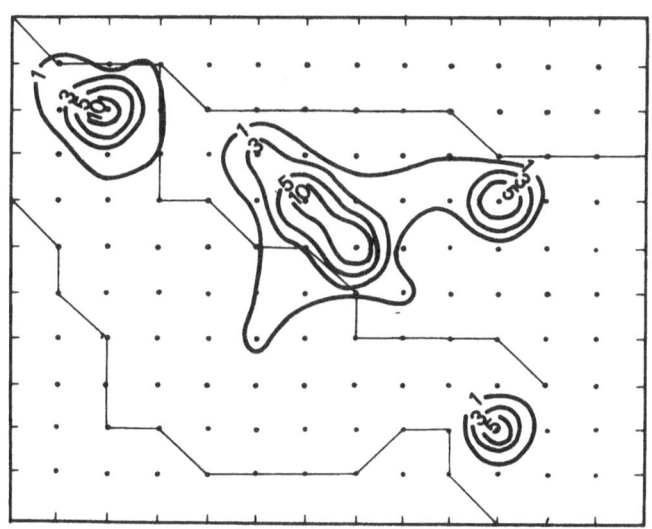

FIG. 6.   THE DIFFERENCE (AS PERCENTAGE) BETWEEN THE CALCULATED
EVAPOTRANSPIRATION FOR SITUATIONS WITHOUT AND WITH
ARTIFICIAL EXTRACTION OF GROUND WATER.

are also taken from these investigations they demonstrate the
decrease of evapotranspiration due to the articifical lowering of
the water table presented in Figure 5. When the land is sued for
agriculture, such a reduction of evapotranspiration means a
smaller crop yield and a financial loss to farmers.

An artificial drop of the water table can affect natural
vegetation as well as agriculture. Species and communities typical
of wet conditions are replaced by those plant species typical of
dry habitats. Generally this means degradation of the vegetation
and reduced variety of plant associations. During this conference,
Van der Maarel will discuss this aspect in more detail. How
sensitive the natural vegetation is to a lowering of the water
table is particularly important, but hard to determine.

Sometimes the influence of the lowering of the water table
can be reduced by artificial ground-water recharge (e.g. by
surface irrigation or sprinkling). Often it is important to use,
for this artificial water supply, water of a similar chemical
composition and temperature to the original water. As discussed
above, an artificial lowering of the water table also affects the
surface water flow (see Table 3). This means a reduced flow
volume and lower water level in the water courses, which may even
dry out completely. There are evident effects on the aquatic
ecosystem and on users of the surface water down-stream of the point
of extraction.

Changes in ground water also affect water quality. In a
recent publication of the Netherlands Committee for Hydrological
Research (Van den Berg (1976) states:

"Pollution of ground-water has been known in the world
for a long time mainly in the extreme form leading to
salinisation of irrigated land. Already more than 4,000
years ago the phenomena have been written down in the chron-
icles of Mesopotamia, though probably it was not known that
the real background of the misery originated from the
rising of saline ground-water because of irrigation."

"Still, nowadays when speaking about pollution we do not
think first of these types of 'natural' pollution. The
attention is much more focussed on the excess of ill-used
matter distributed in the biosphere by human activities ....."

One of the most serious pollutants of ground-water is oil,
easily released by accidents with tank-lorries or storage
reservoirs. Once polluted the ground-water is unsuitable for
domestic use for a long time.

Table 3. DECREASE OF RIVER OUTFLOW DUE TO ARTIFICIAL EXTRACTION OF GROUNDWATER. (after De Laat and Van de Akker (1976))

LEERINK BROOK AREA	RIVER OUTFLOW. NO GROUNDWATER EXTRACTION (mm)*	ACTUAL ARTIFICIAL GROUNDWATER EXTRACTION (mm)*	DECREASE OF RIVER OUTFLOW DUE TO GROUNDWATER EXTRACTION (mm)* (%)		ASSUMED 1.6 TIMES ACTUAL GROUNDWATER EXTRACTION (mm)*	DECREASE OF RIVER OUTFLOW DUE TO 1.6 TIMES ACTUAL GROUNDWATER EXTRACTION (mm)* (%)	
1/IV 1971–5/IV 1972	55	28	19	34.5	46	32	58.2
5/IV 1972–1/IV 1973	207	28	31	15.0	44	51	24.6
1/IV 1973–16/III 1974	162	26	22	13.6	43	37	22.8
	‐‐	‐‐	‐‐	‐‐‐‐	‐‐	‐‐	‐‐‐‐
1/IV 1971–16/III 1974	424	82	72	17.0	133	120	28.3

* In this case 1 mm depth of water equals a volume of 143,000 m³

Another source of ground-water pollution is percolation from
the growing volume of rubbish-dumps. The expanding use of
fertilisers on agriculture may cause such a certain wash-out in
some areas and this pollutes the ground-water as well. The
pollution of ground-water must also be a permanent concern, because
this will ultimately contribute to the pollution of open waters.

One special aspect that must be mentioned briefly is the
growing salt intrusion from the sea by artificial ground-water
extraction. In many coastal areas, a fresh-water body lies upon
the salt water aquifer. By artificial extraction of fresh water
the boundary between fresh and salt water rises rapidly and
salinity sets in. Therefore, in coastal areas it is necessary to
be very careful with ground-water extraction. To form a
barrage against salt intrusion in these areas, surface water from
elsewhere is often recharged.

There are many reports, handbooks, etc., dealing with
ground-water pollution, including Van den Berg (1976) and Fried
(1975).

Soil moisture system. The movement and storage of water in the
unsaturated part of the soil will here be discussed as far as
these are affected by man's activity. The three main processes
involved are the infiltration of water into the surface layer of
the soil; the percolation of water downward through the soil, and
the upward capillary movement of water from the ground-water
(see Figures 2 and 3). These processes are chiefly influenced
by the physical properties of the soil: porosity, grain and
poresize, hydraulic conductivity, etc.

Two more concepts are of importance: "field capacity" and
"wilting point". When a soil has first been saturated with water
and next been allowed to drain under gravity until no more water
moves downward, the soil is said to be holding its field capacity of
water. The wilting point is the quantity of soil moisture below
which plants will be unable to extract further moisture from the
soil, and thus the foliage will wilt. Both characteristics
largely depend on the structure of the soil. A sandy soil has
a relatively low moisture content at field capacity, and a clay
soil a high one.

The main human activities influencing these soil characteristics are observable in the field of agriculture. Land treatment
and such farm practices as levelling, deep ploughing and ter-
racing have a great effect. These activities lower the depression storage and overland flow and increase the infiltration capacity. The porosity, storage capacity and hydraulic conductivity

of the soil are generally increased but compaction by farm tractors has a reverse effect.

The various measures also strongly affect soil erosion and sediment transport. When areas of great conservation value exist in regions mainly used for agriculture, it is likely that these will also be influenced by the changed water management conditions. This aspect should not be neglected, and often additional measures will be necessary to preserve or restore the original conditions.

In summary it can be said that rational management of the sub-surface water system is as complicated as that of the surface system. The relationships and human influences are elucidated in more detail in many reports, handbooks, etc. (see for instance Strahler and Strahler (1973) and Moore and Morgan (1968)).

BALANCING OF INTERESTS

From the foregoing discussion it will have become clear that many interests are intertwined with the water problem. These are brought together in a popular way in Figure 7. Due to the complexity of the problems it will not be easy to arrive at a rational use of all the different hydrological systems. It has been shown that the various subsystems are closely related too. A problem with respect to a certain subsystem cannot be solved in an optimal way by separating it from the related sub-systems. On the other hand, in view of the large number of processes, parameters and variables involved, the problem cannot be solved by optimising the system from the very beginning as a whole.

To solve the problem in an adequate way, the so-called multi-objective optimisation technique seems suitable. This approach is promoted by scientists of the System Research Centre, Cleveland – Ohio (e.g. Mesarovic, Macko and Takahara, 1970 and Haimes, Hall and Friedman, 1975). Mesarovic has introduced the multi-level hierarchical structure. This structure is characterised by (1) vertical arrangements of subsystems, which comprise the overall system; (2) priority of action or right of intervention of the higher level subsystem, and (3) dependencies of the higher level subsystems upon actual performance of the lower levels.

Three types of levels are distinguished:

- the levels of description or abstraction;

- the levels of decision complexity;

- the organisation levels.

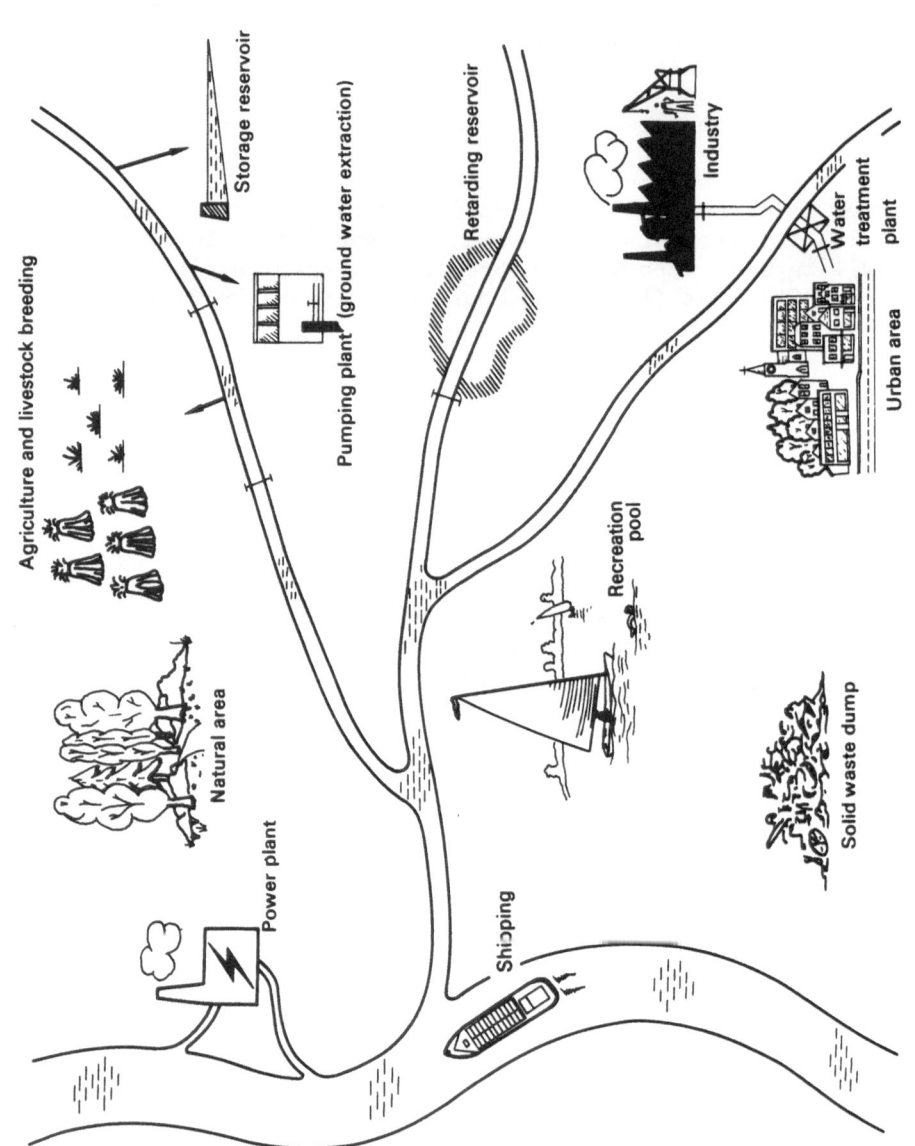

FIG. 7. THE INTERESTS INVOLVED IN WATER RESOURCES MANAGEMENT.

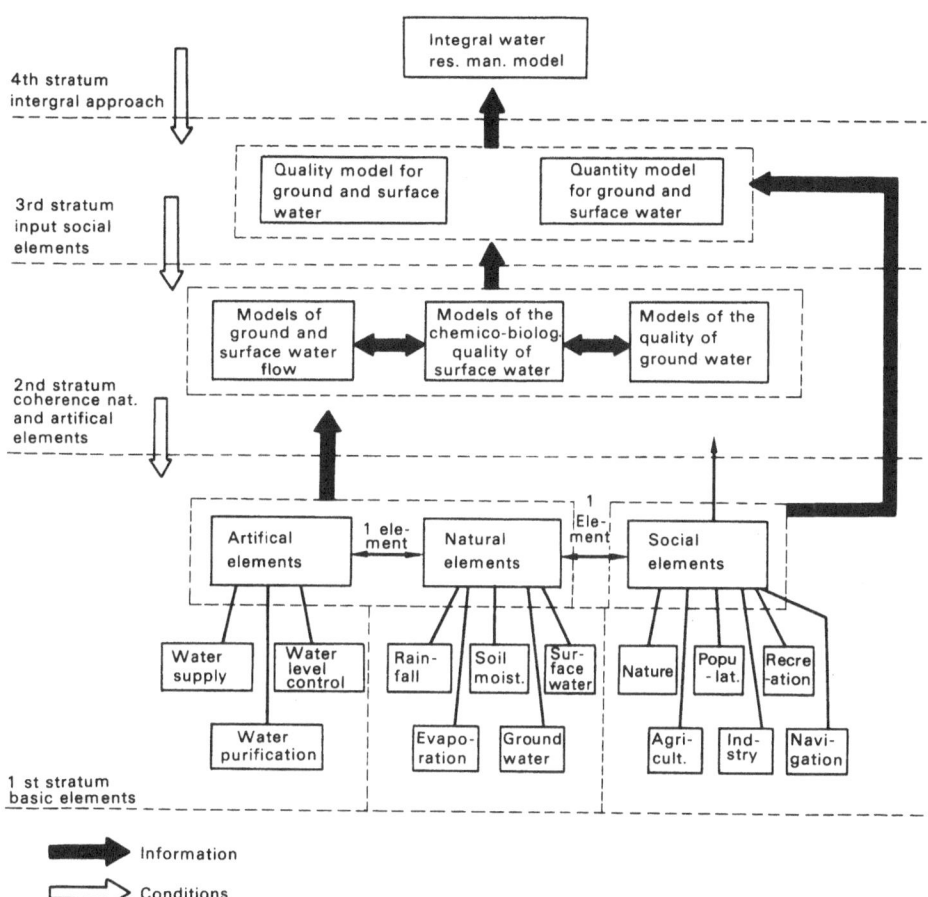

FIG. 8. DESCRIPTION HIERACHY OF THE WATER RESOURCES MANAGEMENT SYSTEM.

Van de Nes (1976) has introduced this technique in a Water
Resources Management study of a part of the province of Gelderland
(Netherlands). In particular the descriptive level has in this
case been analysed.

This level has been subdivided in four strata:

(a)   the basic elements;

(b)   the stratum where the natural and artificial elements are
       combined;

(c)   the stratum where the previously mentioned elements are
       combined with the social elements and

(d)   the stratum of total integration.

At the first stratum, the basic elements have to be described
and the basic relationships established. Three sets of elements
have been distinguished (Figure 8):

- the artificial elements (domestic and industrial water supply;
  surface water management; water quality control);

- the natural elements (precipitation, soil moisture, ground-
  water, etc.);

- the social elements (population, recreation, nature conservation,
  etc.).

A number of these elements have been briefly discussed in
the previous chapters. The relationship between the moisture
condition of the soil (including depth to ground-water) and the
type of natural vegetation is especially hard to establish (see
Van der Maarel, 1976. This symposium). More details concerning
the other relationships can be found in the previously mentioned
report by Van de Nes (1976).

The second stratum comprises the models simulating the various
processes and relationships. There are models for:

- the evapotranspiration, surface and sub-surface water movement;

- the chemical and biological quality of the surface water;

- the chemical quality of the ground-water.

At this stratum the ultimate relationships between the natural
and artificial elements have to be established.

At the third stratum, the social elements have to be combined with the technical results of the models developed. This means that, for example, the real damage caused by human activities in various respects has to be introduced. With respect to agricultural damage, this can nowadays be done with acceptable accuracy. However, with respect to many other interests, this is less easy.

For instance: damage done to terrestrial and aquatic vegetation and fauna cannot be expressed easily in monetary units. Neither can damage done to recreation be expressed in this way. Therefore, an economic criterion alone will not be adequate to balance the various interests: a social criterion will also have to be used. The total set of measures with regard to the management of the hydrological systems (in a technical and legislative sense), must warrant the greatest overall socio-economic benefit.

At the stratum of integration, all relevant aspects have to be brought together. In this respect, different approaches are possible: uni-criteria and multi-criteria ones. These must provide the ultimate results of the descriptive hierarchy. Alternative plans must be prepared at this stratum, which provide the basic information for the decision-making level.

Here politics come in the picture.

This level can be subdivided into several layers too, forming the multi-layer decision hierarchy. This hierarchy deals inter alia with the planning of water resources and water management, the reduction of uncertainties, and with the selection procedure.

As a whole it is a part of the multi-level organisational hierarchy. The relevant decisions are mostly taken at different echelons: local, regional and national. It is essential that the different decision-making echelons co-operate very closely. Only then will it be possible to achieve an optimal overall water management scheme.

A sophisticated technical, socio-economic model is useless without efficient decision-making units, and without an adequate overall organisation. Quite often the problem is, however, even more complicated than so far described, because the surface waters involved may very well be international waters. International water laws are then relevant. These, however, are mostly inadequate. Many of these aspects are discussed by Finkel (1973) and a quotation from his paper forms an appropriate summary of the whole situation:

"During the past thirty or forty years, the world has
witnessed a significant number of water development
projects, which have fallen short of expectations even
though they were designed and constructed to high
technical standards. What went wrong? Hindsight and
closer analysis reveal that the causes for failure in
one after another of these projects were a series of
non-technical human obstacles, which had been ignored
or inadequately solved. An appreciation of human
limitations in hydrological development work, therefore,
is needed if we are to avoid making the same mistakes
in the future."

CONCLUSIONS

1. In view of the increased water-need of modern society, and
   the limited amount of water available, a careful balancing
   of the several interests is indicated.

2. In cost/benefit studies of water projects, the importance of
   nature conservation has often been neglected in the past.

3. Since the various subsystems of the hydrologic cycle are
   closely related, a problem with respect to one of these
   can only be solved adequately by treating the water system
   as a whole.

4. Further to the previous point, the study of interactions between
   the several hydrological subsystems and the human influence on
   these must be clearly given a high priority. In particular
   this is true for the influence of man on ecosystems.

5. Progress to be made in the study of rational management of
   hydrological systems, however, is strongly determined by the
   volume of basic data available. Therefore, much attention
   must be paid to the collection and processing of this type
   of data.

6. The "multi-hierarchical optimisation technique" seems suitable
   to solve the complex problems in the field of water resources
   management. Socio-economic criteria have to be used in this
   procedure.

7. To arrive at rational water resources management 'human'
   problems will be often more difficult to solve than the physical
   and technical ones.

8. A sophisticated technical, socio-economic model has no value
    without a system of efficient decision-making units as a part
    of an adequate overall organisational structure.

REFERENCES

Ackermann, W. C., White, G. F. and Worthington, E. B. 1973.
     Man-made lakes: their problems and environmental effects.
     Geophys. Monogr., 17.

Berg, C. van den. 1976.   Introduction of a technical meeting on
     groundwater pollution. Proc. No. 21. Committee for Hydro-
     logical Research, TNO.

Bogardi, J. 1973. The impact of human activities on hydrologi-
     cal processes in Hungary.

Commissie Bestudering van de Waterbeheofte van de Gelderse Land-
     Bouwgronden. 1962. Hydrological research in the Leerink-
     beek area. Prov. Water Board, Arnhem: (In Dutch with English
     summary).

Dooge, J. C. I. 1973. The nature and components of the hydrolo-
     gical cycle. FAO: Irrigation and draining papers.

Finkel, H. 1973. Human obstacles to the control of the
     hydrological cycle for the benefit of man. FAO: Irrigation
     and drainage papers.

Fried, J. J. 1975. Groundwater pollution. (Developments in
     water science series, no. 4). Amsterdam: Elsevier Sci.

Gardeniers, J. J. P. and Tolkamp, H. H. 1976. Hydrobiological
     quality of channels. Commissie Bestudering Waterhuishouding
     Gelderland. Rapport no. 1. Deel II, section 2.3.3. Prov.
     Water Board - Arnhem (Netherlands). (In Dutch).

Haimes, Y. Y., Hall, W. A. and Friedman, H. T. 1975. Multi-
     objective optimisation in water resources systems. Amsterdam:ɪ
     Elsevier Sci. (Developments in water science series, NO. 3).

Laat, P. J. M. de and Van Den Akker, C. 1976. A model for the
     simultation of actual evapotranspiration and saturated-
     unsaturated flow. Commissie Bestudering Waterhuishouding
     Gelderland. Rapport No. 1. Deel II, sections 3.1. Prov.
     Water Board - Arnhem (Netherlands). (In Dutch).

Laat, P. J. M. de, Van Den Akker, C and Van De Nes, Th. J. 1975.
    Consequences of groundwater extraction on evapotranspiration
    and saturated-unsaturated flow: proceedings of the Bratislava
    symposium, International Assoc. Scientific Hydrology. Publ.
    No. 115.

Lowe - McConnell, R. H. 1966. Man-made lakes: proceedings of a
    symposium held at the Royal Geographical Society, London, 1965.
    London: Academic Pr.

Lvovitch, M. I. 1971. The water balance of the continents of the
    world and the method of studying it. General Assembly of
    Moscow: International Association of Scientific Hydrology.

Mesarovic, M. D., Macko, D. and Takahara, Y. 1970. Theory of
    hierarchical, multi-level systems. London: Academic Pr.
    (Mathematics in science and engineering series, vol. 6).

Moore, W. L. and Morgan, C. W. 1968. Effects of watershed changes
    on streamflow. Univ. of Texas Press.

Naturopa. 1976. Wetlands Campaign 1976. Bulletin of the European
    Information Centre for Nature Conservation, No. 24.

Nes, Th. J. Van De. 1976. The structure of the decision-making
    process within the water resources management system. Proc.Inf.
    No. 22. Committee for Hydrological Research TNO.

Strahler, A. N. and Strahler, A. H. 1973. Environmental geoscience:
    interaction between natural systems and man. New York: Wiley.

DISCUSSION: PAPER 21

M. W. HOLDGATE   Aquatic herbicides are being increasingly
                                 used to control submerged and bank side
                                 vegetation. How valuable are they in main-
                                 taining water flows?

H. J. COLENBRANDER   Labour is expensive in many countries
                                 and this is why the trend to herbicides
                                 has taken place. In parts of the Netherlands
                                 there is now a trend back to mechanical
                                 cutting of the vegetation because of the
                                 side effects of chemicals, but this demands
                                 space along the rivulets and may consequently

also have a considerable impact. There is
also a trend to use chemicals which have
less effect on the environment. Some Water
Boards in fact do not use any chemicals.

P. LEENTVAAR   Chinese grass carp are also being introduced in some Netherland waters and it is
hoped that the quantities of herbicide used
will be reduced.

H. P. BLUME   Your paper demonstrated the role of
water in linking the different biotopes
within the environment. Moving water
carries dissolved or dispersed substances
and links the components of the landscape.
This needs to be borne in mind in landscape
planning, for example when nature reserves
are established in valleys where the hilltops
are used for other purposes and the drainage
brings down water inappropriate to the nature
reserve. In such circumstances, the nature
reserve may need to cover the whole area of
the water catchment.

H. J. COLENBRANDER I agree, but this will be less necessary
in areas where the environment is under less
stress: if the water flowing in the system is
of satisfactory quality it will surely be
acceptable in a nature reserve. The problem
occurs where the water is contaminated
with chemicals different from those in the
natural system. Some very sensitive systems
demand water of the same temperature and
composition as that naturally present, and
this is also a stipulation laid down when
water is recharged into some underground
aquifers.

E. M. NICHOLSON   In your paper you alluded to Finkel's
comments on non-technical human obstacles!
Isn't the problem that water composition can
be determined precisely but social factors are
non-precise? Are we not deceiving ourselves
in trying to balance strictly measurable
hydrological parameters and vaguely measurable human ones?

H. J. COLENBRANDER   Yes. The problems need to be resolved
                            in the technical, economic, and also social
                            sense and this is the difficulty. It will,
                            I think, be even more important in future.

E. VAN DER MAAREL   It is possible to link these components
                            into terms of "relative function fulfilment",
                            determine the maximum possible and measure
                            the short fall. In this way social ecological
                            and economic factors can be brought into a
                            single combination. The problem of course is
                            of weighting the different functions, but it
                            could none the less be an explicit technique.

E. Patterns of Land Use

22. THE BALANCE BETWEEN AGRICULTURE, FORESTRY, URBANISATION AND

CONSERVATION: OPTIMAL PATTERN OF LAND USE

N. Kingo Jacobsen

Geografisk Centralinstitut, Kobenhavns Universitet

Haraldsgade 68, DK-2100 Copenhagen Ø, Denmark

INTRODUCTION

This paper is concerned with studies of ecological systems within a geographically limited area which is also representative of the natural regions of a country or of a larger morphological region.

The selected types can be utilised as model areas for the evaluation of the landscape's 'wearing qualities', that is to say, the amount of human impact it can tolerate. We are concerned here with true landscapes, i.e. integrated natural complexes which have been variously exposed to human influence. To put it in another way, we have selected a variety of cultural landscapes with different sensitivities to encroachments of various sorts.

For Denmark such areas have been chosen in 8 municipalities covering the morphological types of landscapes within Denmark proper (Blavandshuk-Vejle) and in the Wadden Sea area along the southern west coast of Denmark (Blavandshuk-Danish-German frontier; Fig. 1).

Present-day studies of the landscape mean investigations of ecosystems in which man's activities are an important factor. The balances we consider are often not natural, but in many cases influenced by man and sometimes these balances may have shifted radically. Investigations of such disturbance and of the nature of alternative kinds of balance are very important. It is necessary to know how the balance has changed - and consequently the environmental requirements of plants and animals which may again enter man's environment - following such activities as diking, drainage,

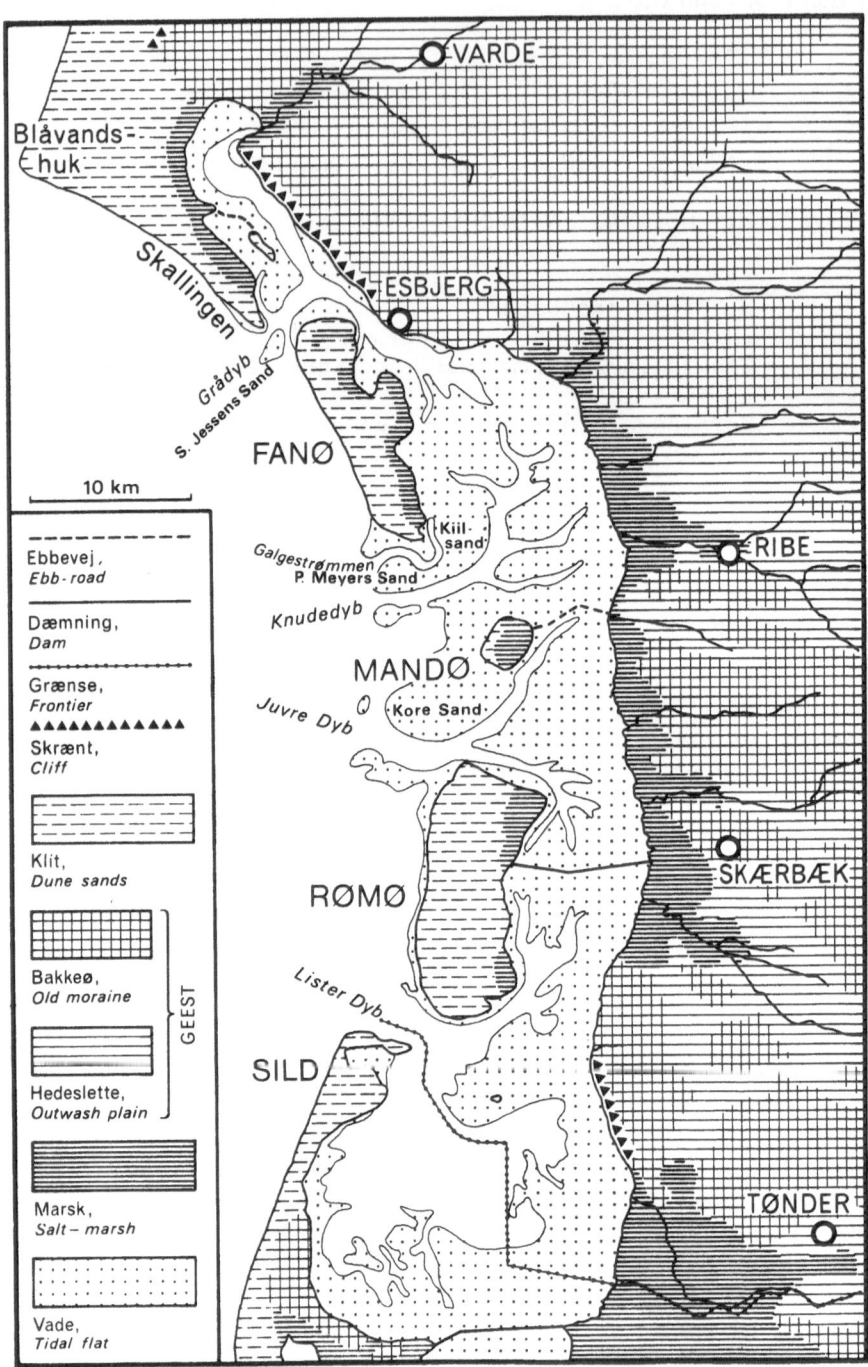

Figure 1

cultivation, industrial development, sewage discharge etc. While
the natural landscape without man's interference is only slowly
changing, can only yield a limited output, can only be exploited
within certain limits, and as an ecological system is very sensitive
to interference, the counterpart of this system - the human society -
is rapidly changing and making new and increasing demands on available resources.

Planners must provide for a natural balance between the environment's potential and society's demand; this will often mean restrictions by protection and conservation measures designed to ensure
that society is optimally adapted to the environment. The word
'optimal' is used here in an ecological sense and the task is to
overcome a landscape problem created by technology and urbanisation
within a limited space. Formerly, planning was exclusively concerned
with solving economic and sociological problems. Today the 'consumption' of nature and thereby the planning of nature is the central
problem.

As to the ecological research involved, it is important that
time and money are not wasted on collecting irrelevant information.
Technological and biological development make new demands, and the
information must be of a character and a quality which makes it
useful as a basis for new investigations. In other words, the data
must be on permanent qualities. A survey must be capable of later
analysis without having to start afresh. A landscape classification
must be repeatedly renewed. It can never be static, but it must be
developed from basic research.

What political goals does one have? There have been frequent
demands for new priorities, i.e. new goals for planning. A primary
intention must be to preserve valuable elements and the unity of the
open land, as well as to prevent ecological threats to the landscape. This primarily requires the mapping or description of landscape resources of the model area. What methods should be used for
the evaluation of a landscape's ecological potential and optimal
use? There are differences here in the needs of the various interest groups concerned with agriculture, forestry, recreation, transportation, urban development or raw material utilisation. One must
determine for an area:

1.   the uses it can tolerate,

2.   the possibilities for improvement,

3.   the care it requires,

4.   the re-establishment of environments which have been
      destroyed.

There is, therefore, a call for an evaluation of land use types and classifications of component regions related to the factors which cause environmental destruction, and indicate alternative forms of use. A time component is important here in order to incorporate a dynamic element.

With regard to the regions, the geographic size should be assessed in relation to practical, administrative and planning units. The description should include natural and economic qualities as well as locational and areal landscape characteristics which point out valuable qualities with regard to nature, traffic and economy.

Examples will be given of environment-disturbing measures in relation to dikes, overgrazing, tourist pressures, industrial pollution, and wind erosion on marginal soils. All of these examples will refer to the model regions. How can we structure relevant investigations of these phenomena and what kind of measures can we take to ensure an optimal exploitation in harmony with the balance of nature?

A WETLAND EXAMPLE

Marsh formation and design of a dike for the protection of the Tønder salt marsh.

The tidal area and salt marshes of South Jutland have always provided an important breeding and foraging area for water fowl, and have thus been of vital importance for the ecological balance, or - to be more precise - the diversity and biological productivity of the North Sea coastlands. The importance of these areas for migratory birds lies in their location along preferred migratory routes and their function as vital stopping points for such species as the Grey duck, the avocet and the godwit.

The marine foreland in the tidal district encompasses areas which are primarily characterised by their high productivity. This is due to tidal exchange, that is the continuous fresh supply of nutrient-rich water from the North Sea to the extensive tidal flats where there is little water cover, thus allowing for high water temperatures and abundant exposure to light. Seventy per cent of the 850 km^2 consist of tidal flats. The tides and wave action carry large amounts of sediment into the tidal flats resulting in a continuous generation of new land, despite the fact that this is a coast where a relative subsidence has occurred. This positive material balance has resulted in the formation of natural salt marshes. The most recent marsh formation of any extent, which has been followed sedimentologically and botanically, is that of the Skalling salt marsh. In 1870 the area was a bare sand flat. Niels Nielsen

commenced his studies of the formation of this marsh and plant
succession in 1932. These studies resulted in Børge Jakobsen's
recognition of the wind-exposed marsh problem along the South
Jutland mainland coast, which, in turn, led to land reclamation
based on modern principles utilising natural processes, whereby,
for example, 1,000 ha of foreland outside Højer (the foreland of
Ny Frederujsjig) could be created in a period of 15 years. Today
this area is an eldorado for water fowl, while at the same time
it provides essential protection for the Højer dike, which is strong,
though low. Furthermore, it provides grazing land for 2,000 sheep
and produces a significant hay harvest.

The basis for these modern land reclamation methods lies in
two important preconditions: (1) the creation of a fascine fence
system which is equipped with "holes" as a wind shelter. This
fascine fence stabilises the high flats by protecting the sedimentation of coarse material – the sand of the flats – which is
deposited at a level near the line of the high tide thus allowing
plants to take root. (2) the creation of a silt ditching system
covering the flats. This hinders salt pan formation and vegetation
is encouraged, thereby creating the requisite conditions for the
sedimentation of fine-grained material and consequently the establishment of a salt marsh. Silt ditching can be done today with
tractors or, where the terrain is soft, as along the Rømø causeway
with special ditching machines.

A great number of questions concerning marsh formation are
raised once man has come upon the scene as the all dominating
element, and the marsh and the higher lying tidal flats thereafter
should properly be called a cultural landscape. One should remember
that man has played this role for over a thousand years, ever since
the Frisians colonised the area about 800 A.D., and introduced dike
building based upon the establishment of sluices which permitted
the drainage of the hinterlands. Dike building has occurred through
all these centuries with the result that today we have a simplified
coastline where only exposed foreland marsh formation occurs. The
lagoons have all been diked. The special conditions for the formation of lagoon or basin marshes cannot be found along our coasts
today, and the flats are largely hard sand flats. All this is the
result of man's work. Another result is due to dike failures, such
as occurred during the storms of 1–2 February 1362, 3–4 October
1634, or 3–4 March 1825. The Rudbøl lake and the polder Magisterkogen are two examples of what nature got out of these catastrophes,
for they are now natural areas of great value.

The central questions which should be considered in connection
with both natural marsh formation and land reclamation have to do
with the quantity of material which is available, where the material
comes from, how it is brought in, how and why deposition occurs,

what ecological balance is brought to bear, and the role of animals
and plants in this mechanism. The answers to these questions are
'preconditions' for the rational development of reclamation methods
as well as for the undertaking of a rational evaluation of nature
itself, and its development possibilities. In this connection, it
is particularly plant growth which captures one's interest, in that
it is specific plant species and specific conditions for growth
which are essential. The role of animal life is just as important,
but here many species function equally effectively, and the productivity of the water surface is decisive in the final analysis.

Common mussel (Mytilus edulis), cockle (Cardium edule), or
the Soft clam (Mya arenaria) can be given as an example of the role
of animal life. During a year these shellfish deposit about one
million tons of clay in the Danish salt marshes. These animal
organisms extract fine particulate matter from the water and concentrate it through their faeces so that it can be deposited in the
plant society's natural sedimentation 'basins'. Diatoms, which
live in enormous numbers in the widespread tidal flats are another
important factor. They are to be found on the flats, where light
conditions are ideal, since they have the ability to grow up through
the sediments, which continually cover them. It is this process
which creates the characteristic slime layer which covers the flats,
and which is grazed by the periwinkle (Littorina), Corophium
volutator and others.

Plant growth, as mentioned, plays a decisive role. Glasswort
(Salicornia herbacea), Sea meadow grass (Puccinellia maritima),
and Rice grass (Spartina townsendii) are totally dominant. Glasswort establishes itself from highwater mark down to about 40 cm
below this line. Sea meadow grass follows next though only down
to about 20 cm below the highwater mark. Rice grass can extend
itself further out, and can likewise extend into the salt marsh,
though Sea meadow grass usually takes over here. The zoning of
plant societies is determined by high tide coverage, cf. table:

The high tide distribution, Højer Sluse, compared to morphology
and vegetation cover

morphology and vegetation cover	m DNN	No. high tide/yr
low water line	÷ 0,50	720
mean sea level	0	700
glasswort starts	+ 0,60	665
sea meadow grass starts	+ 0,80	550
high water line, top of tidal flat	+ 0,96	400
top of foreland cliff, outermost		
salt marsh	+ 1,20	200

normal foreland	+ 1,50	75
high foreland	+ 1,80	30
top foreland	+ 2,00	10
topmost foreland	+ 2,65	1

The coverage of top and topmost forelands occur in connection with rough weather and the consequent deposition of large amounts of coarse sediments (sand). The quieter high tides bring smaller amounts of suspended material, composed of much finer grains.

A varied use plan for the area should be established as soon as possible. It should be a plan which secures both economic and recreational utilisation as well as nature itself. The tidal areas should be preserved as an essential element of nature. This must necessarily include restrictions, such as in casu the prohibition of further land reclamation and of the establishment of advanced dikes, causeway connections to the islands etc. Necessary safety measures and primary drainage must however be allowed. The dikes protecting the Ribe and Tønder marshes should thus be strengthened out of consideration for the fact that security and drainage problems must be solved at the same time (Fig. 1).

For the Ribe salt marsh, the solution naturally must be a strengthening of the existing sea dike, which is placed on high-lying geest (cf. Fig. 1). Under the tidal flats to the west thick layers of peat can be found. The inner marsh west of Ribe is likewise characterised by the presence of peat under the clay. Finally there is no larger foreland in front of the present sea dike. Drainage should be carried out with the help of high water pumps.

In the Tønder salt marsh the situation is different. Here c. 1,100 ha foreland, the Ny Frederikskog foreland, is to be found. It extends from Emmerlev to the boundary at Siltoft. The tidal flats in the area of the Højer Canal are lowlying. All this makes the building of an advanced dike along the western fascine fence possible: a solution which will at the same time create 300-400 ha drainage reservoir behind the new sea lock. Ground conditions are reasonable because a 7-10 m thick layer of tidal flat deposits is found above the geest (old land surface). Peat layers are, in general, missing. One could, of course, strengthen the existing sea dike. This would be a lot cheaper, but such a solution would require, in addition, a high water pump costing about 10 million kr. and a new lock, which would cost as much as the pump. The saving would thus be only about 10 million kr. This solution, furthermore, would involve the excavation of the present Ny Frederikskog foreland. A new foreland would naturally have to be created out of consideration for dike safety, but this would occur in connnection with both solutions. A strengthened dike would mean the loss of 1,100 ha

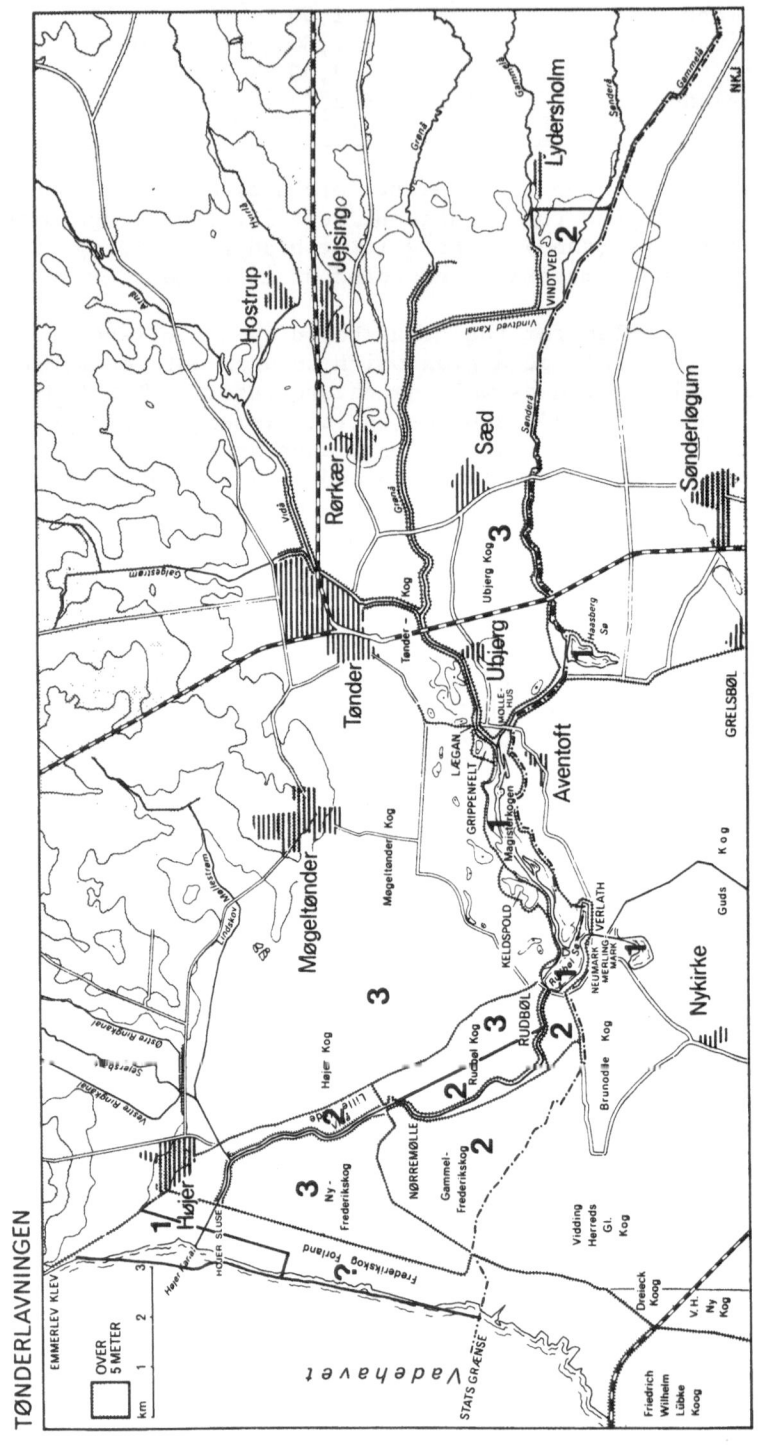

Figure 2

polder land, and neither the area as a whole, the agricultural nor
the environmental interests would be served by this. An advanced
dike would solve the drainage problems and mean the establishment
of a varied sea and shallow terrain. It would, at the same time,
provide 800 ha of new agricultural polder land. This would likewise allow for rounding off of the inner polder thus creating the
possibility of intensive cultivation. It would also allow for the
creation of especially favourable zones for nature which would be
built up around the Vida river, the outer reservoir, Rudbøl lake,
and the polder Magisterkogen along Gudskog and the Hasberger lake
south of the border. All these areas are marked with a 1. on the
map (Fig. 2). The grazing zones are marked with a 2. The remaining areas all marked 3. are used entirely for intensive cultivation. The use of the 800 ha of the new outer polder depends
entirely upon the nature of the technical installations and the
coming political discussions. Only the advanced dike gives all
these possibilities. A strengthened dike will mean great expense -
also in the future due to the high water pumps - and would make it
necessary for intensified agriculture to do without the help of
the possibilities of rounding off and drainage which a projected
dike would allow. This must necessarily result in smaller areas
ending in a condition which, seen from a natural and environmental
viewpoint, must be characterised as primary. At the same time the
blend of natural features and extensive and intensive land use would
be the only desirable pattern, emphasising the character of the
locality and catering for both recreational and leisure interests.

Economic Activities

Besides land reclamation for protection of dikes and for grazing
mainly by sheep, the economic importance of the Danish Wadden area
comes from fishing, harbours, and tourism.

As to economic fishery the Danish Wadden area is only of minor
importance: a little mussel fishery and shrimping within the
Lister Dyb area and Lug-worm digging (some for export) in the Gradyb
and Knude Dyb tidal areas. On the other hand the Danish tidal area
is very important as a nursery area for flounders, and the valuable
sole fishery in that part of the North Sea may be dependent on this
area.

The important west coast harbours of Denmark have always been
located within this area. Today only Esbjerg is worth mentioning.
It is the most important fishing port in Denmark and its importance
as a deep sea harbour is rising. Recently a series of container
traffic lines has been established. Today it is a town of 75,000
inhabitants but it is rapidly growing.

Tourism in the islands is the only economic interest of real
importance within the area itself. This is growing and conflicts
with already established nature conservation areas and with the
planning of future land use of the islands are increasing.

Threats to the Natural Systems

1. Pollution problems in the Danish Wadden Sea. There are three
main sources:

a)   The construction of the harbour and town of Esbjerg. Until
1945 the town had only up to 40,000 inhabitants and minor industry.
In the 1950s, the herring oil industry started and later several
other types have developed rapidly. The harbour itself is expanding as the container traffic develops and since the depth of
Gradyb barrier in 1969 was enlarged to 9 m depth at low tide.

b)   Through the river system Grindsted A-Varde A. A chemical
plant was established in Grindsted in about 1924. It has developed
slowly and expanded considerably since the war as a medical plant
(700 employees). There has been trouble from mercury pollution.
The Skalling-Laboratory has made studies in the river system
because of bed load transport causing great quantities of deposits
at the power station of Karlsgarde, destroying the canals and the
artificial lake. About 1970 pollution problems became acute and
investigations of the balance of mercury stated that at least 5-10
tons had disappeared. Where had it gone? As deposits in the artificial lake or in the Ho Bugt area? In co-operation with Isotop-
centralen we have made investigations showing that parts of this
amount of mercury are to be found in the lake but that considerable
amounts also will be found in the Ho Bugt area. Here it takes part
in the food chain circulation and continues as sedimentation among
the plants in the salt marsh. On Skallingen we have sedimentation
fields where red sand in 1932, 1938 and 1949 was laid out for
studies of the sedimentation rate. This makes time series of samples
possible. Analyses show a close relation between mercury levels
g^{-1} organic matter and time. The amounts of mercury are small in
levels deposited before the war. They declined during and immediately
after the war because of an extension of the artificial lake and
reservoir of Karlsgarde. Since the middle of the 1950s they increased steadily.

c)   Pollution through the North Sea. In 1968 there was a catastrophic mortality among the Lug-worm,(Arenicola marina). On the
tidal flats heaps of dead Lug-worms were found in all small hollows
etc. The situation developed through 4 weeks starting on the tidal
flats close to the tidal inlets, e.g. at Sdr. Ho, North Fanø, and
Langli Sand. From here it was spread, e.g. to the east coast of

Skallingen or a little later to Halen (east coast of Fanø). A lot
of questions can be put forward on such an occasion:

1.   The geographical pattern - is it a local or a general
     phenomenon?

2.   Is a single species involved or a whole animal community?

3.   Are species living on the same type of tidal flat involved,
     or species living in the same food chain?

4.   Is the pollutant effective through one link only or through
     several species of the food chain?

5.   What differences are there between different age groups of
     the species involved?

6.   What is the pollutant itself: biological or chemical?

     Today, however, pollution from the North Sea is not, so far
as we know, an important factor. The Danish Wadden Sea is as a
whole rather pure, except for the Ho Bugt area, and even this area
today is better off than it was a few years ago. The towns of
Varde and Esbjerg now have mechanical and partial chemical waste
water treatment. Locally there will be minor pollution, e.g. in
the Esbjerg area and Mandø.

2.   Land use type, wear and tear at Skallingen. Although the
peninsula of Skallingen is a nature conservation area, during recent
years there have been alarming changes due to the recent develop-
memt of tourism in the Ho-Blavandshuk area. In the last five years
there has been an enormous rise in the number of people going by
car through the area, mostly in the dunes but some in the salt
marsh, destroying the vegetation. The bad economy of the farmers
in this sandy west coast area has also brought about an intensifi-
cation of grazing and land use in Skallingen.

     This has changed the physiognomy of the area partly by the
effect of trampling by fattening cattle and partly by overgrazing.
Formerly 800 sheep and 100 to 200 fattening cattle were raised on
Skallingen each summer. In 1975 it was about 1,000 sheep and 750
head of cattle. There has also been a considerable increase in
shooting, because game licenses for a day's shooting are on sale
and some 50 embrasures have been constructed.

     The only way to control the situation is to protect the area
as a nature conservation area. Grazing e.g. by sheep must continue
or there will be an inevitable change of the vegetation to a

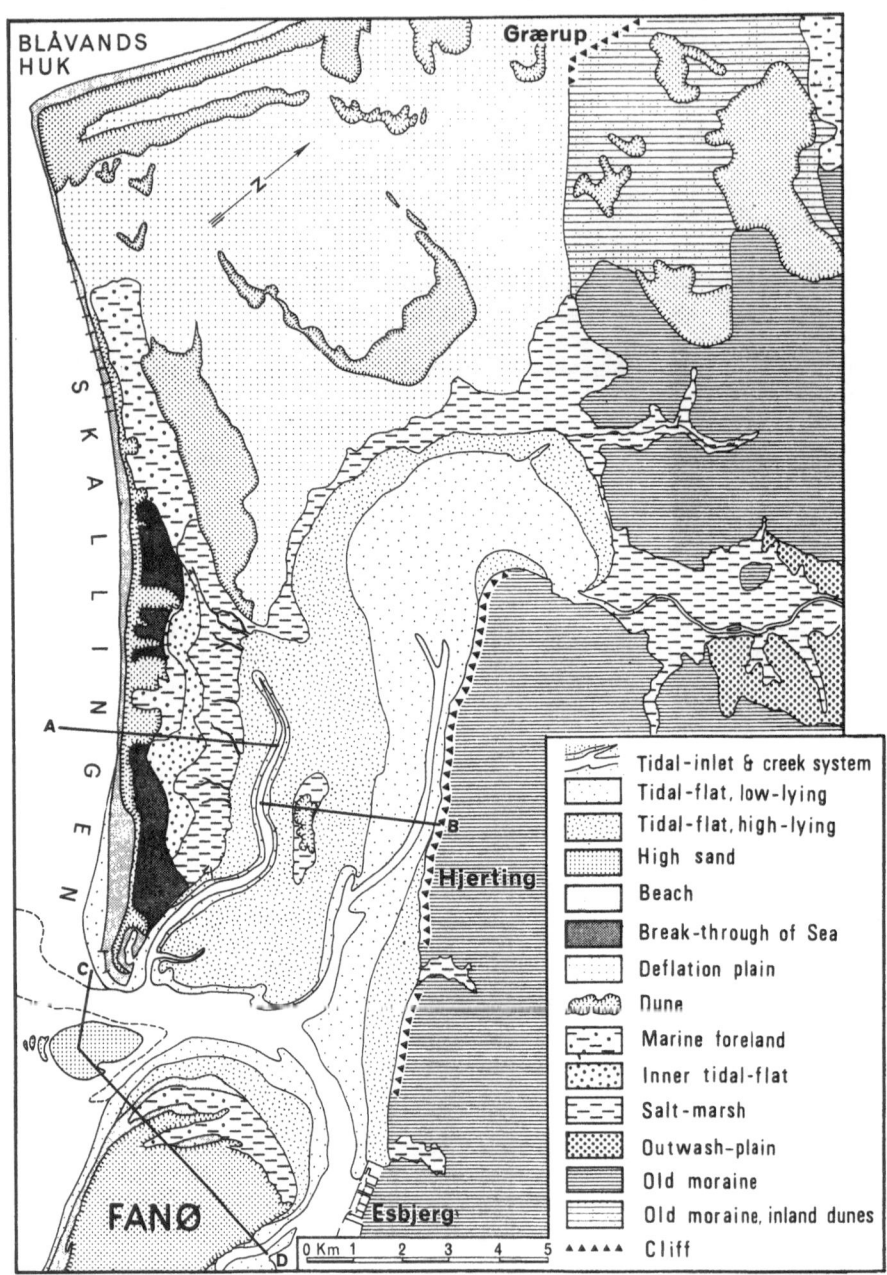

Figure 3

Phragmites marsh. Possible vegetation changes are being investigated at the moment in 8 enclosures in the main landscape types
(Fig. 3).

3. Construction of dams.   A plan for enclosure of the central part
of the Danish Wadden Sea from the Rømø dam in the south to Esbjerg
was put forward by the Vandbygningsvaesen in 1969 (Fig. 4). The
reason is coast protection, but the encroachment would be so large
that the whole nature and physical environment would change.
Politically it was close to reality, but opposition particularly
from nature conservation intests prevented the immediate construction and today it is a sketch only (Fig. 4).

The plan was based on a dike from Fanø to Mandø and Rømø and
a dam with a sluice from Esbjerg to Fanø. This would enclose an
area of about 30,000 ha, the Mandø Vehle, of which one third would
be reclaimed new land. It was furthermore supposed that a new
coastline of 18 km with dune areas behind would appear. The drainage of the salt marshes would be solved, the traffic to Fanø and
Mandø arranged for and the drinking water supply for Manø guaranteed.
The plan would supposedly benefit the development of the harbour of
Esbjerg, the fishing, the hunting possibilities etc.

The Danish Wadden Sea is of primary interest in a national,
Scandinavian, and European context. The advantages which the plan
suggest could be obtained in other ways, perhaps more economically.
The drainage and coast protection problems of the area must be
solved but this is much better done on a local scale. For these
reasons the plan has been re-appraised.

The proposal has nevertheless brought about a discussion of
the environmental conditions, the balance of sediments and economic
conditions concerning a general plan for the area. As to the sediment balance, map studies (1807-1965) indicate an annual contribution of sediments of 200,000 m^3. The total endikement would be
close to a balance of sediments after re-arrangement of the coastal
profile if mean water level is considered as a base level. If the
profile of a normal coastline with a single row of dunes is counted
on, it will mean a deficit of about 100 mill. m^3 of sand or the
amount of sedimentation of 100-500 years.

4. Mandø Ebbevej. The island of Mandø is isolated in the Wadden
Sea, surrounded by tidal flats and a single shallow creek system
which excludes modern shipping. Formerly all supplies to the
island were brought by an evert, a flat-bottomed Wadden Sea sailing
vessel. The island is 6-7 km from the mainland, with a high tidal
flat of sand as connection. At the watershed of this tidal flat an

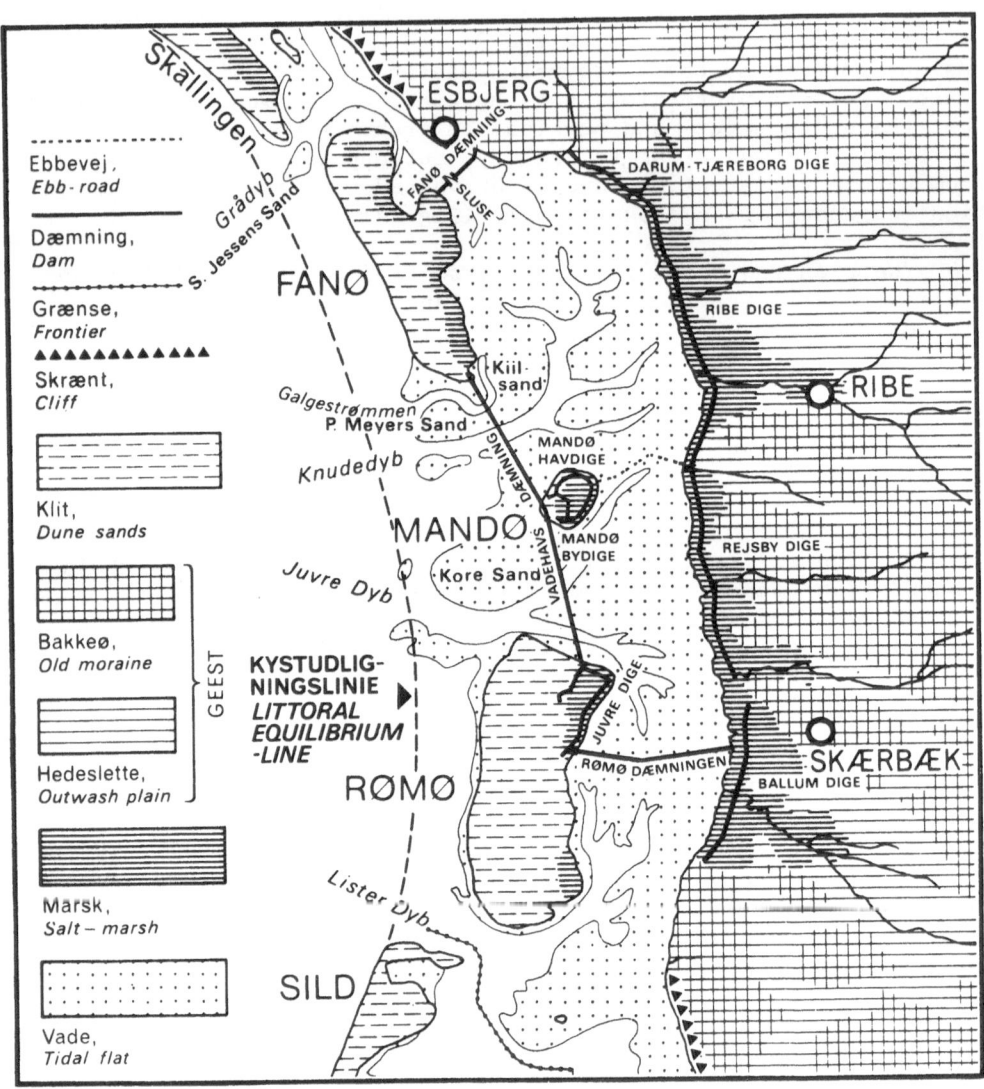

Figure 4

ebb-road was established in 1914 at the same time as the Ribe sea
dike was built. Today the watershed has moved north about 200 to
500 m. The island now has about 150 inhabitants and it is
necessary in some way to find a solution to the traffic problem
(Fig. 5).

This would be an obvious case for an expert committee, but
a single department has begun to build a low dam to Mandø and
started land reclamation along it. This part of the Wadden Sea
represents the only undisturbed, natural part of the Danish tidal
area and the construction works menace the whole environment,
especially since they are quite uncontrolled. It is a single
illustration of the need for co-ordination of environmental interest
on a higher state level.

5. Dikes and drainage. The problems of today along the mainland
coast are partly the sluices i.e. the drainage of the salt marshes,
as the watershed as a whole is placed close to the east coast of
Jutland, and partly storm floods, i.e. a high and safe dike system
along the coast.

Most of the salt marshes need new drainage plans arranged
either by pumping or by a new dike with sluices concentrated outside and around inner reservoirs. Two places along the Danish
mainland coast need such re-organisation: (1) the Tønder area,
where most people prefer a new dike system and the establishment
of inner reservoirs, which could be a fine arrangement also seen
from a nature conservation point of view. (2) the Ribe area, where
reinforcement of the sea dike will be the most reasonable solution.
This arrangement should be considered in relation to the final
solution of Mandø's problems as to traffic, drinking water supply
etc. All this points at the necessity for control and planning.

6. Excavation of gravel, sand etc. Excavation of sand is only
found in connection with Gradyb Barre, the entrance to the harbour
of Esbjerg, which is out of consideration in this respect. Excavation of gravel is going on at Horns Rev, but is dangerous environ-
mentally and should be stopped as soon as possible. Excavation of
shells has stopped because it does not pay, and no oil or gas is
found in the Danish Wadden Sea.

7. Fishing etc. Fishing is mainly going on from Havneby, for
mussels (Mytilus edulis) and shrimps (Crangon vulgaris). It poses
no environmental problems. In the Ho Bugt area and south in the
Knude Dyb area there is a considerable industry digging Lug-worms
(Arenicola marina) for angling, and some are even exported. In

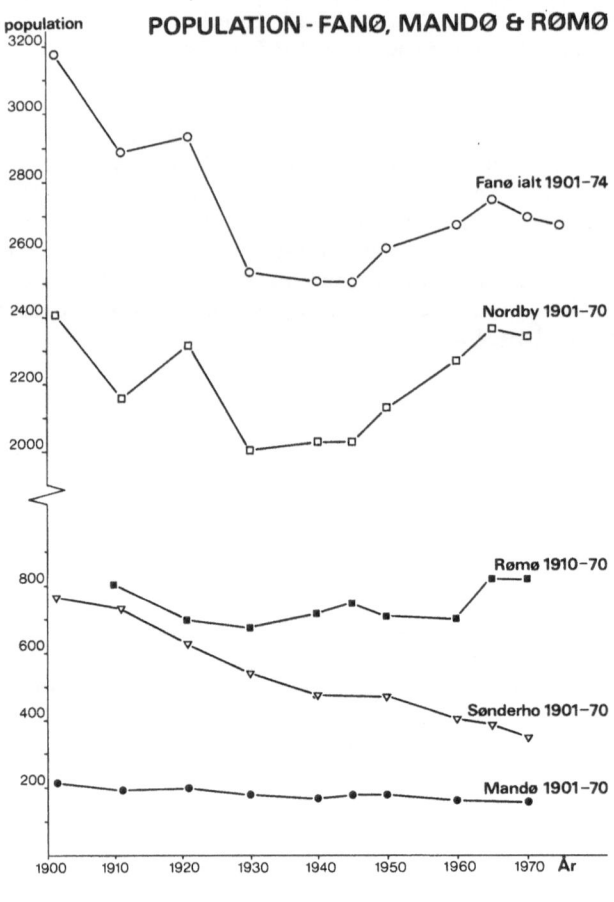

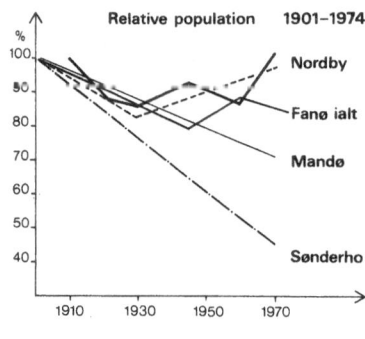

Figure 5

recent times a problem has risen as mechanisation of this activity
has been introduced.

8. Harbour constructions. The harbour of Esbjerg is about 100 years
old and still expanding rapidly. The environment is not at risk
so long as pollution remains under control (cf. 1).

     There exist old harbours at Ribe, Varde, Hjerting, Nordby,
Sønderho, and Jøjer as well as landing places at the sluices. A
new harbour has been built at Havneby, Rømø. There are no problems
today, but it is necessary to consider future problems of motorboat
yachting in the Wadden Sea, caused by building marinas and increased
wildfowling.

9. Recreation and tourism. The island of Rømø provides a good
example of the issues posed by recreation in the Danish Wadden Sea.
This island is 16 km (N-S) and 4 to 6 km (E-W). It is a quite new
element in the landscape, formed as an offshore bar with dunes.
Seven types of landscape element can easily be distinguished:
(1) high sands and foreshore, (2) young dunes, (3) the 1805 coastline, (4) the nucleus of the island, a deflation plain placed about
+ 5 m DNN with parabolic dunes, (5) the salt marsh on the east coast, (6) tidal flats, (7) creeks and channels.

     The agriculture and the settlements are placed in a zone on
the east side of the island. The traffic system is a N-S road
which now is connected with a dam to the mainland and its road
connection E-W to the beach.

     The traffic system, i.e. the accessibility to the area has been
the main controlling factor in the development of a tourist industry.
Within the Danish area we have Fanø as an island with ferryboat
connection, Mandø as an island with connection on an ebb-road, Rømø
with a dam connection to the mainland and Skallingen-Blavandshuk as
a mainland peninsula.

     Because of the geographical position close to the Danish-German
border, and the dam, Rømø has appeared as a magnet attracting as
many tourists on Sundays as the four roadway dam can take. Rømø
is a paradise for one-day tourists, because of the beach 2 km broad,
which can take cars in rows for a length of 2 to 3 km. The road
crossing the island goes through a dune area, the eastern part of
which is a pine plantation. Here parking places flank both sides
of the road. On a summer Sunday Rømø is visited by 100,000 one-day
tourists. The island has about 900 summerhuts and the bed capacity
is in total about 8,800 beds. The traffic on the dam will in July
be about 1 million cars or about 8,000 on a normal summerday. The

one-day tourists do not disturb the recreational demands of the
tourists from the summerhuts except in the Lakolk area: generally
the interests of the tourists appear the same.

The development of population on the Danish Wadden Sea islands
is given in Fig. 5, from which it is obvious that only Rømø and
the parish of Nordby on Fanø have a stable development.   A decline
in population was slowed in 1950 and changed to a rise since 1960
at Rømø, and in 1930 and since 1945 at Nordby.   For the rest of
the area the population has diminished by even as much as 50 per
cent in the last 100 years (Sønderho) (Fig. 5).

All this has certainly changed the whole environment considerably since 1949, when the dam was opened.   The question is how far
will this development go.   Is it possible for the nature as a whole
and for a normal life of the residents to continue on the island
of Rømø or in the Danish Wadden Sea area?   The alternative is
business arranged as a tourist industry.   This is a possibility
which we hope to escape.

DIFFERENTIATION OF THE DANISH WADDEN SEA AREA IN ZONES

To be in advance of misuse and to protect the most valuable parts
of this unique area it is reasonable to try to differentiate the
area in zones according to types of land use and to degree of natural
state.   It is possible to find nuclei of ecotypes which are still
of a high standard and which need protection as far as possible,
(Zone I).   In close connection to this zone there will often exist
areas (Zone II) which need a certain degree of protection.   There
is also a zone of integration in which activities of today can
continue without doing any harm (Zone III).   The rest forms a
'marginal zone' (IV), in which all commercial activities are given
a reasonable priority (Fig. 6).

The solution of the problems in the Danish Wadden Sea area
is possible, if such a zonation is carried out.   The reason for
this can be summarised as follows:

1)   The area is composed of ecosystems of different kinds, each
     in need of its own protection and care.

2)   Different types of people (commercial and/or tourists) have
     distinct interests.

3)   The interests of these different groups can easily conflict.
     Zoning makes reconciliation possible.

4)   International tourism can put quite a pressure on the de-

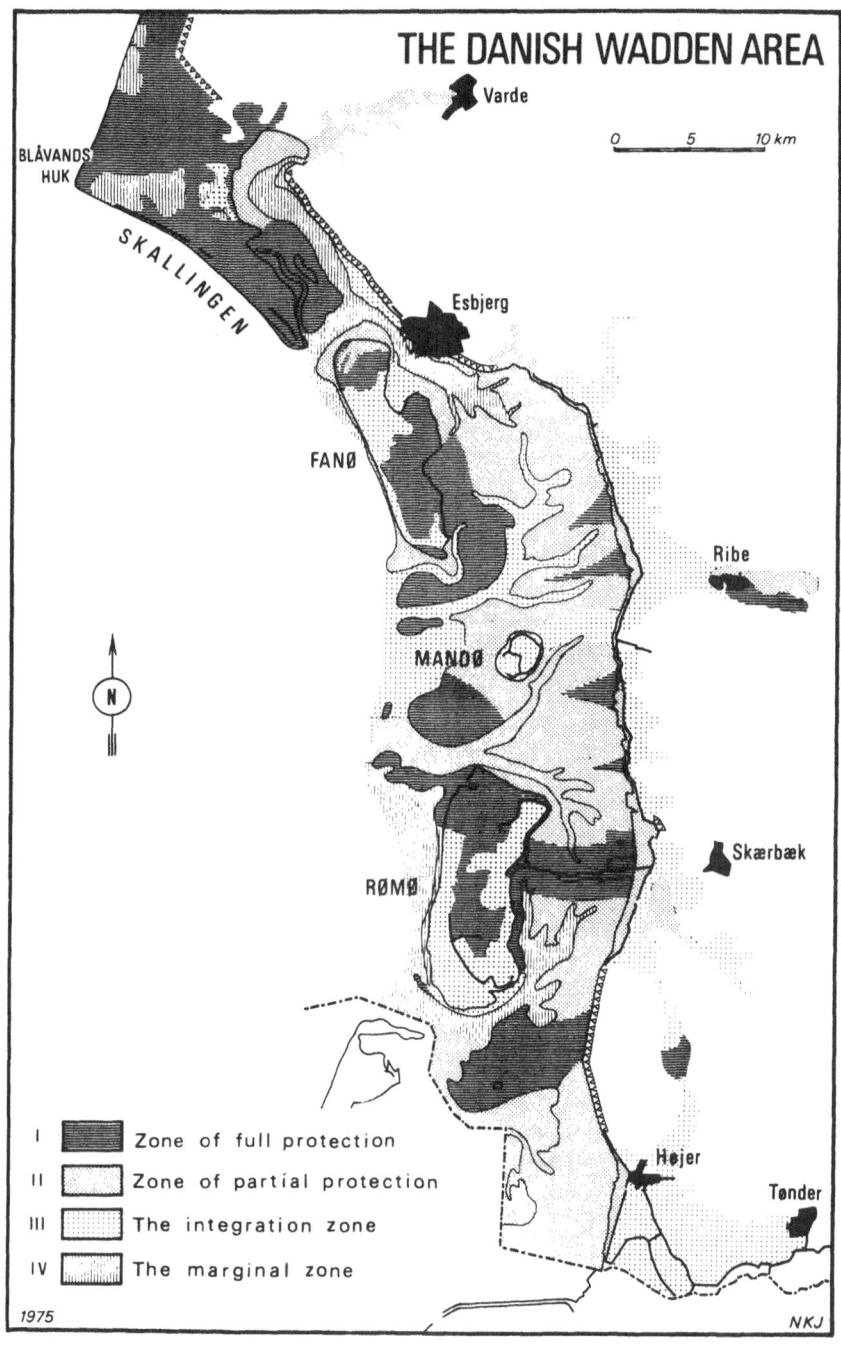

THE DANISH WADDEN AREA

Varde

0   5   10 km

BLÅVANDS
HUK

SKALLINGEN

Esbjerg

FANØ

Ribe

N

MANØ

Skærbæk

RØMØ

Højer

Tønder

I   Zone of full protection

II   Zone of partial protection

III   The integration zone

IV   The marginal zone

1975   NKJ

Figure 6

velopment of the area, which is most easily controlled by
zonation.

If such policies are pursued, there are grounds for hoping
that the environmental character of central parts of the Wadden
Sea area will survive.

DISCUSSION: PAPER 22

H. KÖPP   The example given for the Danish Wadden
                                 Sea has parallels with the Rhine. It illu-
                                 strates problems that can only be solved
                                 through international co-operation involving
                                 the Netherlands, the Federal Republic of
                                 Germany and Denmark. In all these areas
                                 there are threats of environmental change
                                 from the damming of estuarine areas. The
                                 Wadden Sea should be examined from the planning
                                 point of view as a unit.

N. K. JACOBSEN   Co-operation between these countries has
                                 started and will certainly have good results.

E. VAN DER MAAREL   In referring to the physical planning of
                                 this region it is important to recall that we
                                 are dealing with three different Governments,
                                 six to eight state authorities and dozens of
                                 municipalities and it is consequently hard to
                                 link all of them into a common scheme. A
                                 special group has been started in the Nether-
                                 lands and ecologists are beginning to co-operate
                                 on the Wadden Sea region there. A book giving
                                 the basic information about the environment
                                 has been produced and it is clear that the
                                 planners are keen to hear from the ecologists.

A. D. BRADSHAW   I would like to address a question to the
                                 builders of mathematical models. Areas like
                                 that described by Professor Jacobsen demand
                                 a complex analysis of many variables. Is it
                                 possible to make an overall model for so complex
                                 a system or does it have to be broken down into
                                 components?

N. K. JACOBSEN   My own study is of the relationships
                                 between morphogenesis and facies pattern.
                                 I have made a model of sediment distribution
                                 in a tidal regime. This model has been verified

for instance in the Skjern A delta where the
biggest reclamation scheme in Scandinavia
has been worked out.

Maps of the environment in Denmark have
been produced and they show how the nature
conservation interest has influenced physical
planning in advance of other topics. Agriculturalists are at present trying to develop soil
capability maps for which they need a data
bank of soil profile information. At present,
in order to simplify the system and make the
data more quickly obtainable, mapping of clay
content of soil is going on rather than the
investigation of full profile details.

E. VAN DER MAAREL   Perhaps I might respond to Professor
Bradshaw's question. My belief is that a
mathematical model can only be built from the
bottom upwards - that is to say from an
analysis of sub-systems which come together
to describe whole systems. If the term
'model' is defined more broadly it may however
embrace structural models, and models used for
planning optimisation which may have little
mathematical basis. It is useful to build
some kind of a model from a general point of
view because it helps to define the system
more precisely and guide data acquisition.
There is a gap at present between schematic
models used in planning and real models.

J. N. R. JEFFERS   I feel this discussion is becoming almost
dishonest. We all use models, only some are
held in the mind in a form that is almost impossible to transfer from person to person.
Pictures aid the transfer of ideas but they
are coloured by interpretation by the person
seeing them. Such people may believe that
they are working to the same model and drawing
the same conclusions when in fact they are reacting differently. The advantage of a
mathematical model is to ensure that hypotheses
are clearly stated and that the approach is a
strict one. Sometimes mathematical statements
of this kind are helpful: at other times they
may not be. The value of a model in the explicit sense used by George Van Dyne or myself
is that it seeks a way of transferring inform-

ation into a form which allows real and critical
testing of it and this is more certain in the
explicit, mathematical model than the imperfect
verbal one.

G. VAN DYNE   Professor Bradshaw's question was related
to the need for detailed knowledge of changes
from sea to land or land type to land type.
A matrix transition model does not incorporate
details of the workings of the system and this
detailed knowledge is what is most likely to
be transferable to another area. Matrix transition models are valuable in the analysis
within an area. The more widely useful general
types of model take much more building.

23. ECOLOGICAL PRINCIPLES FOR PHYSICAL PLANNING

E. van der Maarel

Division of Geobotany

University of Nijmegen, Nijmegen, The Netherlands

INTRODUCTION

Rehabilitation of damaged ecosystems is a matter of natural resource management and optimal land use, and also of integrated physical planning. This contribution will touch upon ideas ecologists can contribute to physical planning. The formal and theoretical treatment will be extended by examples, mainly from the Netherlands. This small country is still reasonably representative for temperate environments while, because of its small size, conflicts between different types of land use have become manifest earlier than in most larger countries. Moreover, the traditional Dutch passion for regulation has led to a rather full development of physical planning, while the traditional strong emphasis on nature conservation has caused physical planning to be infiltrated by ecological thinking over the last fifteen years.

Within the framework of physical planning we may adopt a rather broad definition of "damage to" or "degradation of" an ecosystem. (cf. Chadwick & Goodman, 1975). We need not restrict ourselves to the breakdown of a productive ecosystem and subsequent attempts at re-establishment. A considerable loss of species would also constitute damage as would the replacement of a mature forest by a grassland. Obviously we are dealing with something akin to "ecological damage" in these two cases and we may distinguish between various types of damage. Generally each ecosystem may be considered to have a number of functions for society (see below) and damage is then to be understood as a loss of function.

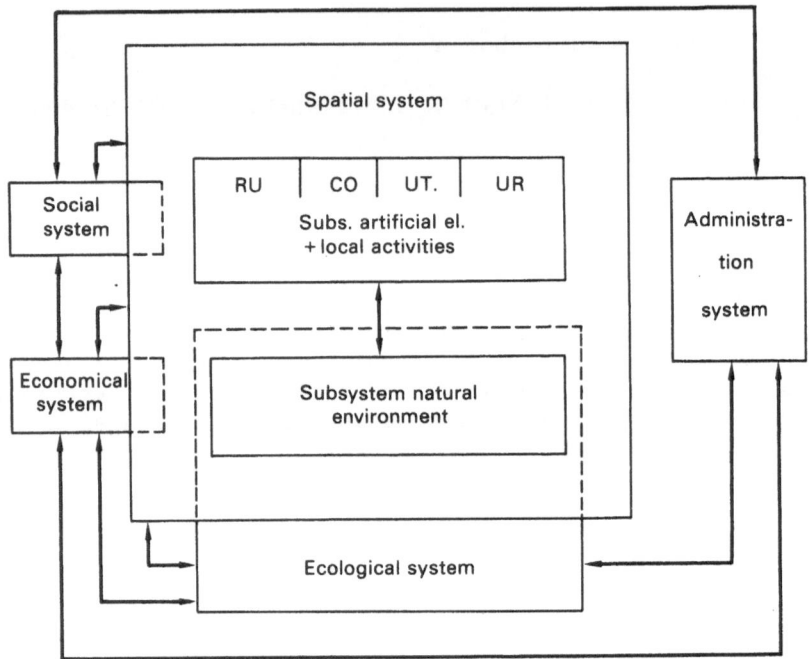

<u>Fig. 1.</u>   Interaction between society and natural environment in a
physical planning model. (adapted from Anon. 1975, cf. Van
der Maarel et al. 1977). See text.

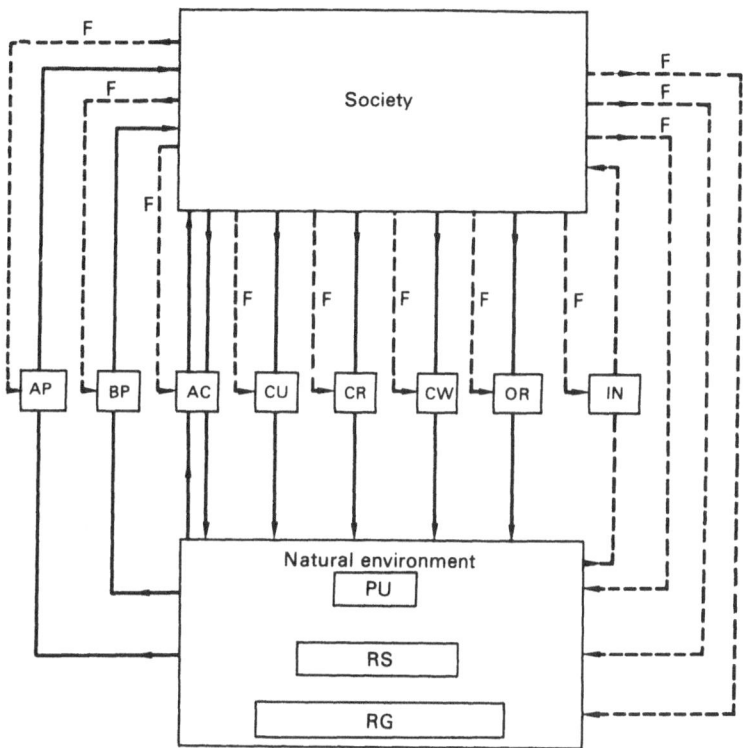

<u>Fig. 2.</u> Interaction between society and natural environment through
functions and feedbacks (adapted from Van der Maarel et al.
1977). AP abiotic production, BP biotic production, AC
agricultural production, CU carrier of urban activities,
CR carrier of rural activities, CW carrier of waste, OR
carrier of outdoor recreation activities, IN information,
PU purification, RS reservoir, RG regulation functions.
withdrawal of matter and energy from natural environment.
addition of matter and energy to the natural environment
withdrawal of information from the natural environment
feedback: addition of information for control of urban and
rural activities, or for control (management) of the natural
environment.

The integration of resource management and land use in
physical planning can be approached through a general ecological
model of interactions between natural environment and society.
The National Physical Planning Agency of the Netherlands has asked
a team of ecologists and planners to contribute to such a model
(cf. van der Maarel and Vellema, 1975) and will incorporate this
in a larger physical planning model. The model includes a number
of sub-models allowing aspect, and sub-systems as well as their
mutual relationships,to be distinguished (Fig. 1, cf. Anon, 1975).

The main model refers to the spatial system which is divided into
the technical and sub-system of artificial elements plus localised
activities and the sub-system of the natural environment. The
technical sub-system is again divided into urban, public utility,
communication and rural components. The social system, comprising
society and its sociocultural relationships and the economic system
are aspect systems overlapping the spatial system which have a
planning of their own: social and economic planning. The ecological
system is also an aspect system, i.e. it includes only the
(ecological) relations between society and natural environment. The
corresponding model is the general ecological model mentioned earlier;
the corresponding control is ecological planning.

FUNCTIONS OF THE NATURAL ENVIRONMENT

To get to grips with the ecological relationships we may distinguish
functions (human uses) of the natural environment which we can place
in four interrelated groups: production, carrier, information and
regulation functions (Table 1, based on data from van der Maarel
et al, 1977). Functions can be measured by estimating a "capacity"
of the natural environment or - particularly for carrier functions -
its potential for use. For most of the functions a maximal possible
value exists. Over-use or over-exploitation may lead to a decrease
in capacity and ultimately to complete exhaustion or decline.
Society has developed feedback mechanisms known as resource management, alternative agriculture (with decreased matter and energy
inputs), land use control, physical planning and nature conservation
and management. Fig. 2 presents a scheme of relationships between
society and natural environment in terms of functions and feedbacks.
In this contribution emphasis will be put on the information and
regulation functions, mainly because of the situation in the Netherlands and related areas where the capacities for these functions have
decreased more than other capacities.

INTERACTION OF FUNCTIONS

Usually a particular area has a potential for many functions, but the fulfilment of one function will reduce the capacity of most other functions. From an ecological viewpoint we may distinguish between functions provided by the more or less unchanged natural environment versus functions whose fulfilment may include a considerable change in the original characteristics.

To specify the expressions "more or less unchanged" and "considerable change" we use a classification of degrees of naturalness. Naturalness is a state of not being influenced by man. One should rather speak of culturalness. Indeed attempts to measure degrees of culturalness, have been made, for example for the vegetation component of terrestrial ecosystems, by estimating the share of neophytic species, i.e. species that form part of a regional flora only since ca. 1500, and have probably been promoted in their establishment there by man (deliberately or not). Sukopp (1972) devised such a system of "hemerobiotic degrees" (hemeros = cultivated) and tried to characterise each degree by a specification of the human influence, and the character and intensity of changes in the soil and the structure of vegetation. He also added an estimation of the loss in nature species. Van der Maarel (1975) summarised this scheme and combined it with the traditional indications of naturalness (Table 2) and pointed out that the present state of naturalness of an ecosystem has to be considered with respect to at least two further criteria: the origin of the ecosystem and its successional state. Inclusion of these two criteria would, of course, greatly complicate the system of Table 2. Instead we adopt a much more simplified system: the ensemble of the first three degrees is characterised by relatively small changes in soil and a largely spontaneous flora and fauna. We may take them together as "more or less natural", or simply natural.

Turning back to the fulfilment of functions we may now specify two types of functions with respect to the degree of naturalness. The first type includes functions bound to more or less natural conditions. They are called ecological functions and what is known as evaluation, valuation assessment of conservation values (cf. Helliwell, 1969, 1971, 1973, Sukopp, 1971, Genuin et al, 1975), (landscape) ecological or biological evaluation (van der Maarel, 1970, van der Maarel & Stumpel, 1975, Tjallingii, 1974) or naturalistic evaluation (Kuyken, 1975) can be considered a measure of these ecological functions, particularly information and regulation functions.

In contrast the other functions, particularly the carrier, storage and most of the production functions, may be called socioeconomic functions. These terms are not satisfactory. Here we

Table 1. Functions of the natural environment.

(a) Production functions supply of matter and energy from natural resources.	1. Production from abiotic resources	supply of light, heat, oxygen, water, energy (wind-power, waterpower, geothermal, nuclear, fuels).
	2. Production from biotic resources, not necessarily managed.	supply of raw materials and natural products (wood, moss, fibres, latex, resin, fungi, sea-weed, fish, ivory, skins etc.
	3. Agricultural production	supply from especially managed natural resources which receive matter and energy from man: fish, oyster and mussel culture, agriculture, silviculture.
(b) Carrier Functions provision of space and surface for human activities.	4. Carrier of urba-industrial activities.	supply of space and surface for artificial elements and localised activities: urban (-industrial), public utilities (water and energy supply, waste processing), communication (roads, railroads, pipelines, harbours, airfields).
	5. Carrier of rural activities.	ibid: rural water management, coastal defence. military defence and training.
	6. Carrier of waste	supply of space for discharge of solid and liquid waste.
	7. Carrier of recreation activities	supply of space for recreational facilities and activities: substrate-bound, landscape-bound.
(c) Information functions	8. Supply of information	supply of information for orientation (incl. aesthetic appreciation of our surroundings, philosophic identification), scientific research, education, and signalling of environmental changes through indicator organisms.
	9. Reservoir	supply of potential information to be used in future.
(d) Regulation functions	10. Purification	regulation by waste assimilation: noise, dust, organic waste ("biological purification").
	11. Regulation	environmental stabilisation through atmospherical filtration of cosmic rays, biospherical damping of climatic fluctuations, retention of water (mainly by soils), soil protection and biotic regulation. (biological control and "biological equilibrium").

Table 2. Degrees of naturalness in ecosystems, hemerobiotic state and some characteristics of vegetation and soil (Van der Maarel 1975, largely after Sukopp 1972).

NATURALNESS	HEMERO-BIOTIC STATE	CHANGES SUBSTRATE	CHANGES VEGETATION STRUCTURE	CHANGES FLORISTIC COMPOSITION	LOSS NATIVES (1000 sq. km)	GAIN NEOPHYTES
NATURAL	A-HEMEROBIOTIC	NO	NO	NO	0	0
NEAR-NATURAL	OLIGO-	FEW	NO	MOST SPECIES SPONTANEOUS	<1%	5%
SEMI (AGRI-) NATURAL	MESO-	SMALL, SUPERFICIAL	OTHER LIFE FORM DOMINATING	MOST SPECIES SPONTANEOUS	1-5%	5-12%
AGRI-CULTURAL	EU-	MODERATE TO DRASTIC	CROPS DOMINATING	FEW SPECIES SPONTANEOUS	6%	13-20%
NEAR-NATURAL	POLY-	DRASTIC ARTIFICIAL SUBSTRATE	OPEN EPHEMERAL	FEW TO NO SPECIES	?	21-80%
CULTURAL	META-HEMEROBIOTIC	IBID	-	-	-	-

prefer the terms urban/industrial, or simply urban, functions.
Obviously some functions have a transitional status: they are
bound to semi-natural and agricultural environments; we may call
them rural functions such as agricultural production, biotic
production, recreation and rural carrying functions. They are
measured by what is usually called land evaluation (Zonneveld, 1972;
Vink, 1975) or land appraisal (Whyte, 1976).

In this contribution we are mainly interested in the interaction between natural functions s.s. and rural + urban functions.
Such interaction studies include:

a)   characterisation of the natural environment, particularly
its ecosystems with respect to the natural functions:
"evaluation";

b)   survey of trends of environmental changes due to fulfilment of rural and urban functions and detection of vulnerability
of natural ecosystems for such changes: "devaluation";

c)   estimation of possible improvement or rehabilitation of
natural functions: "revaluation".

CHARACTERISATION AND SURVEY OF ECOSYSTEMS

Natural functions are linked to the productivity, structure and
species composition of ecosystems. These characteristics are
determined by various abiotic and biotic conditions, notably
geological and geomorphological conditions, soil and (soil) water
conditions, type of vegetation, flora and fauna, landscape and
potential natural vegetation (i.e. "the vegetation that would
develop if all human influences on the site and its immediate
surroundings would stop at once, and if the terminal stage could
be reached at once", Westhoff & van der Maarel, 1973, after Tuxen,
1956). From the systematical description of ecosystems (c.f. Haase
1973, Tjallingii, 1974, Long, 1974, 1975, Seibert, 1975) we may
derive values for a number of evaluation criteria (c.f. Adrian &
van der Maarel, 1968; Helliwell, 1971, van der Maarel, 1970, 1971;
Sukopp, 1971, van der Maarel & Stumpel, 1975). Since the descriptions are often still not complete enough, it may be necessary to
simplify the evaluation by estimating only some general evaluation
indices. Fig. 3 shows the relation between environmental conditions
and natural functions through evaluation criteria and indices (after
van der Maarel et al, 1977, c.f. Zonneveld 1972; Seibert, 1975).

For terrestrial ecosystems the description is preferably based
on a phytosociological analysis of vegetation (c.f. Westhoff & van
der Maarel, 1973, Mueller-Dombois & Ellenberg, 1974 for recent

surveys) and a soil description (often general soil maps are already
available). In addition one or more of the following groups of
organisms are analysed as well: vascular plants, breeding birds,
migrating birds, and mammals. Exceptionally lower plants and herpeto-
and entomofauna are included.

Distribution patterns of soil (and geomorphological) characters
can be integrated with patterns of plant communities and ecotopes
and geotopes may thus be established. An ecotype is the manifestation of a (broadly defined) ecosystem (Tansley, 1939; Troll, 1939)
and it usually covers a local mosaic or zonation of plant communities:
vegetation complex (c.f. van der Maarel & Stumpel, 1975). A geotope
in its form and function is a homogenous and recognisable part of
the geosphere (the earth's surface). Man's impact is an intrinsic
characteristic of geotopes. Geotopes are considered the basic
landscape units (c.f. Haase, 1973; Socava, 1972; van der Maarel et
al, 1977).

So called landscape ecological or environmental surveys are
performed to produce maps (scale 1:25,000) showing the pattern of
geotopes and/or ecotopes or of plant communities and/or vegetation
complexes (see Tjallingii, 1974; Werkgroep GRIM, 1974 and Kalkhoven
et al 1976 for first examples).

Estimation of the various natural function capacities can be
done on the geotope level. Usually estimation comes down to a
ranking in five or ten degrees, ranging from hardly to highly
valuable - or similar categories. Such ranking may be based on
comparative figures, sometimes it is merely the outcome of "a best
professional judgement" (c.f. Helliwell, 1969, 1971, 1973;
Tjallingii, 1974; van der Maarel & Stumpel, 1975; Bugmann, 1975).

Ideally the estimation of a geotope's natural functions should
be to a local potential maximum or rather to an average value for
a much larger area. Turning back to Fig. 3 we may specify this as
follows:

Scarcity index: Rarity or scarcity of geological, geomorphological, pedological, landscape, ecosystem, phytosociological, floral
and faunal components can be estimated through the total number of
occurrences and, preferably as a relative proportion of population
size or ecosystem area within the geotope of a regional, national
or world total. Scarcity is then replaced by the more appropriate
measure "geotope contribution to the maintenance of geo/eco or bio-
components".

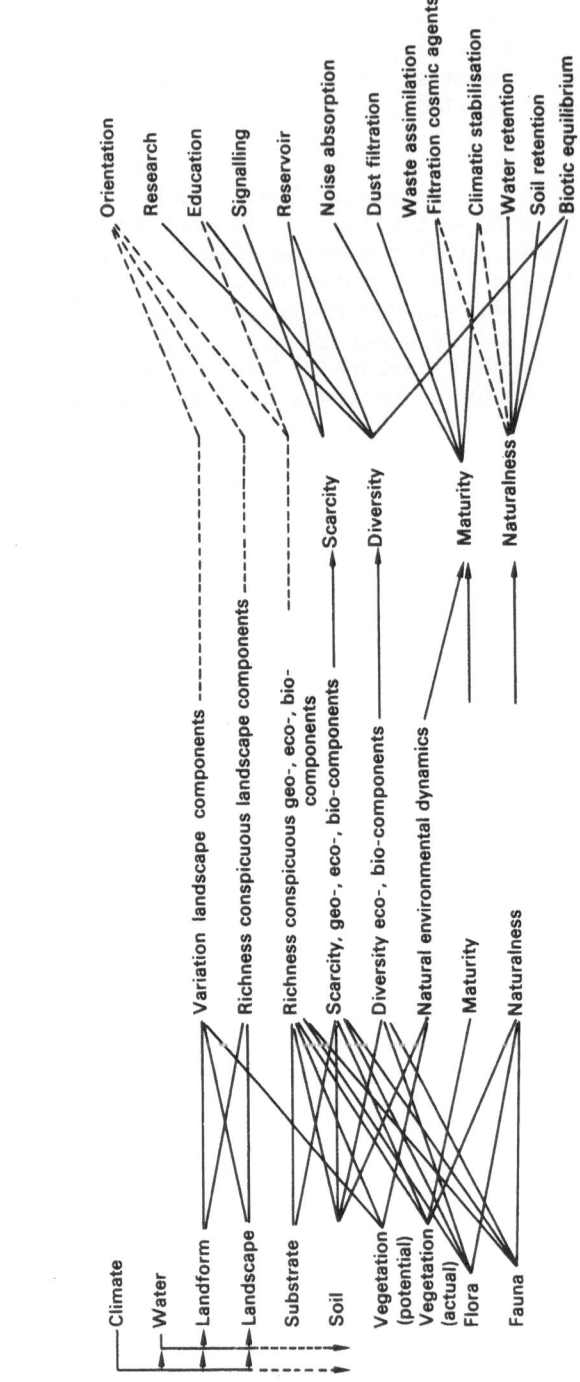

Fig. 3. Natural functions as determined by the abiotic and biotic conditions of the natural environment (Van der Maarel et al. 1977).

Ecotope diversity index: In the concise index system of Fig. 3 only ecotope diversity is measured. This can be done through simple counting of ecotopes per geotope or a larger landscape unit. It is thought that floral and faunal diversity are sufficiently covered by the scarcity and ecotope diversity indices.

Structural differentiation index: This index can be measured through stratification or life-form diversity analysis. It is supposed to relate to both the maturity of the ecosystems and to the natural environmental dynamics.

Naturalness index: This index has been discussed already.

These indices can be used to give a very rough estimate of the capacity a geotope may have with respect to the natural functions distinguished. It is still largely tentative to connect functions with indices as has been done in Fig. 3. Still we may come already a bit further than just saying "high" or "low" value. Our uncertainty is two-fold - we lack the ability to define natural functions precisely and do not have enough analytical data on the environmental conditions.

Information functions relate to all kinds of geotope components. We may say that each component has an information value, which we cannot yet specify per type of component and, particularly, per group of organisms. The larger the number of scarce components and the greater their scarcity, the higher the information will be. The reservoir function referring to a source of potential inform- ation is estimated similarly i.e. through scarcity and diversity indices.

Purification of noise and dust is largely a matter of structural differentiation, as are some of the regulations. Biotic purification seems to be related to natural environmental dynamics, which means that high turn-over speeds may better enable a system to take up waste material. Naturalness is related to various functions, notably those regarding stabilisation. Generally, regulation functions may be estimated through structural different- iation and naturalness indices. Since insufficient is known about the relations between functions and environmental conditions and since it is obvious that some of the indices are "lumping parameters", we may proceed just by trying to measure these - or similar - indices. It would go too far here to describe measurements in detail (c.f. van der Maarel et al, 1977). Generally speaking we may draw two conclusions:

(1) Regulation, upon which the existence of mankind ultimately depends is particularly provided by the natural mature eco-

Table 3. Scheme of influences on the natural environment caused by human activities.

GEOSPHERE COMPONENT	ENVIRONMENTAL INFLUENCE
Substrate	appearance of artifacts, appearance of new substrate, disappearance of substrate.
Soil: Structure	ploughing up, erosion, digging, heightening, compaction, pollution.
Water	lowering phreatic watertable, raising " " inundation, irrigation, warming, pollution, purification.
Nutrition	eutrophication, oligotrophication, acidification, salination, desalination.
Plants	poisoning, removal, introduction.
Vegetation	felling, burning, removal, mowing, grazing, treading, planting.
Animals	poisoning, removal, introduction, disturbance.

systems, both those of relatively constant and those of relatively fluctuating environments.

(2) Information upon which mankind's civilisation is based
in so many ways, is particularly provided by complexes of
natural, near-natural and semi-natural environments. Clearly
we have to give up part of the regulation functions in at least
parts of the world in order to gain information. Of course,
we recognise that in order to provide a physical basis for
the existence of a certain human population (i.e. the fulfilment of so-called primary needs) part of the earth's surface
has to be changed into urban and rural environment. We are
just referring here to the relation between information and
regulation.

SURVEY OF TRENDS OF ENVIRONMENTAL CHANGES AND DETECTION OF ECOSYSTEM VULNERABILITY

Environmental changes are considered here as side-effects of human
activities for the fulfilment of urban and rural functions. The
more or less natural ecosystems on which society depends for the
natural functions may be vulnerable to these environmental changes.
A systematic survey will start with a listing of human activities
together with an indication of the type of environmental influence
involved in these activities, as well as the character, intensity
and pattern of spatial impact. Further side-activities should be
predicted, like suburban or recreational developments after the
construction of new highways.

Van der Maarel et al (1977) listed ca. 140 activities, grouped
per function and including activities in favour of information and
regulation functions, which we may summarise as environmental
management. Table 3 presents 36 influences resulting from the
various activities, arranged according to the geosphere component
they act on.

The sequence of geosphere components in Table 3 is in accord-
ance with the "hierarchy of spheres", an order of abiotic and biotic
spheres according to the total energetic influence each sphere has
on other spheres (Van Leeuwen, 1973, c.f. van der Maarel & Vellema,
1975). This sequence is: cosmosphere> atmosphere > hydrosphere >
lithosphere > pedosphere > phytosphere > zoosphere > noosphere.
In this order the spheres are more and more dependent on energy
from a higher sphere. At the same time each sphere has some counter-
influence on higher spheres, mainly in the form of structural
changes. This influence is roughly increasing in the same order,
with the noosphere as most powerful sphere at the end. The
regulation functions mentioned above are in part the result of the

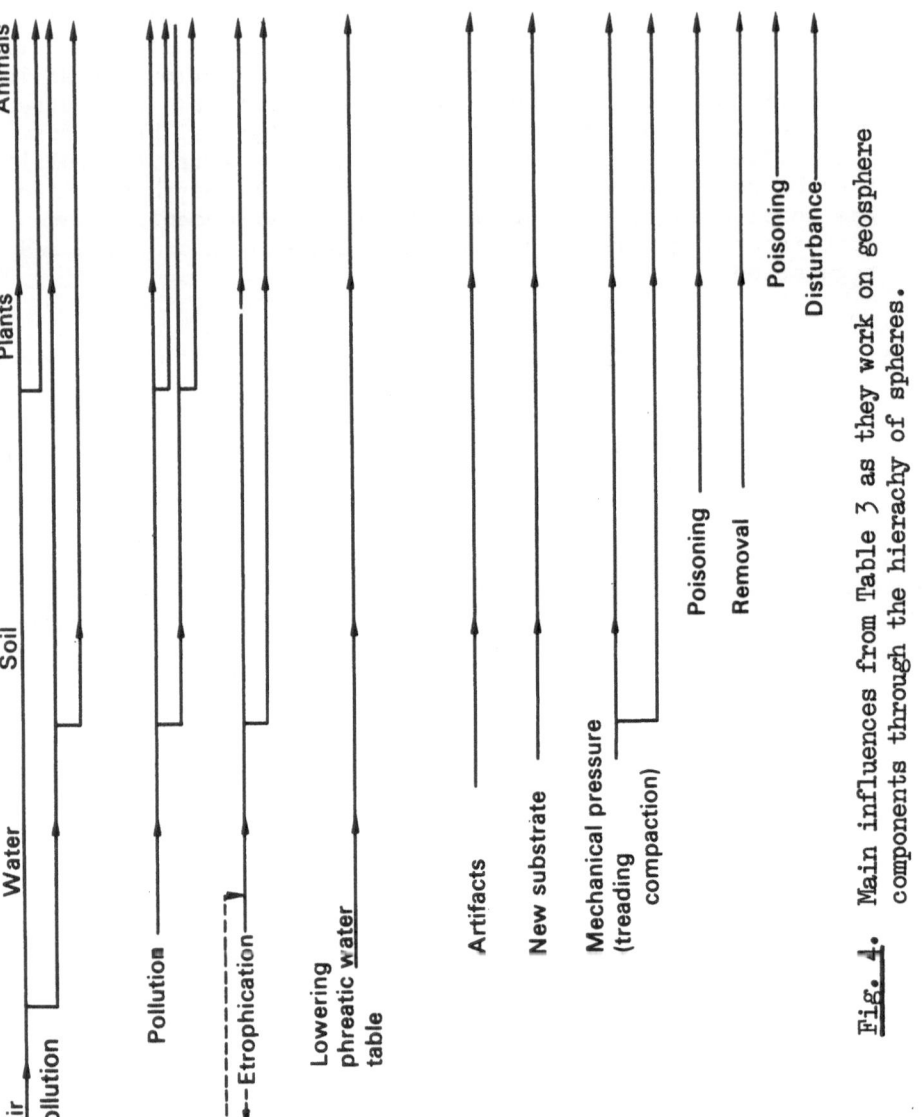

Fig. 4. Main influences from Table 3 as they work on geosphere components through the hierachy of spheres.

damping of atmospheric fluctuations by hydrosphere, and phytosphere. The loss of natural functions of the geosphere, and particularly of the biosphere connected with the damage of ecosystems is the result of noospherical influence, either direct or indirect; i.e. through a higher sphere. Fig. 4 presents the lines along which the major environmental influences from Table 3 work on the various geosphere components.

We may add some general considerations to this scheme:

- The main influences are replacement of the substrate, pollution, eutrophication, lowering of the phreatic watertable, mechanical pressure on the surface.

- Air pollution is an impact high in the hierarchy and thus it has many indirect effects.

- Eutrophication of water and soil water are both impacts relatively high in the hierarchy and profound in effect.

- Urban-industrial and agricultural activities in so far as they involve pollution and eutrophication, urban and road development through spacial impact; and recreation with its direct influences on plants and animals, they all may be considered as the main human activities with ecological side-effects.

- Disturbance only affects animals but it is brought about by many activities.

The next step is the estimation of the susceptibility of all the different ecosystems for all the influences mentioned. This is complicated and difficult to analyse. Only in cases of drastic environmental change such as creating a new substrate or total disturbance of a soil profile do we know that all ecosystems are maximally vulnerable. For most influences however we do not exactly know the susceptibility. For Dutch ecosystems we now have made up tentative susceptibility estimates with 5 point scales, for three influences: eutrophication, desiccation (particularly lowering the phreatic water table) and treading. Generally the susceptibility will depend on the nutritional status, with of course the oligotrophic systems as the most susceptible; the soil moisture conditions and the status of natural environmental dynamics. Here also the disposition may be important: how easily or difficult an influence can really reach a system. The next step is the estimation of regeneration possibilities. Some so-called elastic ecosystems will rapidly return to their pre-disturbance situation if the disturbance does not persist too long. Here we are dealing with the resilient type of stability (cf. Holling, 1973; Patten, 1974). Ellenberg (1972) suggests as a first naturally

0 10 20 30 40 50 km

(a)

Fig. 5.   Changes in the nutrient status of the Dutch terrestrial
          environment. a. Situation ca. 1900 b. Situation ca. 1975.
          Simplified from figures in Van der Maarel et al. 1977,
          which have been prepared by J.H. Smittenberg, mainly from
          evidence collected by the State Research Institute for
          Nature Management.

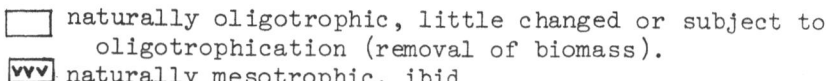

0 10 20 30 40 50 km

(b)

☐ naturally oligotrophic, little changed or subject to
     oligotrophication (removal of biomass).
▨ naturally mesotrophic, ibid.
▤ naturally eutrophic
▥ subject to moderate to strong eutrophication
■ urban-industrial areas

rough approach, a formula to estimate what he called "Belastbarkeit" (load capacity). We may consider vulnerability as an equivalent: it expresses how far and how long an ecosystem (or an ecosystem component) can be burdened before it definitely changes. The formula reads:

$$B = \frac{(100 - D \times L) \times R}{10}$$

where D = disposition, L = susceptibility, R = regeneration, all in 10 point estimation scales. Such values, however rough they must be, may still give some idea about the vulnerability of a system. The vulnerability estimates must finally be used to estimate the change in the various evaluation indices with which we estimate the capacity for information and regulation functions.

The main trend of environmental changes in large parts of the world is deforestation. In most parts of Western Europe the resulting pattern of largely semi-natural ecosystems, with near-natural remnants and agricultural and urban concentrations, was still of considerable value and a good deal of nature conservation has been devoted to semi-natural ecosystems (cf. van Leeuwen, 1965, 1966; Westhoff, 1968, 1970; Duffey & Watt, 1971; Duffey, 1971; van der Maarel, 1971; Westhoff, 1971 and particularly van der Maarel, 1975). In terms of functions the semi-natural ecosystems fulfil most information functions.

Since ca. 1900 the semi-natural landscape has been changed very considerably, particularly through eutrophication and water table regulation, usually in combination with reclamation work. "The whole country on the spade" has been a motto in Dutch land development circles for decades. The result is drastic change from oligotrophic to eutrophic conditions.

This may be illustrated by Fig. 5. Ca. 1900 the Pleistocene eastern part of the Netherlands (nos 1-5 on the map) was almost entirely oligotrophic with only locally some eutrophicated reclaimed areas. The oligotrophic dunes of the NW coast and the West-Frisian islands belong to the same category. The Holocene peat soils in the west and north parts of the country (nr. 6) were mesotrophic with a tendency towards oligotrophication, due to hay making, whilst the mesotrophic to eutrophic calcareous dunes and some inland soils (nr. 7) were also subject to some oligotrophication. Only the marine clay soils in SW Netherlands and the polders of S and N Holland, Friesland and Groningen were eutrophic by origin (cf. Edelman, 1950 for further details).

The 1975 situation is one of eutrophication throughout the country. Only some larger moraine and bog areas as well as most of the dunes remained in the same nutrient status.

ENVIRONMENTAL IMPACT ANALYSIS

Applications of vulnerability analysis usually have an ad hoc
character, when some specific development plan is checked on its
environmental side-effects. First they were rather superficial and
emotional, done by "environmental activists". Now they may be
very profound and part of the governmental system, leading to
official "environmental impact statements". In physical planning
the development plan will have considerable spatial impact. The
obvious examples are urban development and planning of new highways. McHarg (1969) presented impressive visual confrontation
maps of plans versus existing qualities. Such confrontation can
be easily done through computer-based maps (e.g. Steinitz & Rogers,
1970; Anon, 1974; Kiefer & Robbins, 1973; McCarthy et al, 1974;
Koeppel, 1975).

An important step forward would be a more detailed estimation
of "environmental quality" to be confronted with impacts from
various activities. The Netherlands provide some examples how this
could proceed:

(1)   In the north east part of this country the possible impact
of the governmental highway development scheme was tested on
the basis of an ecological evaluation according to an estimation
of geotype naturalness, with the assumption that environmental
influences would be highest in areas of the highest naturalness
class and close to the roadway. (van der Maarel, 1972).

(2)   The urban development scheme for the Arnhem-Nijmegen
area was tested on the basis of an integrated geotope evaluation.
This evaluation included vegetation and some faunal components
and resulted from estimations of the four evaluation indices
mentioned in an earlier section. Through this evaluation the
information capacity within the area was estimated in a 5 point
national scale. The result is presented in Fig. 6. Next the
possible degradation as a result of urban development was estimated as follows. Spatial occupation as such was considered
of the most serious impact in areas of class 5, etc. Side-
activities like further development of infrastructure and outdoor recreation, were considered the more serious according
as the area is closer to the area to be built up and its value
class is higher. The maximum width of the zone of influence
was arbitrarily taken 2 km, with a reduction to 1 km in cases
of barriers like canals and dikes. Accordingly five degrees
of "ecological objectionableness" were established. The
corresponding map is presented in Fig. 7 (from Werkgroep GRAN
1973).

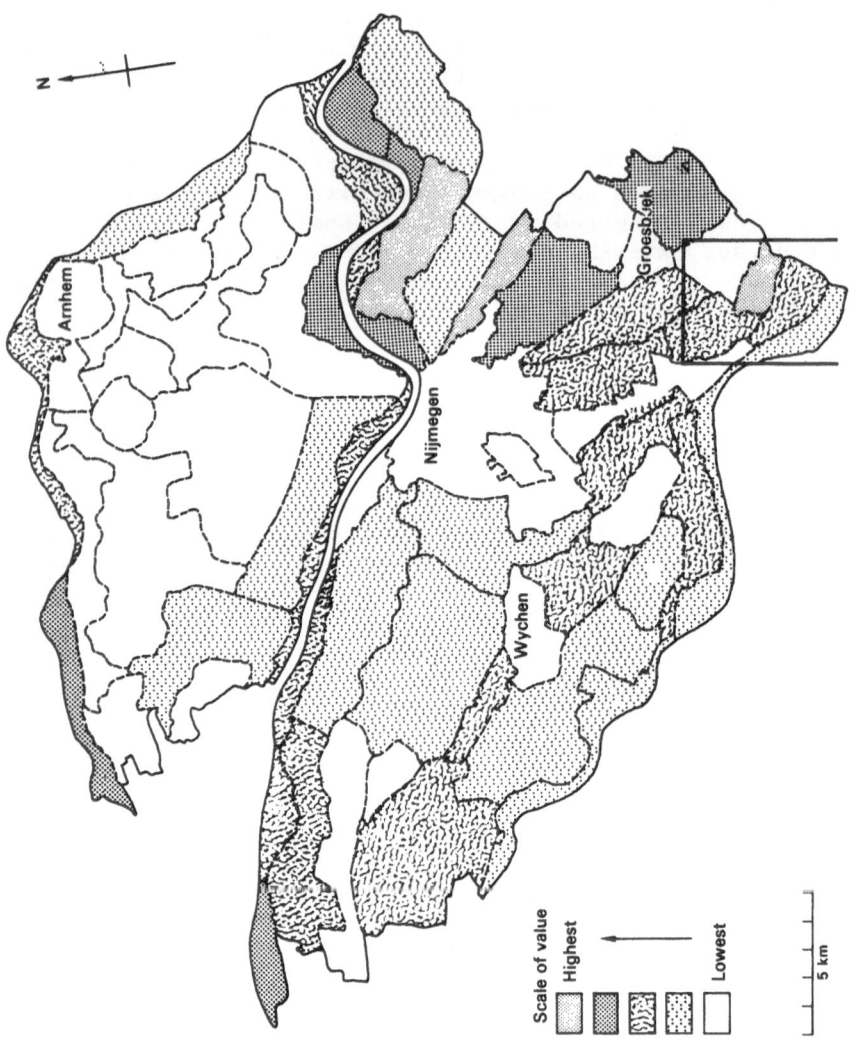

Fig. 6. Integrated ecological evaluation of the Arnhem–Nijmegen region, the Netherlands along a 5 point scale, ranging from lowest value to highest value (from Harms et al. 1976).

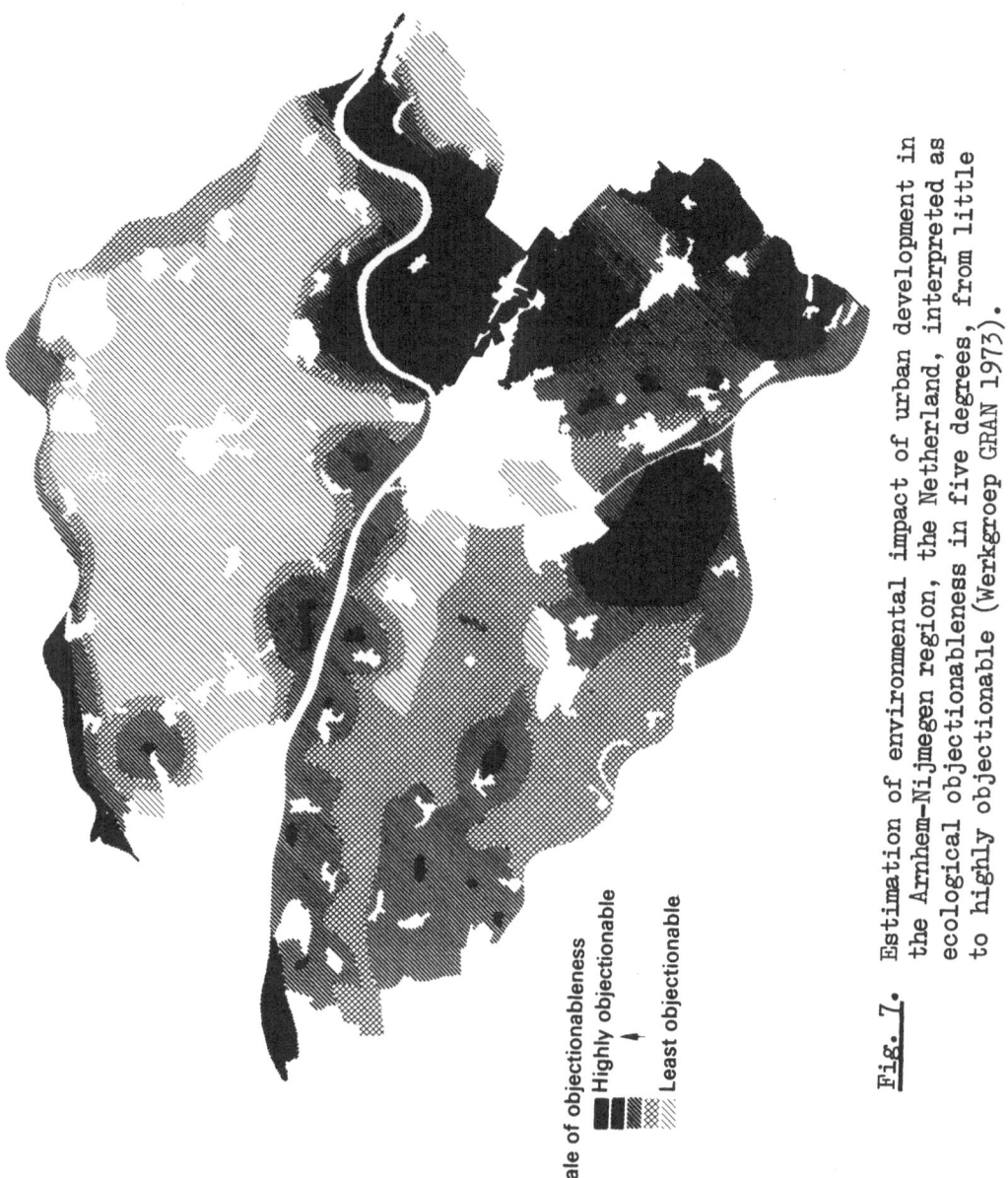

Fig. 7. Estimation of environmental impact of urban development in the Arnhem-Nijmegen region, the Netherland, interpreted as ecological objectionableness in five degrees, from little to highly objectionable (Werkgroep GRAN 1973).

Scale of objectionableness

Highly objectionable

Least objectionable

(3) Within the framework of a water resource management study
in the province of Gelderland (c.f. de Boer et al, 1976;
Colenbrander this volume) an inventory of more or less natural
ecosystems was made and their ecological capacity (overall
estimate of natural functions) along a 10 point scale was
estimated. These estimates were transformed to values for
each sq km. Then the vulnerability for lowering of the phreatic
watertable was estimated and expected losses in natural functions indicated. By combination with the hydrological models,
which predict falls in the phreatic watertable at different
extraction quantities at each place, the ecological effects
of such extractions could be roughly indicated (van der Maarel,
1975a).

ECOSYSTEM RESTORATION POTENTIAL

We are not only concerned with possible degradation but also with
possible restoration of the natural environment. From the present
pattern of ecosystems, or of ecotopes and geotopes we may derive
a pattern of "under-developed", i.e. degraded environments.

To judge the possibilities of restoration we must first know
the potentials of the area under consideration. These are effectively described by the pattern of potential natural vegetation
types. The determination of the potential vegetation may be very
difficult in completely devastated or cultivated areas since it is
based on the knowledge of both the present soil conditions and the
phytosociological structure of remaining (near) natural vegetation
usually a woodland type.

Per type of potential natural vegetation, a list of so-called
"replacement" or substitute plant communities (c.f. Tuxen, 1956) is
derived. Such series of replacement communities ("community complex",
Seibert, 1968, or "vegetation series", Werkgroep GRIM 1974) can be
used as reference. Developments of small-scale maps of potential
natural vegetations can be traced in various European countries,
e.g. Poland (c.f. Falinski, 1968), German Federal Republic (c.f.
Trautmann, 1966, 1972) and also in Japan (c.f. Numata et al 1972;
Miyawaki & Okuda, 1975). For the Netherlands a map of the actual
vegetation, scale 1 : 200.000 has been completed (Kalkhoven et al
1976) in which data on potential natural vegetation and vegetation
series are incorporated per geotope.

In connection with these data on the potentials of an area
a decision has to be made on the kind of development one would
wish in that area and the kind of ecosystem pattern which should
result, given both the prospects and restraints of physical planning.
For that decision we need some indication of the perspectives of

development within the total range of perspectives in terms of ecological success and social needs.

Ecological success, i.e. the effectiveness with which the desired pattern will be attained, will depend on the still existing variation in abiotic (and biotic) conditions and the environmental dynamics within the area.

IMPORTANCE OF GRADIENTS

Environmental variation is particularly important when gradients occur, i.e. gradual transitions in the conditions of one or more environmental factors. With reference to some literature on gradients used in connection with physical planning (Whittaker, 1967, 1973, 1975; MacIntosh, 1967; Van Leeuwen, 1965, 1966a; van der Maarel & Leertouwer, 1967, van der Maarel, 1971; c.f. van der Maarel, 1976 for a recent account), the following information may be presented in summarised form:

(1)   Gradients may occur on various scales, ranging from large gradients between biomes,to microtopographical gradients within some narrow valleys. The type of environmental factor involved in the gradient, changes from the large-scale to the small-scale situation. Following the hierarchy of spheres we may distinguish climatic, hydrological, geomorphological, soil texture, soil chemical, animal influence and human gradients. In the framework of physical planning the gradients from soil texture gradients onwards are of greatest interest.

(2)   Gradients are of great importance for niche different-iation of plant species (and indirectly for animal species as well), leading to local concentrations of different species, including rare ones. Preservation or restoration of gradients will effectively contribute to the fulfilment of information functions.

(3)   Four main single-factor gradients, usually developing along topographical gradients are formed where acid conditions dominate over base-rich; organic soil over mineral soil; dry over wet conditions; and oligotrophic over eutrophic conditions.

(4)   Single factor gradients may occur in spatial combinations, and the combined gradients develop more potentials than the single factor ones.

(5)   Dispersion along environmental gradients is often partly counteracted by concentration effects from environmental dyn-amics occurring within the zone of the gradient (e.g. fluct-

uations of the phreatic water table within a dry-over-moist
gradient zone). Such unstable gradients may be of significance
for particular species and ecosystems.

(6) A gradient will only arise or remain in existence if there
are no directed changes in conditions along the gradient which
tend to diminish the differences between the ends of the gradient. For example, along a topographical gradient under moist
conditions, a gradient from eutrophic to oligotrophic conditions
will not persist since the nutrients will be gradually moved
downwards and hence the oligotrophic lower part will become
eutrophic as well. However, when the oligotrophic condition
is on top then the gradient will be maintained.

Generally a gradient will level out when the more active,
energy-rich expression of the factor is dominating over the
less active, energy-poor expression: base-rich over acid,
mineral over organic, wet over dry, and eutrophic over oligotrophic.

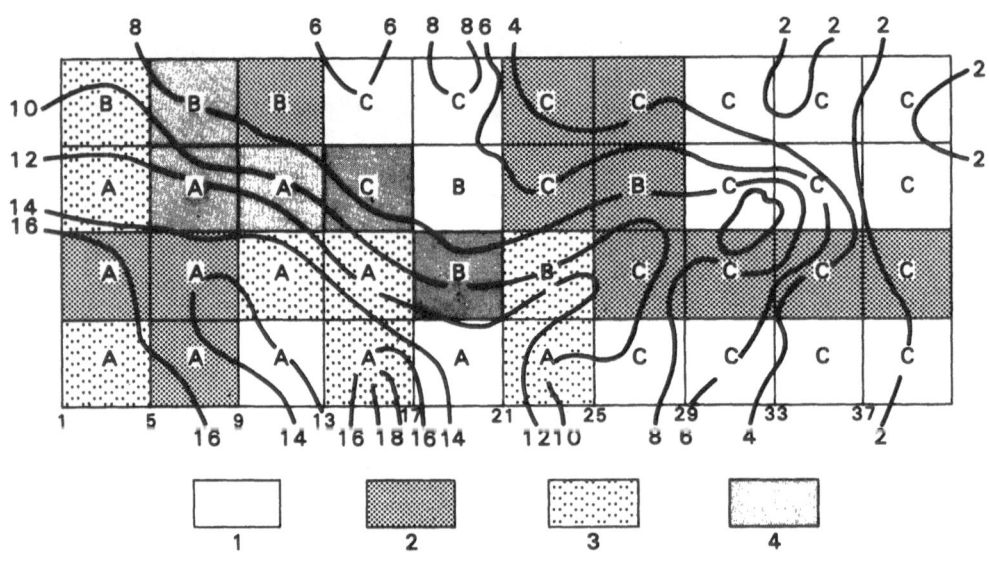

Fig. 8. Distribution of species diversity, vegetation type and height
(2 cm contours) along a transect in a rabbit-grazed duneslack
(Van der Maarel 1971 after Van der Maarel & Leertouwer 1967).
Diversity class 1: 17-25 species/sq.m; 2: 26-30 sp; 3: 31-35
sp.; 4: 36-43 sp. A Radiola linoides community, B Linum
catharticum community, C. Parnassia palustris community.

(7)   Animal influence, notably grazing; human influence,
notably mowing, may be important in addition to an abiotic
gradient. Although the influence as such is a kind of disturbance and a contribution to the environmental dynamics, the
effect may be positive; the vegetation remains low, the nutrient status may be rather poor, even with a tendency towards
oligotrophication (c.f. fig. 5), and small microtopographical
gradients may be fully exploited by niche differentiating
species. Fig. 8 presents an example of such a subtle cor-
relation of species diversity and position along a gradient
in a moderately to heavily rabbit-grazed dune slack area.
Small changes in height within the transition between a dune
slack and a small dune hummock are accompanied by marked changes
in both moisture conditions and pH. In the lower part of the
transect the soil is inundated each winter, pH values are 6-7;
in the upper part inundation hardly occurs, pH values are 4.2-5.
The intermediate zone with pH 5.1-5.7 is restricted to a height
interval of only 4 cm with Linum catharticum L. as a narrow-
range characteristic species. Not only does the species richness in the intermediate zone reach the maximum of 43 species/
sq.m. but in places over 26 species were found in only 1/16
sq.m. (Thalen, 1971).

(8)   Human and also animal influences can easily destroy abiotic
gradients whenever they act from the top of a gradient downwards, or level the entire gradient - e.g. overgrazing, desiccation and particularly eutrophication.

TOWARDS A STRATEGY FOR ECOSYSTEM DEVELOPMENT

The ecological restoration of an area should be based on the present
conditions of the area, i.e. the existing pattern of ecosystems of
various degrees of naturalness with various degrees of environmental
variation and dynamics, with the potential natural ecosystem or ecosystem pattern in mind. As a very tentative general guideline we
may adopt a twofold strategy:

(1)   Whenever the original environmental conditions included
or still includes gradients, and whenever there is any chance
of restoring them, a complex of natural, near-natural and
semi-natural ecosystems should be the aim. When the gradients
are small in extent, particularly small soil gradients, the
semi-natural types with a low vegetation structure should
prevail. Here we are particularly developing information
functions.

(2)   When the original environmental conditions are rather
uniform and/or human impact has levelled down the original
variation so drastically (especially by eutrophication and
desiccation) that restoration is only possible after a heavy
input of energy (and money), one should rather develop ecosystems adapted to uniform and often dynamic conditions i.e.
certain types of woodland as well as savanna-like grasslands
with a possible wild-life stock. Here we are promoting regulation rather than information functions, besides purification
functions.

Examples of both strategies are being developed but no extensive reports have been published so far (see, however, Duffey
& Watt, 1971). An example of the first approach is the (re) introduction of extensive grazing in neglected semi-natural ecosystems
with local eutrophication and wood development.

Experiments are being carried out in several centres, including
the Department of Plant Ecology at Lund, Sweden (N. Malmer, A.
Larsson); the State Institute for Nature Management at Leersum,
the Netherlands (C.G. van Leeuwen, P. Oosterveld) and the Monks Wood
Experimental Station, Abbots Ripton, England (T C E Wells, J P
Dempster). An example of the second approach is the afforestation
of marginal arable land, through planting or through spontaneous
development. The latter case is the standard secondary succession
as described in extenso for American situations (c.f. Waggoner &
Ovington, 1962; Daubenmire, 1968; Odum, 1969; Knapp, 1974.

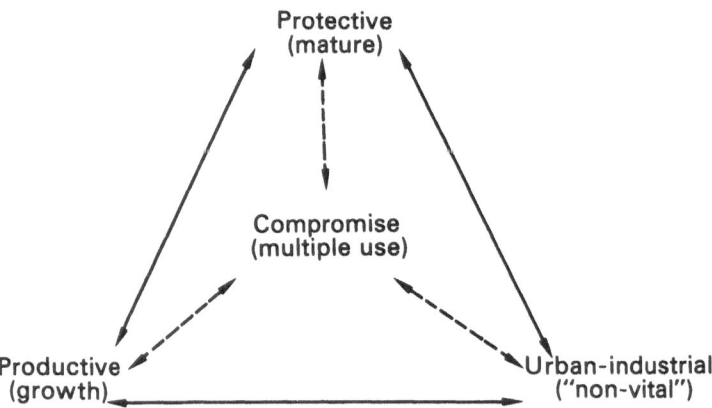

Fig. 9.   Interaction between three main environmental compartments
          (redrawn after Odum 1969).

ECOLOGICAL OPTIMALISATION IN PHYSICAL PLANNING

Rehabilitation of ecosystems, possibly along the lines indicated,
must of course be fitted into plans for the overall management of
an area. This means that reinforcement of natural functions, or
even maintenance of them, is always considered in the realisation
of urban and rural functions. To provide physical planning with
a basis for the balancing of functions we can make up a sort of
compatibility matrix of all natural, rural and urban functions we
are able or wish to distinguish. The ecologist is then mainly
concerned with the mutual relations amongst the natural functions
and between those and the others. We can simplify this interaction
matrix by devising a small number of environmental compartments as
Odum (1969) did. Curiously this attempt is quoted very frequently
but in the last six years not much progress seems to have been made
in refining the model. Fig. 9 present Odum's scheme in a slightly
altered form, i.e. the multiple use compartment is not considered
as a separate but rather as an intermediate compartment between
the others.   For the purpose of this paper a system of 8 compartments has been devised (based on van der Maarel et al, 1977).
They are described as follows (see also Fig. 2).

   (1) INAG "information-regulation" (comparable to Odum's
   mature compartment).

   (2) INBP "information-biotic production", comprising natural
   ecosystems of dynamic environment and semi-natural ecosystems
   with some form of exploitation. Here we could also take up
   purification functions.

   (3) INLR "information-recreation", where a variety of natural
   and semi-natural types can be kept both for purposes of sig-
   nalling, research etc. and for forms of recreation bound to
   landscape and natural history.

   (4) RU "rural". In this compartment we may situate rural
   activities, particularly defence works (against flooding and
   other hazards) and military training.

   (5) AC "agricultural", which is identical to Odum's productive
   compartment.

   (6) SR "recreation bound to substrate" - sporting, leisure
   etc. for which we are developing special environments, such
   as suburban parks and recreation waters and shores.

   (7) UR "urban" i.e. comprising housing and public utilities.
   Compartments 6-8 form Odum's "non-vital" compartment.

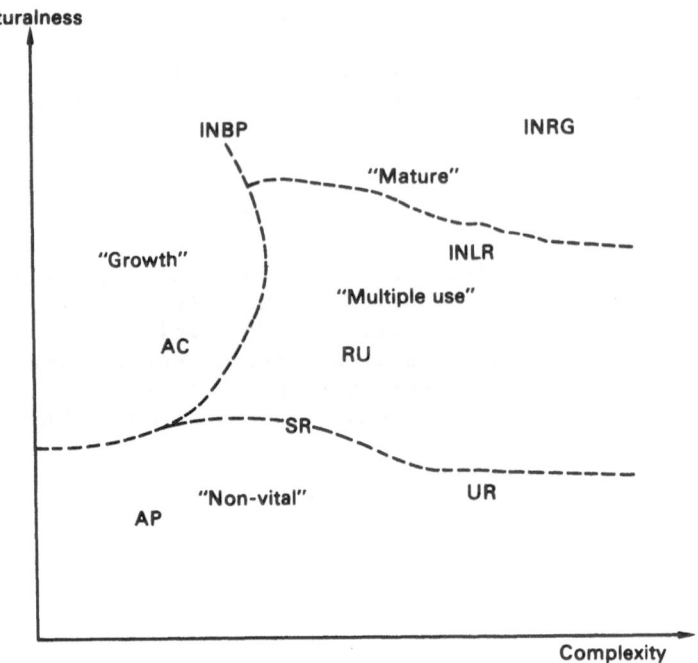

Fig. 10. Relation between environmental compartments as far as their degree of naturalness and their complexity is concerned (after Van der Maarel et al. 1977). The compartments sensu Odum (1969) are indicated as well.

	INRG	INBP	INLR	RU	AC	SR	UR	AP
INRG	●	[+]	[+]	±	−	−	−	−
INBP		●	[+]	±	+	±	±	[±]
INLR			●	±	[±]	±	±	
RU				●	I	±	±	[±]
AC					●	−	−	−
SR						●	[+]	±
UR							●	±
AP								●

Fig. 11. Compatibility matrix for the eight environmental compartments of Fig. 10 + = compatible as neighbouring compartments; − = incompatible; ± compatibility depending on situation; spatial complexes of compartments are possible.

(8) AP "industrial-abiotic production". In this compartment we may include discharge of waste, as well as service and communication systems (road, railroads, waterways, pipelines, wires).

In Figure 10 the position of these compartments is indicated in a two-dimensional scheme, with degree of naturalness and complexity along the axes. This shows that complexity as a valuable characteristic of systems is not restricted to natural ecosystems. Technical ecosystems may be complex as well, particularly the metropolitan system, although no one seems to have compared complexity levels in any quantitative way. This most complex urban environment is thus not merely a "non-vital" compartment as designated by Odum. It is a highly vital complex stable system. The only problem is that its development requires the sacrifice of so much of its natural counterpart, the mature natural environment. The interaction between compartments in terms of matter and energy exchange, as was envisaged by Odum (1969) has not yet been described in great detail (see however H T Odum 1971). Besides, it is not sufficient to have interaction coefficients available for just that type of exchange.

For physical planning it is more important to estimate the spatial, structural impacts of one compartment on the others. This is equally difficult and undeveloped! Fig. 11 presents a so-called compatibility matrix for the eight compartments mentioned above. It is meant as a first approximation. The scheme starts from the idea that a compartment of a certain type will be mainly arranged so as to fulfil the functions by which it is characterised. Two compartment types are considered compatible if they may exist next to each other without considerable changes in either compartment caused by the other e.g. information/regulation and information/biotic production. Such compartments are considered incompatible when at least one compartment may be considerably changed and less suited by influences from the other e.g. industry/abiotic production and information/landscape recreation. In other words: the fulfilment of functions in one compartment will be hindered by the fulfilment of functions in the other. In many cases the compatibility may depend on the local situation, e.g. rural military activities may influence a nearby agricultural compartment but it need not. The ▢ sign means that adjacent compartments may be compatible or not, but joining them in a large spatial complex is desired anyway, either to create transitional zones between the compartments, or to protect a third compartment from side-effects coming from one particular compartment. In this respect the INBP compartment is rather crucial. We should indeed try to develop various kinds of transitional and protecting productive ecosystems between the essential natural and cultural compartments.

a.   INRG >   INBP >   INLR >   AC

b.   INRG >   INLR >   INBP >   AC

c.   INBP >   INLR >   UR > INBP > AP
                      ↘ SR ↙

Fig. 12.   Examples of zonations of environmental compartments leading
to ecological optimilisation. For symbols see Fig. 10 and
text.

The idea of joining compartments into larger complexes may be
elaborated by developing schemes for zonation of compartments. Fig.
12 presents three examples of such zonations. Example A starts
from an agriculture compartment, which is surrounded by a semi-
natural compartment where facilities for landscape bound recreation
exist. This compartment may be joint with a zone of hay meadows
and/or exploited woodland which in its turn joins a mature ecosystem
compartment. In example B the biotic production zone may be one of
extensive grazing, buffering the fully agricultural compartment from
a zoned semi-natural to natural area. Example C indicates a possibility of connecting urban and industrial areas with buffer zones
and recreation areas.

These considerations may be rather obvious to many ecologists,
but they are still rather obscure to physical planners, let alone
social and economical planners, so we need to describe them (cf.
Dasmann et al, 1973; Ovington, 1975).

ECOLOGICAL PLANNING

This brings us to some concluding remarks. So far the ecologists
have been successfully involved, mainly in the preservation of
remnants of our natural environment. However, preservation no
longer means safeguarding alone since there are so many side-
effects arising from human activities elsewhere, or even within
the nature reserves. One only needs to think of pollution and
recreation effects. We are really in need of a much larger
"ecological grip" on human activities in and around nature reserves.
A first aim could be to present guidelines for the real protection
of conservation of those areas society wishes to preserve for
information and regulation (cf. Dasmann, 1972; Duffey & Watt, 1971;
Holling & Clark, 1975; Westhoff, 1971, 1971a).

But that is not enough. We should go one step further and
screen the other functions ecologically, i.e. analyse how efficiently
they are being exploited. This could come down to a balance sheet
of matter and energy input and output (cf. Odum H T, 1972). This

could lead to far-reaching implications and conclusions such as
replacing the entire urban development strategy for the Netherlands,
i.e. "concentrated spread" to the north east and southern parts of
the country by "spread concentration" in the Centre! (Thesis to
be worked out elsewhere). The ecological basis for rehabilitation
of damaged ecosystems should be developed in relation to the "ecology
of functions". Here we should keep in mind or rather bring into
the mind of many, what a Dutch environmental engineer, Peters (1972),
called the "Law of Conservation of Misery". What we try to gain
in order and energy efficiency, we are losing again elsewhere by
the very attempts, so that the overall effect is an increase of
entropy. Our aim should be to reduce this inevitable increase as
much as possible. In terms of planners we should, however, proceed
and devise an ecological planning system, as an aspect-system of
reality with the same level of impact on physical planning as the
two existing aspect-systems, the social and the economical system.
This would lead to a real ecological planning. In addition and
last but not least, ecologists may infiltrate the administration
system with ecological thinking.

REFERENCES

Anon, 1974. The Canada geographic information system. Ottawa:
    Environment Canada. 22pp.

Anon, 1975. Planningmethodiek. Studierapport 5.1. The Hague
    Rijksplanologische Dienst. Ministerie Volkshuisvesting,
    Ruimtelijke Ordening.

Adriani, M. J. & Maarel, E. van der, 1968. Voorne in de branding.
    Oostvoorne: Stichting Wet. Duinonderzoek. 104 pp.

Beguin, C., Hegg, O. & Zoller, H. 1975. Landschaftsökologisch-
    vegetationskundliche Bestandsaufnahme der Schweiz zu
    Naturschutzzwecken. Verhandl. Ges. Ökologie, 4th, Erlangen,
    1974, 245-251. The Hague: Junk.

Boer, B. de, Nes, Th. J. van de, & Romijn, E. (ed). 1976. Model-
    onderzoek 1971-1974 ten behoeve van de waterhuishouding in
    Gelderland. Deel 2: Grondslagen, Arnhem: Comm. Bestudering
    Waterhuishouding Gelderland, 381 pp.

Bugmann, E. 1975. Die formale Umweltqualität. Ein quantitativer
    Ansatz auf geographischökologischer Grundlage. Solothurn:
    Vogt-Schild A.g. 100 pp.

Chadwick, M. J. & Goodman, G. T. (eds). 1975. The ecology of
    resource degradation and renewal. Oxford: Blackwell.

Dasmann, R. F. 1972. Environmental conservation. 3rd ed., New
    York: Wiley.

Dasmann, R. F.? Milton, J. P. & Freeman, P. H. 1973. Ecological
    principles for economic development. London: Wiley.

Daubenmire, R. 1968. Plant communities: a textbook of plant
    syn ecology. New York: Harper & Row.

Duffey, E. 1971. The management of Woodwalton Fen: a multi-
    disciplinary approach. In: The scientific management of
    animal and plant communities for conservation, edited by E.
    Duffey and A. S. Watt, 581-597. Oxford: Blackwell.

Duffey, E. & Watt, A. S. (eds.) 1971. The scientific management
    of animal and plant communities for conservation. Oxford:
    Blackwell.

Edelman, C. H. 1950. Soils of the Netherlands. Amsterdam:
    North Holland.

Ellenberg, H. 1972. Belastung und Belastbarkeit von Ökosystemen.
    TagBer. Ges. Ökologie Giessen 1972m k9-26.

Falinski, J. B. 1968. Methodical basis for map of potential
    natural vegetation of Poland. Acta Soc. Bot. Pol., 40, 209-221.

Haase, G. (ed.) 1973. Beitrage zur Klärung der Terminologie in der
    Landschaftsforschung. Leipzig: Geogr. Inst. Akad. Wiss. D.D.R.

Harms, W. B., Wittgen, A. & Reijnen, M. 1976. De groene ruimte
    tussen Arnhem en Nijmegen, een biologisch onderzoek. Natuur
    Landsch. 30, 70-79.

Helliwell, D. R. 1969. Valuation of wildlife resources. Reg.
    Stud. 3, 41-47.

Helliwell, D. R. 1971. A methodology for the assessment of
    priorities and values in nature conservation. Merlewood Res.
    Dev. Paper, no. 28.

Helliwell, D. R. 1973. Priorities and values in nature conservation.
    J. Environ. Manage. 1, 85-127.

Holling, C. S. 1973. Resilience and stability in ecological
    systems. Annu. Rev. Ecol. & Syst., 4, 1-23.

Holling, C. S. and Clark, W. C. 1975. Notes towards a science of ecological management. In: Unifying concepts in ecology, edited by W. H. van Dobben, R. H. Lowe-McConnell, 247-252. The Hague: Junk.

Kalkhoven, J.T.R., Stumpel, A.H.P. & Stumpel-Rienks, S.E. 1976. Landelijke milieukartering van het natuurlijk milieu in Nederland ten behoeve van de ruimtelijke planning op nationaal niveau. The Hague: R.I.N. Leersum, R.P.D. (in press).

Kiefer, R. W. and Robbins, M. L. 1973. Computer-based land use suitability maps. J. Surv. Mapp. Div. Am. Soc. civ. Engrs, 99 (SU1) 39-62.

Knapp, R. (ed.) 1974. Vegetation dynamics. (Handbook of vegetation science, Part 8). The Hague: Junk.

Koeppel, H. W. 1975. Erfahrungen mit dem Einsatz von Computern. Verh. Ges. Okologie, 4th, Erlangen, 1974, The Hague, Junk.

Kuyken, E. 1975. Landscape ecology and spatial planning in W. Belgium. Verh. Ges. Ökologie, 4th, Erlangen, 1974, 253-256. The Hague, Junk.

Leeuwen, C. G. van, 1965. Het verband tussen natuurlijke en anthropogene landschapsvormen, bezien vanuit de betrekkingen in grensmilieu's (with a summary). Gorteria, 2, 93-105.

Leeuwen, C. G. van, 1966. Het botanisch beheer van natuurreservaten op structuur-oecologische grondslag. (with a summary). Gorteria, 3, 16-28.

Leeuwen, C. G. van, 1966. A relation theoretical approach to pattern and process in vegetation. Wentia, 15, 25-46.

Leeuwen, C. G. van, 1973. Edologie. Collegediktaat HB20A, Delft, Technische Hogeschool, Afd. Bowkunde.

Long, G. 1974. Diagnostic phyto-ecologique et amenagement du territoire. I. Principes generaux et methodes. Paris: Masson et Cie.

Long, G. 1975. Diagnostic phyto-ecologique et amenagement du territoire. II. Application du diagnostic phyto-ecologique. Paris: Masson et Cie.

Maarel, E. van der. 1970. De Ooypolder, biologische evaluatie van natuur en landschap. Natuur Landsch. 23, 201-223.

Maarel, E. van der. 1971. Plant species diversity in relation to management. In: The scientific management of plant and animal communities for conservation, edited by A.S. Watt and E. Duffey, 45-63. Oxford: Blackwell.

Maarel, E. van der, 1972. De invloed van het zich ontwikkelende hoofdwegennet op natuur en landschap. Stedebouw & Volkshuisvesting, 53, extra nr., 3-18.

Maarel, E. van der. 1975. Man-made natural ecosystems in environmental management and planning. In: Unifying concepts in ecology, edited by W. H. van Dobben and R. H. Lowe-McConnell, 263-274. The Hague: Junk; Wageningen: Pudoc.

Maarel, E. van der. 1975a. Ecologische aspecten van de grondwaterhuishouding in Oost-Gelderland. Rapp. Comm. Bestudering Waterhuishouding Gelderland. Prov. Waterstaat, Arnhem. 30pp.

Maarel, E. van der. 1976. On the establishment of plant community boundaries. Ber. dt. bot. Ges. (in press).

Maarel, E. van der et al., 1977. Naar een globaal ecologisch model voor de ruimtelijke ontwikkeling van Nederland. Ministerie Volkshuisvesting Buimtelijke Ordening. The Hague: (in press).

Maarel, E. van der and Leertouwer, J. 1967. Variation in vegetation and species diversity along a local environmental gradient. Acta bot. neerl., 16, 211-221.

Maarel, E. van der and Stumpel, A.H.P. 1975. Landschaftsökologische Kartierung und Bewertung in den Niederlanden. Verh. Ges. Ökologie, 4th, Erlangen, 1974, 231-240. The Hague, Junk.

Maarel, E. van der and Vellema, K. 1975. Towards an ecological model for physical planning in the Netherlands. In: Ecological aspects of economic development planning. Report Seminar U.N. Economic Commission for Europe, Rotterdam 1975; 128-143. Geneva: E.C.E.

McCarthy, M. M. et al 1974. Regional environmental systems analysis: an approach for management. Proc. Int. Congress Ecology, 1st, 130-135. Wageningen: Centre for Agricultural Publishing and Documentation.

MacHarg, I. L. 1969. Design with Nature. New York: Natural History Press.

MacIntosh, R. P. 1967. The continuum concept of vegetation. Bot. Rev. 33, 131-187.

Miyawaki, A.S. Okuda (eds.), 1975. Potential natural vegetation map of Japan in "Landscape of Japan". Tokyo.

Mueller-Dombois, D. and Ellenberg, H, 1974. Aims and methods of vegetation ecology. New York: Wiley.

Numata, M., Miyawaki, and Itow, D. 1972. Natural and semi-natural vegetation in Japan. Blumea, 20, 435-481.

Odum, E. P. 1969. The strategy of ecosystem development. Science, N.Y. 164, 262-270.

Odum, H. T. 1971. Environment, power and society. New York: Wiley, IX + 331 pp.

Ovington, J. D. 1975. Strategies for management of natural and man-made ecosystems. In: Unifying concepts in ecology edited by W. H. van Dobben and R. H. Lowe-McConnell, 239-247. The Hague: Junk; Wageningen: Pudoc.

Patten, B. C. 1974. The zero state and ecosystem stability. Proc. 1st. Int. Congr. Ecology. The Hague App: Wageningen: Pudoc.

Peters. H. 1973. De wet van behoud van ellende: de natuurlijke grenzen van de welvaart. Amsterdam: Wetenschappelijke Uitgeverij.

Seibert, P. 1968. Gesellschaftsring und Gesellschaftskomplex in der Randschaftsgliederung. In: Pflanzensoziologie und Landschaftosökologie, edited by T. Tuxen, 48-60. Ber. Int. Symposium, Stolzenau 1963. The Hague.

Seibert, P. 1975. Versuch einer synoptischen Eignungsbewertung von Ökosystemen und Landschaftseingeiten. Forstarchiv, 46, (5) 89-97.

Socava, V. B. 1972. Geographie und Olilogie. Petermanns geogr. Mitt., 116, 89-98.

Steinitz, C and Rogers, P. 1970. A systems analysis model of urbanisation and change. MITE Report 20, Cambridge, Mass: MIT Press.

Sukopp, H. 1971. Vewertung und Auswahl von Naturschutzgebieten. Schriftenr. fur Landschaftspflege und Naturschutz, 6, 183-194.

Sukopp, H. 1972. Wandel von Flora und Vegetation in Mitteleuropa unter dem Einfluss des Menschen. Ber. Landw. 50, 112-139.

Tansley, A. G. 1939-1953. The British islands and their vegetation. Cambridge University Press.

Thalen, D. C. P. 1971. Variation in some saltmarsh and dune vegetation in the Netherlands with special reference to gradient situations. Acta bot. neerl., 20, 327-342.

Tjallingii, S. R. 1974. Unity and diversity in Landscape. Landscape Planning, 1, 7-34.

Trautmann, W. 1966. Erläuterungen zur Karte der potentiellen natürlichen Vegetation der Bundesrepublik Deutschland 1:200.000. 85 leaves. Schriftenr. Veg., 1, 1-38.

Trautmann, W. 1972. Erlauterungen zur Karte "Vegetation" (Potentielle naturliche Vegetation). Deutscher Planungsatlas 1: Nordrhein-Westfalen. Lief. 3: 1-29. Hannover.

Troll, C. 1939. Luftbildplan und ökologische Bodenforschung. Z. ges. Erdkunde, (718), 297.

Tuxen, R. 1956. Die heutige potentielle natürliche Vegetation als Gegenstand der Vegetationskartierung. Angew. PflSoziol., 13, 5-42.

Vink, A. P. A. 1975. Land use in advancing agriculture. Berlin: Springer.

Waggoner, P. E. and Ovington, J.D. 1962. Proceedings of the Lockwood Conference on the suburban forest and ecology. Bull. Conn. agric. Exp. Stn., no. 652, 1-102.

Werkgroep Gran. 1973. Biologische kartering en evaluatie van de Groene ruimte in het gebied van de stadsgewesten Arnhem en Migmegen. (with a summary). Rapport Afd. Geobotanie Nijmegen.

Werkgroep Grim. 1974. Landschapsecologische basisstudie voor het streekplangebied Midden-Gelderland. Arnhem: Prov. Planologische Dienst.

Westhoff, V. 1968. Die ausgeräumte Landschaft. Biologische Verarmung und Bereicherung der Kulturlandschaften. In: Handbuch fur Landschaftspflege und Naturschutz, 2, ·p 1-10. Munich; Vienna.

Westhoff, V. 1970. New criteria for nature reserves. New Scient., 46, 108-113.

Westhoff, V. 1971. Choice and management of nature reserves in the
    Netherlands. Bull. Jard. bot. nat. Belg. 41, 231-245.

Westhoff, V. 1971. The dynamic structure of plant communities
    in relation to the objectives of conservation. In: The
    scientific management of animal and plant communities for
    conservation, edited by E. Duffey and A.S. Watt, 3-14.
    Oxford: Blackwell.

Westhoff, V. and Maarel, E. van der, 1973. The Braun-Blanquet
    approach. In: Handbook of vegetation science. Part V.
    Ordination and classification of communities, edited by
    R. H. Whittaker, 617-726. The Hague: Junk.

Whittaker, R. H. 1967. Gradient analysis of vegetation. Biol.
    Rev., 42, 207-264.

Whittaker, R. H. (ed.) 1973. Ordination and classification.
    (Handbook of vegetation science, edited by R. Tuxen. Part V.
    The Hague: Junk.

Whittaker, R. H. 1975. Communities and ecosystems. 2nd Ed.
    New York: MacMillan.

Whyte, R. O. 1976. Land and land appraisal. The Hague: Junk.

Zonneveld, I. S. 1972. Land evaluation and land(scape) science.
    Enschede, ITC Textbook of Phot-Interpretation. Vol. 7,
    Ch. VII-4. 106 pp.

DISCUSSION: PAPER 23

E. M. NICHOLSON   How are we to integrate ecological
                       thinking and conservation within the
                       "technosphere" - that is reconcile these
                       activities with urban systems? Experience
                       in London suggests that there is no linear
                       relationship between biological richness and
                       degree of urbanisation: the city is not much
                       more artificial as a habitat than intensively
                       managed agricultural areas and species
                       diversity may be higher in the city than in
                       intensively managed farm lands. What do
                       you feel about the principles in this
                       situation?

E. VAN DER MAAREL   Some physical planners are waiting for
                          ecologists to join their teams and joint
                          efforts are being made in developing regional
                          plans. There is a search for a recreational
                          environment that is compatible with the
                          natural environment and for an urban environ-
                          ment which does not force people to go out-
                          side it for their enjoyment.

H. A. REGIER   I have reservations about the roles of
                          indices of air quality or water quality. Many
                          of them are simple linear or multiplicative
                          functions and are very arbitary in nature.
                          They can agglomerate empirical relationships
                          or be stripped down models. Those who use
                          them tend to give them more weight than the
                          indices deserve. What led you to choose your
                          four indices and what properties do you attach
                          to them?

E. VAN DER MAAREL   The four indices I illustrated in my
                          paper are pragmatic choices and are the only
                          ones for which we have the data. Some regional
                          studies have tested their validity and found
                          that they do allow us to make statements about
                          the quality of the environment: they consequ-
                          ently simplify the approach to environmental
                          evaluation. Calibration is of course required
                          and most maps are still rougher than we would
                          like. Generally speaking, however, there is
                          a considerable use of empirical professional
                          judgement in planning, and society is only
                          asking in the surveys undertaken for rough
                          figures about general vulnerability that can
                          be provided if the results of evaluations using
                          all the indices are assembled together.

24. THE APPLICATION OF ECOLOGICAL KNOWLEDGE TO LAND USE PLANNING

M. W. Holdgate

Institute of Terrestrial Ecology

Cambridge, England

INTRODUCTION

There is an English rhyme in the style called "Clerihew" which
goes:

"The science of Geography is different from Biography. Geography
 is about maps. Biography is about chaps."

In the context of the present volume and remembering Jeffers'
instruction to define and bound the problem, it is important to
emphasise that we are concerned with both components of the
environmental situation.

The basic system of interaction between man and environment,
leading to the degradation of ecosystems, is shown in Figure 1.
Human actions impinge upon both the physical and biological components of the environment, modifying these and altering the nature or rate of their interactions. The system does not, however,
stop there for these effects provoke reactions within the community
and these social effects in turn determine how far the human actions
which initially perturbed the system are modified. Scientists (and
the papers in this volume confirm this fact) are liable to bound
their discussion of this interacting system in a fashion that omits
many components of the social effect and much of the social response which in turn modifies the actions. Such an approach is
valid if concern is confined to technical questions about how to
rehabilitate ecosystems, while leaving the major decision about
whether to rehabilitate them to others. But if we are seeking to

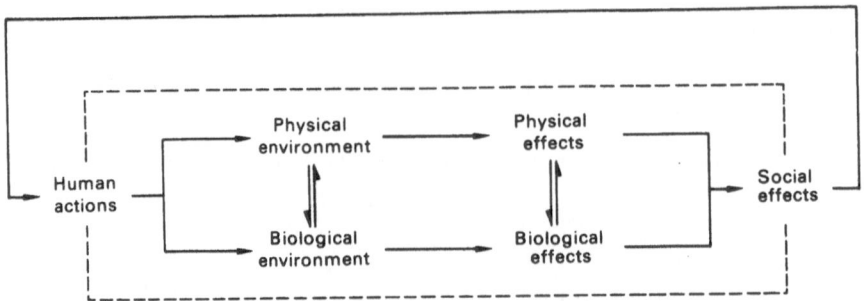

Fig. 1. The interaction between man and environment. The dotted line delimits the portion of the system on which most scientific analyses concentrate.

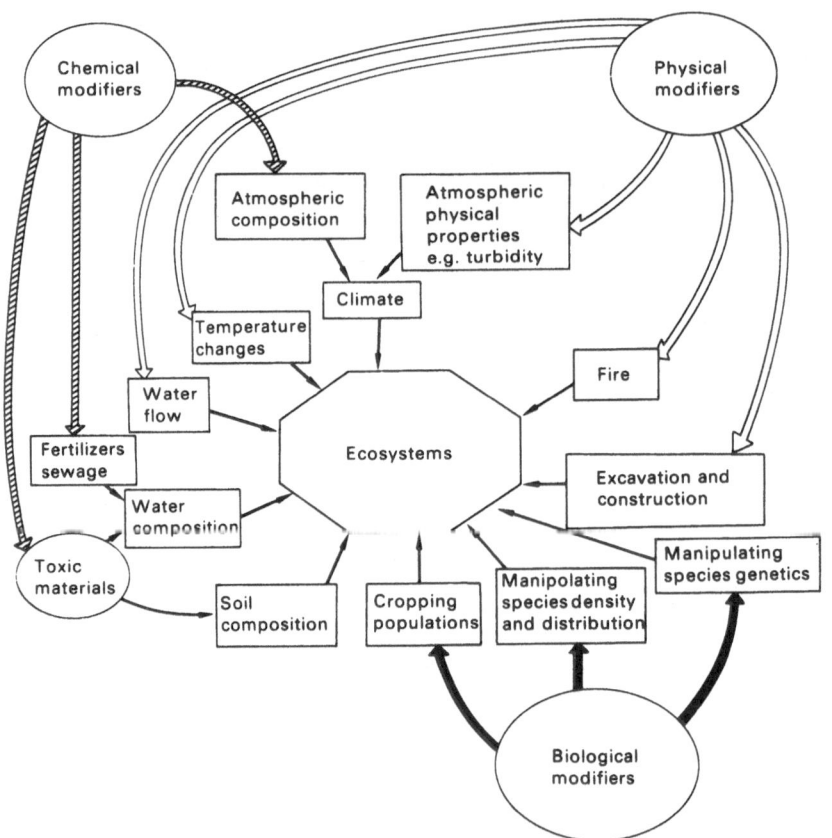

Fig. 2. Some direct and indirect modifiers of ecosystems.

guide the larger decisions, we must understand the social context
at least well enough to explain the scientific options so that
those considering whether to act know what can and cannot be done,
where and how and at what cost in terms of money and human effort.

INTERACTIONS WITH THE ENVIRONMENT

The term "physical environment" in Figure 1 is used to describe
the earth's structure - its geology, topography, climate, soils,
and the hydrological and geochemical cycles that move substances
through these compartments. Similarly, the term "biological
environment" is taken to mean the series of living components of
ecosystems, including plant primary producers, herbivorous animal
consumers, predators, decomposers, and the saprovores and micro-
bivores dependant on the decomposer system.

Man's action can be classified crudely by the compartments
affected (Figure 2), the types of action and the proximate targets,
which in turn influence particular components of the social system.
Other schemes could readily be devised (see Van der Maarel, this
volume, for examples): the important point is that in probing such
interactions the need is to have a classification of the overall
system into sub-systems that provide a meaningful structure, and
to be able to measure the rates of change or of interactive
processes within the system resulting from specified effects.

Such an analysis should provide a kind of dose:effect relationship for various human actions and environmental responses, thereby
ranking the factors prone to create degradation or to resist its
reversal. Of the five main physical modifiers recognised in
Figure 2, alteration of water flow by man is one of the major
impacts leading to ecological changes, because the drainage of
wetlands leads to the modification of the carbon cycle through the
oxidation of organic soils, reductions in the emission of methane
(a sink for atmospheric oxygen), and direct changes in the
distribution of species and composition of ecosystems. Irrigation
similarly leads to major changes in the pattern of primary producers, and affects rates and directions of processes of leaching or
salinisation. Another major physical modifier is certainly fire,
changing soil stability, water run-off and ecosystem composition,
and both these processes have had impacts over large areas of the
earth's surface. In contrast excavation and construction are a
localised group of impacts in urban zones, and heat emissions
similarly have only local effect in the air of cities, and in the
waters of rivers and some estuaries and seas. The impact of human
activities on atmospheric turbidity and radiation transmission is
uncertain, but probably very much smaller in scale than the natural
effects of volcanic eruptions.

Chemical modifiers also vary in their relative importance. Increased nutrient levels (whether through the run-off of fertilisers from agricultural land or the discharge of sewage) have certainly led to substantial changes in freshwater systems, leading to enhanced primary production, enhanced decomposer activity, oxygen depletion and consequential changes in ecosystem composition and function. In contrast the impact of pesticides, biocides and toxic materials has been localised and generally below the point at which widespread ecosystem changes result, and the scale of effect of substances altering the physical system (like carbon dioxide modifying climate), while more extensive spatially is probably not major in quantitative terms at the present time.

The direct biotic modifiers have undoubtedly had profound impact over the world. All three of the activities indicated have had major effects. By and large they have led to the replacement of large, long-lived organisms (K-Strategists) like large herbivores and predators, or large forest trees, by species with more of an r-strategy characteristic of earlier successional stages. In this process of modification, nutrients which are locked up in the biomass of these large long-lived organisms have been liberated and nitrogen, phosphorus and carbon in particular released back to the environmental section of their cycles.

Such an analysis is relevant to our present discussion only if it leads us on to recognise certain especially vulnerable components of environmental systems. It is important to seek to identify targets prone to respond to human interference in an especially sensitive way. At the same time it must be recognised that these will differ from one part of the world to the other because the world supports a spatial mosaic of ecosystems.

Three kinds of system are suggested as being particularly prone to degradation:

1. Systems whose physical integrity depends critically upon the maintenance of the biota

These systems may be exemplified by:

(a) permafrost systems in which the maintenance of vegetation cover is essential to provide insulation of the frozen ground beneath: if vegetation is disrupted thaw lakes appear and there may be a dramatic break-down of the whole system.

(b)   deep peat mire systems, in which the maintenance of the
        organic soil depends on waterlogging, which in turn,
        especially in soligenous mires, depends on the maintenance
        of the unbroken plant cover if the soil is not to be
        disected by erosion.

(c)   sand dune systems or any other systems over loose sub
        strata like the volcanic loessic soils in Iceland. These
        depend for their integrity on plants binding the surface
        soil and protecting it from wind and water erosion.

(d)   steep lands, generally dependant on vegetation cover to
        control water run-off and protect the soils from erosion.

        These four examples are only extremes. All soils to some
extent share the property of vulnerability to erosion. This
vulnerability is greatest where the physical stability of the
soil is potentially low because of freeze-thaw cycling, a low
clay fraction, a steep topographic gradient, or unusually high
or low rainfalls or high exposures to wind.

2. Systems whose biotic function depends on physical integrity

        Such systems may be exemplified by:

(a)   oligotrophic lakes, dependant upon a low nutrient inflow.

(b)   wetland systems dependant upon the maintenance of water-
        logging of the organic soils.

(c)   systems dependant upon high mineral nutrient levels, for
        example in calcareous flush situations associated with
        springs that maintain openness of the habitat through
        continued water erosion.

        All ecological systems, of course, show this basic
property, depending for their stability on the containment of
environmental fluctuation within certain outer limits. The
examples quoted are of systems that in various ways occupy
habitats at the extremes in the range of environmental variation.
This may be one indication of vulnerability, but the other must
come from the amplitude of environmental fluctuations normally
experienced by such systems, since those whose persistence
depends on the maintenance of a relatively narrow range of
conditions are likely to be more vulnerable than those adapted
to withstand a wider variation.

3. Systems with biological features related to abnormal
    ecological functioning

        This category of system may be exemplified by:

    (a) oceanic island ecosystems, which are generally character-
         ised by a relatively low level of herbivory, including the
         absence of vertebrate herbivores, and also a low level
         of predation, in part due to the absence of mammalian
         predators. These systems may be subject to unusual stress
         and break-down if these characteristics are modified by
         man. For example, many of the vegetation types, such as
         the perennial evergreen tussock grassland of southern
         oceanic islands, are not adapted to grazing, and break
         down rapidly leading to consequential soil erosion if
         herbivores are imported. Similarly such islands commonly
         support large populations of breeding sea birds which
         are extremely vulnerable to the sudden importation of
         ground predators.

    (b) peatland systems, with a low decomposer component, liable
         to major changes in this section of the ecosystem if the
         soils are rendered aerobic.

    (c) systems in which a single species is unusually abundant
         or dominant. In the marine field, outside the scope of
         the present conference, a good example may be provided
         by the krill (Euphausia superba) of the Antarctic, which
         alone accounts for perhaps 50% of the zooplankton biomass
         and supports much of the higher trophic level of whales,
         seals and seabirds. This species, now being considered
         for exploitation and with a suggested sustainable yield
         of between 50 and 100 million tons per year, clearly
         occupies an unusual position in the ecosystem and for
         this reason needs management with especial care.

    It may be a general conclusion that systems with unusual
proportions of components or unusual rates of transfer between
components may be particularly susceptible.

                SOCIAL EFFECTS OR RESPONSES TO ENVIRONMENTAL CHANGE

There are four main kinds of social response to environmental effects, including the degradation of ecosystems through human impact:

(a) Blindness, in which the change is ignored. Two contemporary
     examples, both from the United Kingdom, are the lack of
     response to the overpopulation of the Scottish Highlands with
     Red deer (Cervus elaphus), now probably at about twice the
     population level optimal in land management terms, and despite
     consistent scientific advice, consistently undercropped by
     Highland landowners and managers, and the inaction in the
     face of extensive erosion of the blanket peatlands of the
     English North Pennines.

(b) Acceptance, that is a failure to respond because it is
     considered that the effect needs to be tolerated because of
     other social factors. One of the best and clearest examples
     is the acceptance over many past decades of industrial
     dereliction caused by pollution, summed up in the nineteenth
     century Northern English saying "Where there's muck there's
     brass" ! Another example is the acceptance by many develop-
     ing country governments of unwise deforestation, creating low
     grade pasture over substantial areas of land, even though it
     is clear that scientific knowlege could guide such winning
     of land for agricultural purposes in a much more effective
     fashion with much less wastage of nutrients.

(c) Calculated action, in which the response of the community is
     related to an evaluation of the impact and effects, and some
     notion of the benefits and costs related to both.

(d) Over-reaction, in which there is a social response to the
     effects in the environment which is not, however, related
     so closely to a critical evaluation and where the demand to
     be seen to be responding is more important than the scale
     of either the effect or of the response.

     This symposium is concerned with the third of these, and the
whole justification for the conference rested on the assumption
that a calculated action to prevent the unnecessary degradation
of ecosystems and restore those damaged by mismanagement in the
past is both possible and socially desirable.

     Some notion of costs and benefits is inseparable from such a
rational approach. Figure 3 is the familiar curve of cost/benefit
economics relating the degree of an impact (in this case degradation) on an arbitrary scale from 0-100 to the costs of the associated
damage and the costs of rehabilitation or prevention. It must be
emphasised that for these purposes "100 per cent degradation" is
no more than the change that will occur if nothing is done to
prevent or rehabilitate. Conversely zero degradation simply
indicates the situation before the specific action under consider-

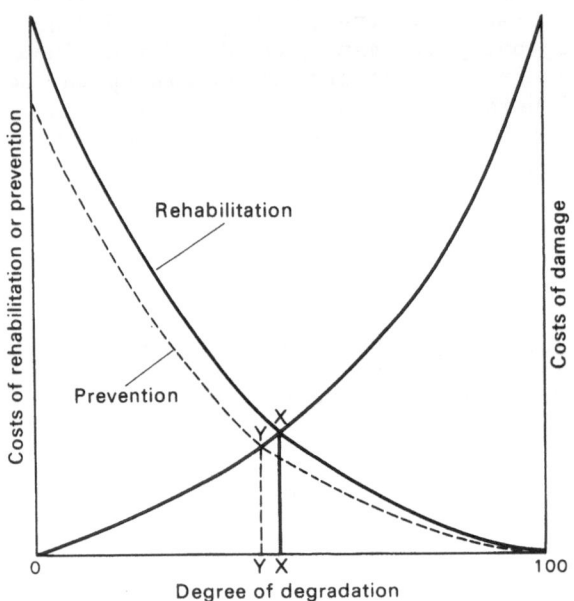

Fig. 3.   Interaction between ecosystem damage, damage costs and
          costs of rehabilitation or prevention.

ation and leading to ecological change takes place: it could mean
the pre-industrial agricultural environment, subsequently damaged
by pollution from a factory nearby, rather than the original pre-
human environment. The intersection point at which the marginal
damage costs balance the marginal costs of rehabilitation or
prevention becomes in social terms the optimum, and it will be noted
that this diagram makes the assumption that some degradation (i.e.
some change from the starting point), is socially the most rational
thing to accept. The diagram also caters for the general theory
that prevention is better than cure, since this must rest on the
assumption that the curve of prevention costs (the dotted line in
Figure 3), is lower than the rehabilitation cost curve. As a
result the optimum following a prevention approach is at point Y
and gives one a lower degree of degradation than the rehabilitation
curve which balances damage costs at point X.

Such curves must not be taken too seriously. They are very
prone to variation according to social judgement. For example,
the rehabilitation or prevention costs depend on the methods
used and these are capable of continuing improvement as new technology comes into use and better understanding is gained of
ecological processes, which may in turn lead to the development of
strains of plant which with minimal human effort will recolonise
contaminated or damaged soil surfaces. On the other scale the
damage costs naturally include a major component related to the
value put by society on the previous environmental condition or
the injury caused in social terms by its present features. Such
injury may be aesthetic or result from a loss of recreational potential. It is probably a general rule that the more affluent the
society the greater the value placed on aesthetic qualities and
recreation and consequently the more "economic" rehabilitation
will become. The important point for the ecologist is, however,
that ecological information is vital to the assessment both of
restoration costs and of damage costs for it is the ecologist's
statement of the potential land use options applied to a damaged
site that will help determine the damage costs and his evaluation
of the most effective way of restoring a more desirable vegetation
cover and associated fauna that is fundamental to the assessment
of the rehabilitation costs. It is necessary for ecologists to
be prepared to communicate their judgements within this kind of
cost-benefit economic framework even though they may be concerned
about the limitations of the approach.

REHABILITATION

There will clearly be several options that might be chosen
in deciding the precise kind of rehabiltation. One option that
is unlikely to be open is the restoration of the precise condition

that prevails before human impact, because of the physical environmental changes and genetic changes in the biological system that
are likely to have taken place following that impact.

The first question to ask is why a system is being rehabilitated (or alternatively protected from devastation). There are
perhaps two types of objective:

(a)   to restore or to sustain certain "outer limits" which deter-
        mine the integrity of the system, for example in terms of
        soil profile or hydrological regime. If this is the object-
        ive it may involve no more than the creation of, or mainte-
        nance of, a self-sustaining producer/consumer/decomposer
        system. Often this is done so as to prevent losses from a
        system which at present is not being used by man, for example
        when forest is left to regulate water run-off and maintain
        soils on steep hill systems, and this kind of activity also
        has the advantage of preventing damage to an adjacent used
        system, for example through uncontrolled floods. This kind
        of restoration or protection is conservationist in approach.

(b)   to sustain a desired functional system, manipulated to provide
        a particular human benefit: an exploitation approach.

In the latter case it is the proportions and the make-up of
the producer/consumer/predator/decomposer components with which
we are concerned rather than the retention of just some kind of
functioning total system. The precise balance and composition
depends on the kind of use for which we are managing or restoring
the area of environment. Agriculture, for example, is concerned
with a narrow range of producers and herbivores, manipulated by
selective cropping (especially of weeds and pests), chemical protection, and chemical promotion of growth. For forestry the range
is still narrower because it is concerned with producers only. In
both agriculture and forestry there has been little deliberate
manipulation of the decomposer system, but this is now beginning
with the development of management techniques for fungi (e.g.
mycorrhiza) or bacteria (especially nitrogen fixers). In contrast
to both these purposes, wildlife and genetic conservation is
concerned with whole systems which are not necessarily managed
to retain all their details, but within which certain key organisms
may have been selected for especial attention. Different kinds of
system again will be required when land is managed for water supply,
or for visual aspect, and in many cases mosaics of all these will
be sought when a land area is either protected or restored. The
important point is to evaluate the objectives and assess the ease
or difficulty of attaining them in each case so that the costs of
producing various combinations of use can be assessed alongside the
anticipated benefits.

THE CONTROL OF ECOSYSTEM DEGRADATION AND RESTORATION

In many countries the social organisation to prevent ecosystem degradation or cure land that has already been damaged is fragmented between central and local Government, and different agencies are often involved in the processes of prevention and cure.

Physical planning is in most countries the context within which the prevention of environmental degradation is furthered. It commonly involves two stages: the strategic zonation of water and land use over wide areas (in the United Kingdom this is done by the development of "structure plans" for counties) and the tactical development control applied to individual applications for changes of land use (in the United Kingdom done at the district level).

Strategic planning clearly depends in the first instance upon surveys which characterise the national environment and its ecological land-use patterns at some starting time, and allow the superimposition of evaluations of potential land use patterns that might meet various social and economic goals. These initial surveys must clearly define the range of variation in the environment in rigorous, objective terms, and consequently involve methodology which is compatible over the whole region that is being considered as a unit (commonly the whole country, but sometimes possibly a whole continent).

Such surveys are of two kinds. The first deals with actual systems - soil, geology, vegetation, fauna, climate, or present rural and urban land use. Generally, in defining the ecological background to planning, soil and vegetation data are of first priority among these (Holdgate and Woodman, 1976). The second type of survey combines parameters to give indices such as those describing the diversity of flora and fauna or the rarity of species in an area, or particular landscape qualities. Biotic diversity indices have been used as indicators of freshwater quality and combined with rarity to help select areas important for nature conservation (Goldsmith, 1975); landscape quality - indices have been elaborated by Helliwell (1973) and are discussed by van der Maarel (this volume). Where such indices are used, it is essential that the methods of evaluation are objective and are distinguished from and do not obscure the basic data.

Such surveys should provide a foundation for the evaluation of options for land use policy. They should indicate the potential of a site for agriculture, forestry, mineral extraction, nature conservation, recreation or other uses. They should also identify (a) the areas prone to degradation because of their special sensitivities, (b) areas of high value for particular forms of use and likely to need especial safe-guarding, and (c) areas already

functioning below their optimum use and possibly meriting restoration. A time component must clearly be built into such surveys,
predicting the rates of change and the rates at which rehabilitation of land that has been damaged may be attainable. The surveys
must equally clearly be followed by an analytic phase which looks
at the resources, costs and needs for different kinds of development even if a strict cost benefit analysis is not achievable.

At the more tactical level, physical planning usually acts
at the margin of man's encroachment on the countryside and steers
new interference to the "right" places. Thus the broad land use
potential surveys need to be anticipatory of such specific development control, providing the context within which individual cases
are judged. The latter may need supplementary surveys or environmental impact assessments, case by case, so that their detailed
features are defined.

It is rare for planning authorities at central or local
Government level to be the only agencies involved in the prevention of environmental damage. The approach needs also to be built
into the activities of agencies controlling aspects of public
policy, for example nature conservation agencies, forestry commissions, or agricultural authorities. These agencies have
optimisation of land use among their main tasks, and provide a
counterpoise to the urban planning control agencies. In most
countries the responsibility for the cure of already damaged land
is also fragmented. If the land is in forest or farm use, but
functioning below its potential, then its improvement may well
rest with national forestry or agriculture agencies, and the same
is true of land which is being taken and developed for wildlife
conservation or rural recreation in those states which have
specific agencies for these purposes. Very commonly, however,
land which is devastated through past industrial exploitation and
pollution is the concern of the Local Government or Authority
which may, however, be aided by Central Government funds and
encouraged by Central Government strategies (this is the case in
the United Kingdom). The planning of restoration of individual
areas which have suffered degradation, like development control,
is naturally a piece-meal activity done on a site-by-site basis
following a critical evaluation of the options.

Overall, the fragmented nature of the picture may be emphasised by the fact that very few countries, if any, represented at the
conference giving rise to this book have overall national environmental characterisation surveys or potential land use surveys,
despite the fact that these are fundamental to the rational
analysis of options for land management, and statistics of the
area of degraded and devastated land are far from complete.

THE ROLE OF THE ECOLOGIST

The professional environmental scientist can contribute both to
the protection and rehabilitation of ecosystems by five main
classes of activity:

1. The conduct of surveys, whether of actual systems, describing
    existing soils, vegetation or land use, or of potential syst-
    ems, defining the configurations that could be attained by
    various routes.

2. Evaluation. The ecologist has an important part to play in
    improving the evalution of the alternative options presented
    to society whether this is done through improved understanding
    of the potential value of the land and the living systems it
    supports or a better definition of the costs of restoration.
    For this, research explaining the cause of the patterns
    defined by the surveys is essential.

3. The statement of issues for the public. Ecological analysis
    should be set out in terms that are understood by the public
    at large, since social attitudes are extremely important if
    land preservation or rehabilitation is to be carried forward.

4. The design and supervision of environmental protection or
    rehabilitation schemes. Once a decision has been taken to go
    ahead with restoration or to formulate a national or regional
    strategy for site protection, ecological information will be
    needed in the detailed design of the programme. The rehabil-
    itation process in particular will also need monitoring.

5. Monitoring generally is an important contribution from
    ecologists. The changes in the environment need to be kept
    under continuing scrutiny and fed back into land management
    policies for only through monitoring will the success or
    failure of such policies become apparent. Monitoring of
    change is also needed to feed into the progressive design of
    research aimed at understanding the processes in the environ-
    ment and the progressive improvement of models.

    There has in the past been some confusion over what ecologists
stand for (at least this has been so in the United Kingdom). Commonly ecologists have been regarded as only concerned with wildlife
protection. It is quite clear that this function is only one
application of ecological sciences. In the present context,
environmental science begins by stating objectively the dimensions
of variation of environmental patterns, in space and time. It should go on to aid in the definition of options, through projecting
trends, explaining causes, and evaluating the capacities for mani-

pulating environmental systems to particular end points. The
resources and effort involved in attaining the different options
can be stated and the yield (in a wide sense) of different end
products can be stated. These are scientifically neutral tasks
standing quite aside from the specific advocacy of wildlife conservation as one of several policies to be considered. It is
absolutely essential in all this for the environmental scientist
to communicate with the non-scientist, especially in the forum
of policy definition. The special skills of the ecologist
need to be set alongside those of the economist, the lawyer and
the administrator in determining just what the best approach to
the environment is. This is unlikely to be achieved simply through
the statement of results of scientific research in the scientific
literature, which is neither read nor readily understood by people
who have not had the specialist professional training of the
scientist.

REFERENCE

Holdgate, M. W. and Woodman, M. J. 1976. Ecology and Planning.
    Report of a workshop. Bull. Br. Ecol. Soc. 6: 4, 5-14.

Goldsmith, F. G. 1975. The evaluation of ecological resources
    in the countryside for conservation purposes. Biol. Cons.,
    8, 89-96.

Helliwell, D. R. 1973. Priorities and values in nature conservation.
    J. Environ. Manage., 1, 85-127.

Part IV: Final Discussion

In introducing the final discussion, Dr. Holdgate suggested that the field covered by the conference might be summarised as follows:

(1) The properties that determine the vulnerability of environmental systems.

    (a) Physical features.
    (b) Biological features.
    (c) Biological processes.

(2) The nature of the ecological principles governing environmental rehabilitation.

    (a) Restoration of the physical structure, chemical composition and biological functioning of the soil, and especially the restoration of nitrogen levels and commencement of an effective nitrogen cycle.

    (b) The role of successional processes leading to restoration: e.g. from sown grass and legume mixtures to multi-layered woodland, or podsolised Callunetum to birch woodland in which soil fauna and a brown soil are restored.

    (c) The need to commence with a producer/decomposer system and control consumers critically in the early stages.

    (d) The restoration of physical and chemical features of fresh-water environments.

(3) The principles for a social strategy in preventing degradation or in restoring damaged lands and freshwaters.

    (a) Types of survey to define the present pattern, including the pattern of actual and potential dereliction.

    (b) Definition of land and water use options in terms of environmental capability, including capability for restoration. Assessment of associated effort and costs.

    (c) The types of communication required to determine action.

The discussion was necessarily curtailed by shortage of time, and could not deal with all these aspects. Most of the contributions addressed themselves to three broad issues: first, how the structure and processes of ecosystems should be analysed and modelled, and what features made ecosystems vulnerable, second, how land use options should be stated, and third, how the findings of ecologists should be communicated to policy makers.

ANALYSIS OF THE STRUCTURE, PROCESSES AND VULNERABILITY OF ECOSYSTEMS

B. ULRICH   Do we agree that before we can discuss the restoration of a system, we have to define both our objectives and the features of the present situation?

M. A. REGIER   The converse of asking what makes a system vulnerable to degradation is the question "what makes a particular system resistant to what kinds of challenge?".

J. CHRISTIE   I was left unsatisfied by Dr. May's observation that we cannot model multiple species systems, because this observation carries the implication that we have no way of proceeding. It also leaves open the question of which is the best direction to take in further studies.

                                    The nature of ecosystem development in particular interests me. Is development continuous towards some consistent, stable level, or can stability be achieved at a variety of maturity plateaus? If the former is true, simple optimisation in fisheries and wildlife management may not only be intolerably expensive; it may be altogether untenable.

                                    My suggestion would be that such questions cannot be answered solely in the realm of mathematics, or by traditional natural history. There is a need for improved interaction between those who model systems and those who study their parts. It may be the best approach to achieving this is to become more inductive in approach – to study the systems where possible by deliberately perturbing them.

This is consistent with Mr Jeffers' re-
commendations here and proposes that we
concentrate more on models which lead to
hypotheses that can be tested in the field.

G. VAN DYNE

There are a number of issues to clarify.
Ecologists may concern themselves with long
term successional changes, short term changes
in nutrients, or short term political changes.
Survey, prediction, and monitoring can take a
long time: does this mean the the ecologist
might be part of the long-term system? Should
we focus on how ecologists of different
affinities can maintain their professional
independence and not be part of the system
judging their evaluations – even at a risk of
being accused of inhabiting ivory towers?

L. M. TALBOT

One approach is to analyse the simplicity
of a system. If predators are removed from
some systems, the balance of the system
changes, simple systems being more vulnerable
than complex ones, partly because more components have to be influenced in the latter
before there is significant impact.

M. W. HOLDGATE

Some complex systems, like tropical rain
forest, are vulnerable if stressed beyond
certain limits: can we identify the limits or
the vulnerable components?

L. M. TALBOT

The impacts that destroy such systems are
often simple, and by their impact they certainly
simplify the ecological situation.

A. D. BRADSHAW

Dr Talbot suggested vulnerability might
be related to simplicity. I would suggest
that the correct argument is that communities
in extreme conditions are likely to be
vulnerable, and that these often have a simple
structure because of the extreme conditions.
Thus the determination of physical external
factors may give a useful indication of
vulnerability.

H. A. REGIER

If a challenge relaxes the constraints
on a system, it is not a threat: if it adds
to the constraint, it is a threat.

J.N.R. JEFFERS   Time scales are also important. In
examining threats to ecological systems, we
may be working within the context of a long
term cycle. For example, recent concern over
the threat from Dutch Elm disease might be
better viewed within the longer time scale of
fluctuations in elm population, caused by many
factors. Our concern should not be with
change, but with how far the system is varying
outside the previous range. Foresters, for
example, are often criticised because their
policies are long term, and get out of step
with short term economic fluctuations.

E. M. NICHOLSON   Many systems are vulnerable to intensive
economic exploitation even though they are in
no way extreme. Wetlands are an example.

L. M. TALBOT   Our concern must be over our objective.
What makes a system vulnerable to what?

M. GODRON   To say a system is vulnerable implies
that it may be 'ill', and that we know what
kind of illness is involved.

M. W. HOLDGATE   No. A degraded or devastated system is
one whose productivity or value to society in
some other way has been impaired. It has been
modified in a way that is unwelcome, because
it is of diminished use and is imposing a
cost. Vulnerability is an expression of a
tendency to behave in this way readily, in
contrast to "robust" systems that do not
easily change.

R.W.J. KEAY   I would like to stress that in considering
the vulnerability of an ecosystem, we must take
into account soil and geomorphology. For
instance, tropical forests can look much the
same on sand and on clays derived from gneiss,
but may differ markedly in their resilience
to disturbance.

M. W. HOLDGATE   A. G. Tansley defined an ecosystem as
embracing an assemblage of plants, the physical
environment with which they interacted directly,
and the animals closely associated with them:
his definition would thus discriminate between
such forest types.

N. POLUNIN   May I insist that we stick to this
                            definition of ecosystems laid down by Tansley?
                            It is just nonsense to use terms like "plant
                            ecosystems" - as a recent A.I.B.S. symposium
                            did.

D. H. SUKOPP   It is important to be clear about
                            definities and classifications. These may
                            be of two kinds - genetic, relating to the
                            origin of systems, and practical, relating
                            to the use of a region. Dansereau, for
                            example, has provided the second type of
                            classification. The objectives of our con-
                            ference do not encompass the classification
                            of populations, communities and ecosystems
                            but land use systems.

                                    It is not useful to try to estimate de-
                            gradation on a numerical scale from 0 to 100,
                            as in many regions we cannot define 0, all
                            regions being altered. But new equilibria
                            are being sought. In urban ecology we clearly
                            cannot study urban systems in comparison with
                            others that display no human influence - we
                            can relate the city to an agglomeration of
                            villages. In the table I showed at the start
                            of Session II, I did put forward a scale of
                            impact that began with zero interference. It
                            is relatively easy to define land use objectives
                            if there is only one system to consider. It
                            becomes more difficult if a range of options
                            is to be stated.

G. VAN DYNE   I would like to enunciate three hypotheses
                            derived from attending this conference:

                            (a) problems of restoration, reclamation and
                            resource management depend for solution on
                            the manipulation of the rates of nutrient
                            cycles;

                            (b) energy in ecological systems is simply a
                            pump for nutrients;

                            (c) we do not understand ecosystems and sub-
                            systems well enough to construct a fully
                            comprehensive model at present.

THE STATEMENT OF OPTIONS

E. M. NICHOLSON   Land use is not in conflict with the
                          aims of this conference in preventing or
                          curing degradation. Management of land for
                          sustained yield is a long-standing tradition.
                          We are not seeking a system for intervening
                          against use, but a framework which minimises
                          conflicts and allows intervention to correct
                          wrong actions.

B. ULRICH   Options may be available. The ecologist's
                          role is to advise the land user or planner
                          which option to adopt.

M. A. REGIER   There are a number of tacit assumptions
                          about the inference we can draw from ecology -
                          some of them mythical rather than real. For
                          example, there are three prevalent myths in
                          fisheries - that eutrophication is a natural
                          process (whereas the reverse is true because
                          oligotrophication is the natural trend in
                          many lakes); that the trophic pyramid is
                          basic to ecology, and that the upper, but not
                          the lower, stages can therefore be sacrificed;
                          and that ecologists always need to use taxo-
                          nomically defined species as their units.
                          None of these myths is valid in fisheries
                          studies.

L. M. TALBOT   Professor Sukopp and Professor Ulrich
                          have both stressed the central role of options
                          now and in the future, and the need to keep
                          future options open. One possible consequential
                          requirement in management and planning is to
                          ensure that a land use mosaic is retained,
                          along lines proposed by Dasmann and others,
                          reserving pieces of little-modified ecosystems
                          as genetic pools and ecological reference
                          points. It would have helped in managing the
                          environment in Iceland, if areas of unaltered
                          landscape had been available as reference
                          points.

                          There is an implication here for comm-
                          unications. If we speak of "reserved eco-
                          systems" we imply a preservationist approach,
                          but the objective here is functional in
                          development. While preservation of certain

representative areas is involved, the
values of these areas can be functionally
basic to development or rehabilitation of the
surrounding lands. Therefore we can make
the proposal more positive, in terms of the
value of reference points and genetic pools to
decision makers and society.

A. D. BRADSHAW   It is important not to overstress
preservation,for this option rarely arises
when dealing with degraded systems. Here we
should think creatively of the new kinds of
species assemblage or ecosystem that can be
put in place of a degraded area. I live in
a totally degraded area - and see how often
planners, seeking to restore such regions,
are motivated by a half truth: they want to
put back what was there, not what could be
there. We should press for analysis of what
could be there in preference to what was there.

V. GEIST   May I point out that in Canada we are
still able to consider large mammals, or
the predators as discussed by Dr. Talbot -
and insist that they be made part of a reconstructed ecosystem where appropriate.

G. L. LUCAS   One thing that comes out of all this is
our lack of knowledge despite the amount of
information already available in the literature.
This is very disturbing. It is all very well
to blame scientists for always wanting to do
more research, but research is clearly essential
in almost all areas of environmental studies
and their component parts.

THE NEED FOR COMMUNICATION

L. M. TALBOT   Another need in our discussion is to
highlight the need for communication between
scientist and decision-makers. Scientists
must be taught how to communicate with policy
makers. In Europe and North America, the
orientation of ecological training is at odds
with this communication, for decision-makers
seek to define issues in black and white, while
scientists are trained to avoid black and white
decisions, qualifying issues in various shades
of grey! It is essential that we, and our

students, understand the needs of decision-
makers and train ourselves to respond.

M. W. HOLDGATE

One problem is that the scientific paper,
which scientists are trained to use as their
main means of communication, is virtually
useless as a method of putting scientific ideas
to non-scientific policy makers. We have to
teach people a new communication style
appropriate to different users.

A. D. BRADSHAW

The ecologist must sustain his position
by showing that he understands ecological
systems and how they work - he must explain
enough of his scientific approach to carry
conviction!

M. W. HOLDGATE

A recent symposium at our Institute's
Monks Wood Experimental Station, on Ecology
and Planning, concluded that ecological
advice must be simple enough to be understood,
but scientific enough to be credible!

V. GEIST

After five years working in an inter-
disciplinary faculty, I am pessimistic. The
viewpoints of different groups just do not
translate readily. We have to get into the
professional schools - law schools, medical
schools and engineering schools - and instil
ecological understanding.

R. W. J. KEAY

Some of the blame for poor communication
rests with ecologists. They claim to understand environmental systems - but the forestry
literature demonstrates that foresters have
had a good measure of understanding at the
practical level for many years. Agriculturalists have likewise had some understanding of
how their systems work. The ecologist should
not claim a new and sudden revelation, they
can learn much from agriculturalists and
foresters and should improve their communication
with these ecologically based industries.

J. BALFOUR

I suggest that ecologists wanting to
influence decision makers should start with
land managers and land owners. These people
are influential with policy makers as well as
directly on the ground and may, from practical
experience, have at least some understanding

of ecological thinking.

B. ULRICH

We must not stay at the level of pointing out the likely nature of environmental impacts, but help in translating these into economic terms. For example, we can state to an economist that if a trend in nutrient level continues, there will be an ecological breakdown - but we can also state the cost of fertiliser to halt or reverse the trend, and so inject an ecological point into economic theory.

J.N.R. JEFFERS

The erection of a one-dimensional system will not get us anywhere: it is a common fault of ecological thinking. There must be a cross-classification of the decision-making impacts in the human system as well as the components of the ecosystem. A wide variety of solutions can usually be generated and this complexity must be communicated. Alternative treatments may succeed because they fit the wider system. Generally, the restoration of degraded ecosystems will need a longer time-scale than the normal political cycle, and may demand the reversal of political decisions. There is a cost implicit in over-turning a decision, and options are foreclosed when policies are fixed - this is one reason why alternative solutions need to be stated.

E. M. NICHOLSON

The essential point to make is that either ecological thinking will permeate and affect the whole community, including those who take economic decisions, or ecologists will become extinct! The whole problem is that the ways of analysing systems and impacts differ in the two groups, and one or other of these ways must become dominant, unless we can evolve a common approach. We should look at the ways of linking the approach of ecologists and decision-makers.

Part V: Conclusions

M. W. Holdgate

From the papers presented at this Conference, certain broad conclusions do emerge.

First, it is now possible to use mathematical language to describe many of the characteristics of plant and animal populations (May), and the interactions of two-component systems, and even relatively simple statements can have challenging implications – for example for the stability of exploited populations or concerning the likelihood of alternative stable states rather than one unique point of balance in a population. It is also possible to make general statements about how the survival of a population in an area is governed by the interplay of immigration, extinction and extent of habitat (Diamond). On the other hand, while multi-component models of ecosystems have been built (Van Dyne, Joyce, and Williams) and proved valuable in posing hypotheses for exploration, we are not at the point where the practical manager of ecosystems can rely on these to the exclusion of the real world (Van Dyne, final discussion: May).

One great asset of a mathematical approach to ecology is that it forces us to define the systems with which we are concerned and the processes we are considering precisely (Jeffers). It can lead to practical analysis of vital components of ecosystems, including the flux and levels of soil nutrients (Ulrich) – which this symposium emphasises as being the key to rehabilitating derelict or degraded ecosystems. Early man's deforestation (Dimbleby) and use of fire (Wein) undoubtedly had its greatest impact especially where phosphorus and nitrogen levels fell so low that replenishment by rock weathering or rainfall were inadequate to re-start properly functioning cycles in the degraded soils. There is a close parallel here with the effects of contemporary degradation caused by mining and quarrying, yielding spoil that is often porous, unconsolidated, and low in N and P, and where the key to revegetation is the establishment of self-sufficiency in these nutrients, in the former case especially through the introduction of leguminous plants (Bradshaw, Humphries, Johnson and Roberts).

Patterns of degradation in the past show regional variations, as would be expected. In the Mediterranean regions scrub woodland is still able to restore nutrient cycles and bind soil on land which has suffered millennia of fire and grazing (Godron). In the oceanic fringes of Britain deforestation appears, in contrast, to have been accompanied by progressive nutrient loss and a replacement of forest by peatland (Dimbleby): the acidity of soils under such conditions may have affected nutrient availability and hampered re-vegetation (Ulrich, discussion). In Iceland the initial ecosystem lacked mammalian herbivores and may have been slow-growing, while the loessic volcanic soils depended for stability on the birch woodlands: forest clearance and grazing caused extensive damage in consequence (Fridriksson, Runolfsson). The lesson of history appears to be that terrestrial ecosystems are especially vulnerable where the stability of soil and retention of nutrients depends critically on a particular vegetation type, and where nutrient replenishment is slow or impossible under the conditions likely to prevail if that vegetation is destroyed by man (Holdgate). It follows that "prevention is better than cure" and that wherever possible, surveys of ecological pattern and analyses of their causes should lead on to deliberate planning and management of human impact; the ecologist should be anticipatory rather than responsive in his actions (van der Maarel, Jacobsen, Holdgate).

To a certain extent, this symposium was on two interwoven themes. The first was this general evaluation of the properties and processes of ecosystems and of methods for the wise use of environmental resources over broad tracts of land and water. The second was a more practical analysis of how to handle land and water that had been devastated in various ways. The papers dealing with freshwater problems illustrated this well. The story of the Rhine (Wolff) eloquently demonstrated how a great and productive river could sustain severe ecological damage through pollution with sewage and toxic substances, losing its fisheries and declining in amenity – and how, although the scientific basis for restoration might be obvious, the need for international agreement, resources and political determination might delay action. On a smaller scale, where resources and policy decisions present fewer problems, lakes can be manipulated to more oligotrophic, preferred conditions by exclusion of sewage and removal of contaminated sediments (Gelin): even in such cases, however, there can be controversy over objectives and over the effectiveness of techniques, and there is an evident need properly to analyse the system, state the options, and test the methods in advance (Rorslett). Short of the point at which such major remedial action is considered, there are well established techniques for fisheries management which have been applied successfully, for example to benefit the tourist industry in Iceland, (Gudjonsson).

On land, successful rehabilitation is very largely based on
successful restoration of the physical stability and nutrient
cycles of the soil. In Iceland, this demands exclosure of grazing
animals, fertilisation, restoration of vegetation (including
nitrogen fixers like Alaska lupin), and careful subsequent control
of grazing, especially on coastal dunes (Runolfsson). Re-afforestation is an important part of the rehabilitation process, re-
establishing nutrient cycles and curbing erosion and the selection
of non-indigenous species of a provenance matching the Iceland
climate has proved the key, overcoming the effects of the ocean
barrier in excluding many well-adapted plants from the island
(Bjarnason). In Germany, forest policy is partly directed to
sustaining the stability and fertility of land surplus to agriculture, but recent research is being used to select the optimum
species and harvesting cycles, sometimes in a way that deliberately
takes a long-term view and sets aside some contemporary economic
approaches (Kopp). Woodland, used for recreation and amenity may
also be the best option for land previously disturbed by mineral
working, where levels of toxic metals prohibit agriculture (Bradshaw, Humphries, Johnson and Roberts) or where old quarry sites
can provide an attractive feature and draw off pressure from
farmland. Such restored sites can also be of value in providing
refuges for native species, and conserving genetic diversity - an
important overall objective (Heslop-Harrison and Lucas).

It is virtually impossible to restore the original ecosystem
as it was prior to many centuries of disturbance by man. Indeed,
it is often difficult to be sure what that ecosystem was (Godron).
What can be done is restore a system that includes all the main
functional components. Just how this is done depends on objectives
for the use of the land. If the aim is essentially conservationist -
to retain soil, control water run-off, preserve genetic diversity
and sustain amenity - a complete system is needed, but there may
be less concern over its precise comparison and room for unchecked
invasion by native species (Holdgate). If the area is large enough
and conflicts with man unlikely, it may be desirable to restore or
retain large predators as integral components of such a system
(Talbot). In contrast, where rehabilitated sites are to be cropped
by man for plant or animal food or for particular recreational use,
the precise structure and composition of the ecosystem is naturally
of concern.

The overall lesson is that it is scientifically possible now
to develop prescriptions for the adjustment of degraded ecosystems
to a more satisfactory state even if more research is needed to
develop more economical ways of doing it, improve productivity and
expand our options. Among the areas of practical research this
symposium points to as needing continuing effort are studies on how
to establish nitrogen fixing plants in barren soil, how to establish

nitrogen fixing nodules in non-leguminous species, and how to
innoculate trees with symbiotic fungi chosen to boost their
growth and productivity on particular sites. It seems to emerge
that research on soil microbiology and nutrient cycling (especially
of N and P) remains the most productive area in improving derelict
land rehabilitation, especially on the severely disturbed sites
(Sukopp, Bradshaw, Ulrich, Blume in discussion).

But the scientific understanding needs to be set in the
context of policy. We are concerned with land use strategies.
These need to be based on a survey and evaluation of contemporary
ecological patterns, an explanation of their cause and a prediction of their likely trends under various systems of human management (van der Maarel, Colenbrander, Jacobsen, Holdgate). This may
point to potential disasters, for example through unchecked river
pollution (Wolff), interference with coastal patterns of tidal
flow and sedimentation (Jacobsen), unchecked grazing (Bjarnason,
Fridriksson, Runolfsson), bad hydrological management (Colenbrander),
or excessive fire (Wein). It should also lead to zonation of land
and water for particular types of use, and a recognition of the
management needs in each area (van der Maarel, Jacobsen). Embedded
in such broad strategies will be specific action to cure past
dereliction (Bradshaw, Humphries, Johnson, Roberts and Gelin). But
all these plans will be of little practical use unless they are
built on collaboration between ecologist, economist, professional
planner and administrator. The ecologist cannot remain a voice
crying in the wilderness - if he is to be heard and understood.
He must come into the arena where policies are fought out and see
that his understanding of the natural world is communicated to
those who shape the world in which men live (Talbot, Nicholson,
in final discussion).

List of Participants

ADALSTEINSSON, H.   National Energy Authority, Iceland.

ALEXANDERSDOTTIR, M., Mrs.   Institute of Freshwater Fisheries, Iceland.

ALLAN, T. D., Dr.   Executive Officer, NATO EcoSciences Panel, NATO, Scientific Affairs Division, 1110 Brussels, Belgium.

BALFOUR, J., Mrs.   Nature Conservancy Council, Kirk Forthar House, Markinch, Fife, KY7 6LS, Scotland.

BARSTAD, J., Dr.   Department of Environmental Toxicology, Statens Institutt for Folkhelse, Postuttak Oslo 1, Norway. (NATO EcoSciences Panel).

BENEDIKZ, T.   Iceland Forest Service.

BERGMANN, S.   Icelandic Association of Nature Conservation.

BJARNASON, A. H.   Iceland Forest Service.

BJARNASON, H.   Director, (Iceland Forest Service), Skograekt Rikisins, Ranagarta 18, Reykjavik, Iceland.

BJORNSSON, J.   Director, National Energy Authority, Iceland.

BLUME, H. P. Prof. Dr.   Institut für Ökologie der Technischen, Universitat Berlin, I. Berlin 10, Franklin Strasse 29, Federal Republic of Germany.

BOERSET, O., Prof.   Institute of Sylviculture, Norge Landbruks Høgskola 14322 Vollebekk, Ås, Norway.

BRADSHAW, A. D. Prof.   Department of Botany, University of Liverpool, P.O. Box 147, Liverpool, England, L69 3BX.

BURGER, R., Fraulein

Kårtehause Strasse 9, Freiburg i Br.
Federal Republic of Germany.

CHRISTIE, W. J.

Glenora Fisheries Station, Ontario
Ministry of Natural Resources, RRH4,
Picton, Ontario, Canada, KOK 2TO.

COLENBRANDER, H. J.

Head Office of the Committee for
Hydrological Research, TNO, Juliana
van Stolberglaan 148, Postbus 297,
den Haag 2076, The Netherlands.

DEMOLENAAR, J. Dr.

Rijksinstituut voor Naturbeheer,
Kasteel Broekhuizen, Leersum,
The Netherlands.

DIAMOND, J. M. Prof.

Department of Physiology, School of
Medicine, Center for the Health
Sciences, Los Angeles, California
90024, USA.

DIMBLEBY, E. W. Prof.

Department of Human Environment,
University of London, Institute of
Archaeology, 31-34 Gordon Square,
London, WC1H OPY, England.
(Chairman Session III of Conference).

FRIDRIKSSON, S. Dr.

Erdafraedinefnd Haskolans Islands
Ingolfstraeti 5, III.h, Reykjavik,
(Administrative Director of Conference).

GEIST, V. Dr.

Faculty of Environmental Design,
University of Calgary, 2920 24 Ave.
N.W., Calgary, Alberta, Canada, T2N 1N4.

GELIN, C. Dr.

Institute of Limnology, University
of Lund, Fack, S-220 03 Lund 3, Sweden.

GODRON, M. Prof.

Directeur, Centre Nationale de la
Recherche Scientifique, Centre d'Etudes
Phytosociologiques et Ecologiques Louis
Emberger, Route de Mende, BP 5051,
34033 Montpellier-Cedex, France.

GOSSEN, R. Dr.

Canadian Arctic Gas Study Ltd.,
1270 Calgary House, 550 6th Avenue SW.,
Calgary, Alberta, Canada, T2P 052.

GRÖNVOLD, K. Dr.	Nordic Volcanological Institute, Iceland.
GUDBERGSSON, G.	Agricultural Research Institute, Iceland.
GUDJONSSON, T.	Director, Institute of Freshwater Fisheries, Veidimalastofnun, Vedurstufus, V/Bustadaveg, Reykjavik, Iceland. (Chairman, Session IV of Conference).
HINDS, W. T. Dr.	Environmental Programs, Division of Biomedical and Environmental Research, U.S. Energy Research and Development Administration, Washington D.C.20545. USA.
HOLDGATE, M. W. Dr.	Director, Institute of Terrestrial Ecology, 68 Hills Road, Cambridge, England, CB2 1LA. (Scientific Director of Conference).
HUTNIK, R. J. Prof.	Forest Resources Laboratory, Pennsylvania State University, College of Agriculture, University Park, Pennsylvania 16802, USA.
ISAKSSON, A.	Institute of Freshwater Fisheries, Iceland.
JACOBSEN, N. K. Prof. Dr.	Geografisk Centralinstitut, Kobenhavns Universitet, Haraldsgade 68, DK-2100, Kobenhavn Ø, Denmark.
JEFFERS J. N. R.	Institute of Terrestrial Ecology, Merlewood Research Station, Grange-over-Sands, Cumbria, England, LA11 6JU.
JONSSON, J.	Editor, Agricultural Society, Iceland.
KEAY, R. W. J. Dr.	The Royal Society, 6 Carlton House Terrace, London, England.
KØIE, M. Prof.	Institut for Økologisk Botanik, Københavns Universitet, Øster Farimagsgade 2 D, 1353 København K, Denmark.

KÖPP, H. Dr.   Institut für Forstpolitik, D-3400
                                  Göttingen, Busgenweg 5, Federal
                                  Republic of Germany.

KRISTJANSSON, J.   Institute of Freshwater Fisheries,
                                  Iceland.

LEBRUN, P. Prof.   Directeur, Ecologie animale,
                                  Laboratoire d'écologie générale, Plea
                                  Croix du Sud 5, 1348, Louvain-la-Neuve,
                                  Belgium.

LEENTVAAR, P. Dr.   Ryksinstituut voor Naturbeheer,
                                  Kasteel Broekhuizen, Leersum, The
                                  Netherlands.

LUCAS, G.   Royal Botanic Gardens, Kew, Richmond,
                                  Surrey, England.

MARGARIS, N. S. Dr.   Institute of General Botany, University
                                  of Athens, Panepistimiopolis, Athens
                                  621, Greece.

MAY, R. M. Prof.   Department of Biology, Princeton
                                  University, Princeton, New Jersey
                                  08540, USA.

MELLQUIST, I.   Norges Vussdrags-Og Elektrisitetsvesen,
                                  Vassdragsdirektoratet Boks 5091,
                                  Majorstua, Oslo 3, Norway.

NICHOLSON, E. M.   Land Use Consultants, 731 Fulham Road,
                                  London, England.

OLSCHOWY, Prof. Dr.   Der Leitende Direktor, Bundesanstalt
                                  für Vegetationskunde, Naturschutz und
                                  Landschaftpflege, 53 Bonn-Bad Godesberg
                                  1, den, Heerstrasse 110, Federal
                                  Republic of Germany.

PARKINSON, D. Prof.   Head, Department of Biology, University
                                  of Calgary, 2920 24 Ave. NW., Calgary,
                                  Alberta, Canada, T2N 1N4.

POLUNIN, N. Prof.   15, Chemin F-Lehmann, 1218 Grand
                                  Saconnex, Geneva, Switzerland.

RAGNARSSON, H.   Iceland Forest Service.

REGIER, H. Dr.   Department of Zoology, University
                                   of Toronto, Toronto, Ontario, Canada.

RORSLETT, B.   Norskinstitutt for Vandforskning,
                                   Ganstadalléen 25, P O Box 260, Blindern,
                                   Oslo 3, Norway.

RUNOLFSSON, S. Dr.   Director, Institute of Soil Conser-
                                   vation, Landgraedsla Rikisins,
                                   Gunnarsholt, Rangarvollum, Iceland.

SEVASTOS, C. G.   Director of Forest Service, 27 Antinoros
                                   Street, Athens, 516A, Greece.

SIGURBJORNSSON, B. Dr.   Director, Agricultural Research
                                   Institute, Iceland.

SIGVALDASON, G. Dr.   Director, Nordic Volcanological
                                   Institute, Iceland.

SKULADOTTIR,V. Mrs.   Icelandic Association of Nature
                                   Conservation.

SUKOPP, H. Prof. Dr.   TU Berlin-Institut für Ökologie,
                                   1 Berlin 33, Albrecht-Thaer Weg 4,
                                   Federal Republic of Germany.
                                   (Chairman, Session II of Conference).

TALBOT, L. M. Dr.   Executive Office of the President,
                                   Council on Environmental Quality,
                                   722, Jackson Place, NW., Washington
                                   DC. 20006, USA.

THAMDRUP, H. M. Dr.   Director, Naturhistorisk Museum
                                   Universitetsparken, DK-8000, Aarhus,
                                   Denmark.

THORARINSSON, S. Prof.   Dean, Department of Geology,
                                   University of Iceland.

THORSTEINSSON, I.   Agricultural Research Institute,
                                   Iceland.

ULRICH, B. Prof. Dr.   Direktor, Institut für BodenKunde v
                                   Waldernährung der Universität Göttingen,
                                   34 Göttingen, Büsgenweg 2, Federal
                                   Republic of Germany.

VAN DER MAAREL, E. Dr.   Botanisch Laboratorium, Afdeling
                                Geobotanie, Toernooiveld, Nijmegen,
                                The Netherlands.

VAN DYNE, G. M. Prof.   Department of Range Science, Colorado
                                State University, Fort Collins,
                                Colorado 80523, USA.

VAN MIEGROET, M. Prof.   Institut de Recherche de Sylviculture
                                de l'Université, Coupure Gauche 533,
                                9000 Gand, Belgium.

VISSER, S. Miss.   Department of Biology, University of
                                Calgary, 2920 24 Avenue NW, Calgary,
                                Canada, T2N 1N4.

WEIN, R. W. Assoc. Prof.   Department of Biology, University of
                                New Brunswick, Fredericton, New
                                Brunswick, Canada, E3B 5A3.

WOLFF, W. J. Dr.   Nederlands Instituut voor Onderzoek
                                de Zee (NIOZ), Postbus 59, Den Hoorn,
                                Texel, The Netherlands.

WOODMAN, M. J.   Institute of Terrestrial Ecology,
                                68 Hills Road, Cambridge, England,
                                CB2 1LA.
                                (Assistant Director of Conference).

The names and addresses given above are correct to the best of our
knowledge as at July 1976. Any general enquiries should be made to
Mr Woodman, Institute of Terrestrial Ecology, UK, and those
specifically concerned with Iceland should be addressed to Dr.
Fridriksson in Reykjavik.

Index